Introduction to Geographic Information Systems

Falguni Mukherjee

Sam Houston State University

ROWMAN & LITTLEFIELD
Lanham • Boulder • New York • London

Executive Editor: Michael Kerns
Associate Acquisitions Editor: Elizabeth Von Buhr
Sales and Marketing Inquiries: RLtextbooks@bloomsbury.com

Credits and acknowledgments for material borrowed from other sources, and reproduced with permission, appear on the appropriate pages within the text.

Published by Rowman & Littlefield
An imprint of The Rowman & Littlefield Publishing Group, Inc.
4501 Forbes Boulevard, Suite 200, Lanham, Maryland 20706
www.rowman.com

86-90 Paul Street, London EC2A 4NE

British Library Cataloguing in Publication Information Available

Library of Congress Cataloging-in-Publication Data
Names: Mukherjee, Falguni, 1975– author.
Title: Introduction to geographic information systems / Falguni Mukherjee.
Description: Lanham, Maryland : Rowman & Littlefield, 2025. |
Series: Exploring geography | Includes bibliographical references and index.
Identifiers: LCCN 2024038009 (print) | LCCN 2024038010 (ebook) | ISBN 9781538152003 (cloth) |
ISBN 9781538152010 (paperback) | ISBN 9781538152027 (epub)
Subjects: LCSH: Geographic information systems.
Classification: LCC G70.212 .M85 2025 (print) | LCC G70.212 (ebook) |
DDC 910.285–dc23/eng/20240911
LC record available at https://lccn.loc.gov/2024038009
LC ebook record available at https://lccn.loc.gov/2024038010

This book is dedicated to my parents, Shyamal Kumar Mukherjee and Itu Mukherjee, who taught me the importance of hard work and determination and most importantly instilled in me a love for lifelong learning.

The book would not have been possible without the support of my husband Kaushik and my daughter Riya, who checked my progress every morning when I started writing the first chapter and ensured that I was able to meet my daily goals. Both of you have been a constant source of support in my life. I don't know what I would do without the two of you.

Contents

Preface

When I am asked what I teach, my typical response is "geographic information systems or GIS." The usual response to my answer is, "What is that?" So, to help others understand what I teach, over time, I changed my response to that question by referencing GPS and satellite imagery in my answer. And immediately I would see an expression in their faces that said, "Wow! That is interesting." What surprised me was that most of the individuals I was speaking to were using GIS every day without knowing that they were using it.

I have been teaching geographic information systems and science to undergraduate and graduate students for over a decade now. And in my interactions with students who were from a non-technical background I felt the same kind of unawareness. They understood that they need to take this class on GIS because, "it will help them get a job." But what exactly is GIS? Why will it help you to get a job? What can it do for you? And how is it possible for GIS to do these things? These are the kinds of questions I wanted them to understand when they took their first introductory class on GIS. However, when I would look for a book to teach this course, I found it hard to find any book that would cater to students who were from a non-geography, non-technical, non-cartography, or non-quantitative background—a book that would lay down the concepts and theories in a very simple language. Additionally, what I was looking for was a book that would cover the concepts of GIS and spatial thinking, while also providing hands-on activities and exercises on those concepts, which I believe is not only the most attractive part about GIS but also helps students better understand the theories through real-world applications. I had to rely on multiple sources and textbooks to accomplish this in my course. It is not difficult to find a book that covers the theories and concepts of GIS if you are teaching a graduate level course or an advanced undergraduate course. But, what about those who are learning about GIS for the first time? How do you get them interested in pursuing a future in GIS or taking more GIS courses? How do you get students who have no idea what GIS is to see its remarkable applications? How do you show them the career trajectories in GIS and geospatial technology? These are the kinds of questions that motivated me to write this book, one that instructors like me can adopt for an introductory GIS course for their lecture and lab sections. I wanted a book that would cover theories *and* hands-on exercises, while written in very simple and non-technical language.

The goal of this book is driven by my motivation to write it: a book that will be all encompassing, can be used by instructors to teach an introductory GIS class, and will make the complex and sophisticated field of GIS and spatial thinking comprehensible for first time users who are new to the world of GIS and geographic knowledge. The book is written in a language that assumes no prior knowledge of GIS. Instead, the goal is to break down the various nuances and facets of GIS and spatial thinking. Covering different aspects that makes a Geographic Information System and Science, this book provides a detailed coverage of the topic. The objective for the textbook is to provide an overview of the topic at the basic level, to provide foundational concepts that students can build upon through additional coursework. Today, GIS is used by professionals in many different disciplines. This text is a one-stop book to introduce someone to the breadth and depth of this complex field in all of its applications.

The book reflects my personal thinking developed from teaching GIS over a decade. GIS is best learned by practicing what you learn, doing hands-on activities and labs. You have to understand the concepts. However, real learning happens when you apply those theories and concepts to analyze data, create a database, or create a visualization using the database. Hands-on experience is important and key to learning GIS and to connecting it to real world. Each chapter in the book reflects this thinking.

Organization of the Book

The book contains ten chapters. Material in each chapter can be covered in a week for a typical semester-length course:

- ***Chapter 1*** is an introduction to the history of GIS, its interdisciplinary nature, different definitions that have emerged for GIS from varying disciplines, and components of GIS, while setting the discussion for all the other chapters in the book.
- ***Chapter 2*** focusses on the importance of location and cartography in GIS. It discusses different projection systems, the importance of projection systems in mapping, projected and geographic coordinate systems, different coordinate systems used in the United States, and how GIS softwares work with them.
- ***Chapter 3*** introduces GIS data models, vector and raster data models, how GIS softwares work with different data models, and different uses of vector and raster data.
- ***Chapter 4*** discusses another important GIS concept – database management systems. This chapter discusses GIS's roots in database management systems, what is a database, components of a database, different types of database management systems, attribute tables, how GIS softwares work with attribute tables, and metadata.
- ***Chapter 5*** carries the discussion forward into different types of geospatial data. The chapter focuses on both vector and raster data, where to look for different geospatial datasets, agencies that create and disseminate geospatial data, how geospatial data are created for fieldwork, and techniques using vector and raster datasets. The chapter also includes discussion on the latest tools and field methods for collecting different geospatial data, namely, unmanned aerial vehicles (UAV's), GPS collection devices, smartphone apps, and remote sensing data collection methods.
- The next chapter, ***Chapter 6*** goes into another important topic, cartography. This chapter focuses on principles of cartography covering cartographic basics such as map design, fonts, color, symbols, and how these cartographic principles are applicable in GIS. The importance of cartographic visualization in GIS is also discussed.
- ***Chapter 7*** discusses different GIS softwares that we have in the market today, softwares that work with vector data, softwares that work with rater data, cloud-based GIS software, open-source GIS softwares, desktop versus cloud-based GIS software, and advantages and disadvantages of each type of software.
- ***Chapter 8*** focuses on the ethics of GIS use and its implications on society. It delves into the meaning of ethics in terms of GIS, how ethics impact every aspect of GIS (software use, GIS applications, data creation, and data use), and discusses real-world tips on ethical practices. The chapter also discusses participatory GIS practices and how GIS influences different aspects of society and is influenced by them. This chapter also includes detailed discussion on GIS project management principles.
- ***Chapter 9*** discusses current and future trends in GIS which includes a discussion of the role of digital transformation on GIS, types of cloud services—software as a service (SaaS), infrastructure as a service (IaaS), platform as a service (PaaS), big data, APIs, and IoT—and how they relate to GIS. This chapter also discusses

different cloud services available for geospatial technology and how digital transformation, cloud computing, big data, and IoT have changed the mechanisms used for geospatial data capture, storage, delivery, and visualization. Some of the latest geospatial applications based on these concepts such as ArcGIS Online and Story Maps are discussed in detail. This chapter also includes a discussion on GIS career trajectories.
- ***Chapter 10***, the final chapter, includes discussion on geospatial data analysis and manipulation. Discussion focuses on both vector and raster data and covers some of the basic spatial analytical operations that can be performed in a GIS.

Organization of Each Chapter

Each chapter includes the following components.

- ***Learning objectives***: Each chapter's content is summarized into a bulleted list. Instructors can use this list as chapter summaries. The learning objectives list will also be helpful to create learning outcomes for the course.
- ***Chapter discussions***: This section is the bulk of the chapter with content that includes detailed discussion of geospatial topics, identified by title.
- ***Art program***: Each chapter contains full-color images to reinforce the topics being discussed including geospatial concepts, applications, geospatial software interfaces, and data collection methods, to engage students and visually explain the topics.
- ***Key terms***: At the end of each chapter, key terms have been compiled for easy reference. Key terms are bolded throughout the chapter and are also compiled in a detailed glossary.
- ***Questions and activities***: Accompanying each chapter at the end are review questions and hands-on activities and applications to reinforce the concepts covered in each chapter. The hands-on activities use freely available geospatial software that can be downloaded or accessed using a web browser. Some of the activities use GIS software that most institutions will have access to through a license agreement with the software vendor. The goal of the hands-on activities is to help the reader understand real world applications of the concepts covered in the chapter. The activities in each chapter use one or more of the following software packages or applications – ArcMap, ArcGIS Online, Story Maps, QGIS, Google Earth Pro, and Earth Explorer. All hands-on activities have step-by-step instructions for students to follow and contain color images of screenshots of the software for each step.

  The hands-on activities have questions integrated with step-by-step instructions that students must answer as they proceed through the activity. The questions are designed to reinforce the concepts covered in the chapter and also to help students work through the activities and connect the dots of geospatial thinking.

  Data for the hands-on activities using GIS software are freely available and can be downloaded by students and instructors. Instructions for the hands-on activities provide a link to the websites from where you can download data and step-by-step instructions on how to download the data.

The questions and activities section at the end of each chapter also contains hands-on applications using freely available resources on the internet that students can use to understand the concepts covered in each chapter and explore the world of geospatial thinking.

Ancillaries

Instructor Resources for this book, available on this book's companion website (at https://bloomsbury.pub/mukherjee-intro-to-gis), includes additional hands-on activities using GIS softwares. Instructions for the activities following each chapter and in the Instructor Resources include brief explanation about software access and availability.

- ***Summary and study guide:*** Instructor Resources on this book's companion website contain a test bank for each chapter that includes individual chapter summary and study guide in question-answer format.
- ***People in Action:*** Chapters 1 and 9 include *People in Action* boxes highlighting interviews with seasoned GIS professionals. These interviews help students understand various career pathways using geospatial technologies. Chapters 1 and 9 in the textbook have abridged versions of these interviews. The Student Resources for this book, available on this book's companion website, contains the full interviews.

Acknowledgments

This book was not supported or endorsed by owners of any of the software products and applications that are discussed herein. The names, logos, designs, trademarks, and graphical user interfaces are the intellectual property of their respective owners: ArcGIS Desktop 10.8 (Esri), ArcGIS Online (Esri), QGIS, Google Earth Pro (Google), Google Maps (Google). Data for many of the hands-on activities and maps in this book were derived from ArcGIS Living Atlas of the World. Maps in many chapters in this book were created using ArcGIS® software by Esri. ArcGIS® and ArcMap™ are the intellectual property of Esri and are used herein under license. Copyright © Esri. All rights reserved. For more information about Esri® software, please visit www.esri.com.

When I started writing this book I wanted to create screenshots to demonstrate the concepts and theories I was discussing in the text. And this would not have been possible if I had not had access to geospatial data on ArcGIS Living Atlas of the World. For those of us who teach or conduct research using GIS this is very helpful and for that I am grateful. I would like to thank the anonymous reviewers who provided valuable feedback which helped to shape the final text. I would also like to thank Rowman & Littlefield for working with me on this book during different stages and for their feedback as it took shape.

What Is GIS? 1

Learning Objectives

- Understand and explain why GIS is important in today's digital world
- Understand and discuss what data, information, and information systems are
- Describe what GIS is and compare the different definitions of GIS from different disciplinary perspectives
- Explain the different components of GIS and their role
- Describe location information
- Discuss the different processes performed by a GIS
- Understand the history of GIS

Why GIS?

If you are reading this chapter, you must have asked yourself, what is GIS? Why do I need to know about GIS? Everything we see around us and don't see around us—a person, an object, an event, an activity—everything that exists is located or occurs somewhere. Information about the "somewhere" is crucial. It is the very foundation of the quest that has driven explorers for centuries. And it is this key information that is at the crux of geographic research. We humans have explored every corner of the earth for ages looking for resources, people, or habitable space. And this pursuit continues even today. With time the tools used for exploration and investigation have evolved. For instance, a magnetic compass that was used as early as the 11th or 12th century has evolved from a magnetic needle attached to a piece of wood to a highly advanced and sophisticated device capable of complex calculations used for navigation purposes.

As tools available for exploration have evolved into sophisticated devices, it has allowed humans to ask and investigate more-complex questions. It has also enabled explorers and researchers to reach and access parts of the earth that were previously considered elusive and inaccessible. Research and exploration have come a long way from traveling to a faraway location and describing native people and indigenous resources. Researchers now focus on how, why, what, and where, investigating distributions and patterns of activities and events, examining why the patterns exist and their impacts on society and ecosystems, and predicting future patterns. An essential component of any such investigation is the question about where. This lies at the heart of any geographic research. Location is the root of all geographic investigation.

With technology and advancements in technology, we have more investigative tools at our disposal today than early explorers. Fieldwork in the wilderness today consists of several other technological devices such as a GPS unit, in addition to a magnetic compass. Moreover, investigations about the earth and its processes and patterns and their impacts can now be performed using modern technological devices and applications. Paper maps are being replaced by digital maps that we view on a computer screen or mobile device. Today we live in a digital era, the era of Facebook, Instagram, and X. Most of us cannot imagine life without a smartphone.

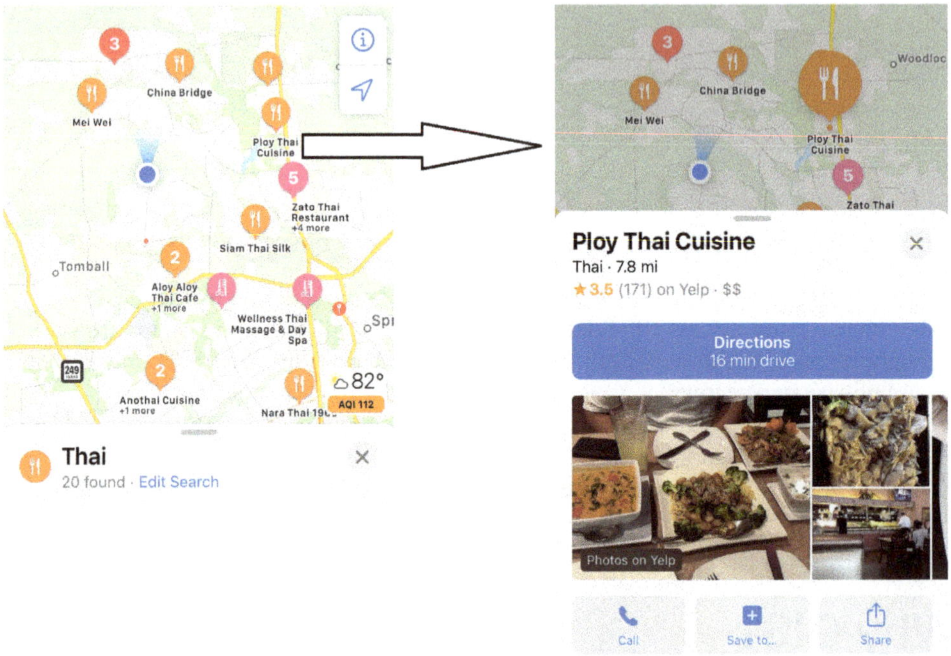

FIGURE 1.1 A smartphone's map application to search for nearby restaurants.

At a time when information is predominantly digital, GIS is a key component of the vast collection of tools, and its knowledge is paramount for exploring the earth, its processes and patterns, and their impacts.

Today if you wish to go to a restaurant for lunch, let's say a Thai restaurant, all you need to do is take out your smartphone and search for the nearest Thai restaurant on your phone's map application. What you will see next on your map are point markers displaying the locations of Thai restaurants near *your* location (figure 1.1). You can click on each of the markers to get detailed information about each restaurant represented by that particular marker, including the website, location, hours, and customer reviews in some cases. Once you have selected a restaurant, your next step would be to look for directions to navigate to the restaurant. Your phone's map application not only provides directions to the restaurant, but also displays it on your map. Would you believe that this map of Thai restaurants on your smartphone's map application with the associated information was possible because of GIS?

What Is GIS?

Consider a situation where you are part of a team working for a national fleet management company. One of your primary job responsibilities is resource management wherein your duties include maintaining an up-to-date inventory of the fleet of vehicles owned by the company and lease information for each vehicle; tracking vehicular fuel consumption and vehicle routes; managing on-time vehicular maintenance; tracking fraud alerts, health of vehicles, driver behavior, vehicle recalls, and total cost of ownership of a vehicle; and recording historical information from acquisition to termination of vehicle use. Additionally, you must also assess and make decisions about when to replace a vehicle to maintain a healthy fleet and when to change vehicle routes or drivers, and evaluate if there is a need or shortage of certain vehicle types in a certain region and if vehicles are being utilized optimally in every region. You must maintain this information and make your evaluations so that your superiors can ensure that the company has a management plan in place for an adequate number of healthy vehicles of all makes and models available for any company in any region to lease at any time. You must also ensure that there are no violations on any vehicle and all local, regional, and national policies and legislations are followed.

There is no doubt that the tasks outlined for you will require vast amounts of data collection, collation, compilation, management, evaluation, and analysis. It will require you to assemble a variety of databases and maps that provide different sets of information for the fleet, and you then must extract the information from the databases and maps and put it all together in a manner that can be visualized seamlessly at once. Manually producing such an interface would involve several hundred hours of work and manpower and a huge investment. An efficient way to avoid these consequences and complete the tasks involves the use of an automated system that has the capability to work with both tabular and map data. Such a system can extract information from tables and maps to help you display to your superiors all the different datasets and parameters, relations and distributions, and whatever patterns emerge from your data analysis seamlessly in one location at any given time.

The need to efficiently and effectively perform the kinds of tasks outlined above does not exist in this scenario only. You will come across such scenarios in a variety of industries and disciplines. Emergency management and fire departments have the need for such an automated system that can assist the department in efficient allocation of resources and provide optimal routing capabilities to ensure rapid response during emergencies. Similarly, forestry departments would need such an automated system to assemble and manage an inventory of timber, wildlife, biodiversity, and other such resources for effective daily operation. Utility companies would need such an automated system to evaluate and forecast future demands based on current coverage. They would also need an automated system to determine where to place the next cell network tower or fiber network cable. An energy company would need such an automated system to maintain and manage its grid of equipment, resources, and customers and to manage and evaluate service needs for new customers. Local governments would utilize such an automated system to evaluate future infrastructure and resource needs and create development plans based on those needs. Similarly, police departments, retail industries, real estate companies, and defense industries would all utilize this automated system for effective management and maintenance of their resources and to create future plans.

So, what is this *automated system*? In order to answer that question, let's first briefly talk about data and information. The scenarios discussed above all involve data of some sort. More importantly, they involve examining, analyzing, and working with information and making important decisions based on information generated or extracted from your data. If you look around, everything is driven by data and information. No matter the industry and discipline, applications and business processes are all driven by data. This is more relevant in today's digital age where technology drives everything and value of data is paramount.

Data can be thought of as any material that is used to extract information after careful evaluation. Data is available in different formats. It can be plain facts, observations, texts, measurements, characters, symbols, images, numbers, tables, maps, figures, or diagrams—anything that can be collected and analyzed. In order to make sense for meaningful use, raw data must first be processed. Data may be processed by a human or a machine. A data ultimately represents a phenomenon, an event, an occurrence, or a measurement of something. However, it may or may not be a satisfactory or good representation of the phenomenon. How well your data represents something will depend on how, what, where, and when the data was put together. Data can exist in a physical format or in a digital format. In the current digital age where technology seems to drive everything, data is represented, stored, and analyzed in a digital format for its use by computers. Digital data can be processed, transferred, and transmitted more efficiently and rapidly than data in physical format. We will discuss data and data formats in chapter 4.

Once data has been processed, analyzed, and organized in a way that makes sense and can be used, what you get is information. **Information** is data that has been prepared and structured in a meaningful manner that can be utilized to make informed decisions. The process is based on context, relevance, and objectives. Data is the building block of information.

Consider the location of all elementary schools within a school district and all the houses within a neighborhood that are zoned to this school district. The location of elementary schools and houses is data. When this data is used to answer questions like "What is the fastest way to get to the elementary school from a bus stop in the neighborhood?" or "How many houses in the neighborhood have elementary-school-aged children?" the data is processed and transformed into information. Similarly, a list of precipitation readings by itself will have no significance until the precipitation data is organized and analyzed to extract information about trends and pattern. One can then use that information to see if there are any regional trends or certain patterns exhibited by this data or if the data indicates some kind of a natural disaster. One can ask and seek answers to such questions from information that is extracted by processing and analyzing data.

As one seeks answers to questions that become more complex, involving large amounts and different types of data, it becomes more difficult to analyze the data in an effective and timely manner by using data in physical formats such as paper maps, tables, lists, and catalogs. Data can be reused and reanalyzed several times to extract information. However, data must be up to date in order to be reused and analyzed to provide any meaningful information. In the current digital age, information is a commodity, and the ease of access to technology has made it easier to produce information from data, making it a major industry.

So how do you process data and extract information? **Information systems** are automated systems used to gather, store, organize, process, and analyze data to extract information. An information system is a combination of hardware, software, data, people, and methods to transform data into information. Organizations and businesses use information systems to convert data into information and transform information into organizational knowledge. Every business, small or large, makes use of an information system. Businesses like Walmart, Amazon, Experian, Esri, and Netflix all make use of some kind of an information system or systems.

Now, let's get back to our original question. So what is GIS? GIS stands for **geographic information system**, which is an automated system to process, analyze, and extract geographic information from geographic data. It is a combination of hardware, software, data, people, and methods used for processing spatial or geographic data to obtain geographic information and make decisions. There is no one agreed-upon definition of GIS because GIS represents the integration of many disciplinary concepts and subject areas. The term has evolved over time. A search on Google for "geographic information systems" yields 367,000,000 results and several different ways to define the term. The very first search result page from Esri defines GIS as "a framework for gathering, managing, and analyzing data." Wikipedia defines GIS as "a conceptualized framework that provides the ability to capture and analyze spatial and geographic data."

So why is it difficult to arrive at an agreed-upon definition of GIS? To answer that question you need to answer another question: If you were asked to name a discipline GIS belongs to, what comes to mind? It would be difficult to name one discipline that GIS exclusively belongs to, because it is inherently multidisciplinary in nature. The foundation of GIS was built by integrating theoretical concepts from many different disciplines, including cartography, information sciences, computer science, and geography. The field of GIS technology integrates several other fields and technologies. As we cover different chapters in this book, the multidisciplinary nature of GIS will become clear. Hence, several definitions of GIS have evolved over time. The definitions allude to the different disciplinary backgrounds to the field of GIS and the evolution of the field.

Four decades ago, Dueker defined GIS as "a special case of information systems where the database consists of observations on spatially distributed features, activities or events, which are definable in space as points, lines, or areas. A geographic information system manipulates data about these points, lines, and areas to retrieve data for ad hoc queries and analyses" (Dueker 1979, 106). This is one of the oldest definitions of GIS and alludes to the context of GIS being an information system.

Other definitions underscore the role of GIS as an information technology or an automated system. For instance, Ozernoy, Smith, and Sicherman (1981, 92) defined GIS as "an automated set of functions that provides professionals with advanced capabilities for the storage, retrieval, manipulation and display of geographically located data." Similarly, Aronoff (1989, 39) defined GIS as "any manual or *computer* based set of procedures used to store and manipulate geographically referenced data." Others have defined GIS by emphasizing the use of GIS as a tool for the purpose of transforming data. For example, Burrough (1986) defined GIS as "a powerful set of *tools* for storing and retrieving at will, transforming and displaying spatial data from the real world for a particular set of purposes."

A few have defined GIS using a *systems* approach where the emphasis is on the *S* of the GIS and the *S* could be a computer system, a decision-support system, a management information system, or a system of hardware, software, data, people, organizations, and institutional arrangements. Finally, a definition that was coined in the 1990s and marked a turning point in the field of GIS was an approach that underscored the importance of science in a GIS. This approach brought to light an important message: There is more to GIS than pushing buttons. As Goodchild (2009, 9) stated, "GIScience is the science behind the systems, the scientific knowledge on which GIS is based."

Geographic information science is the theoretical foundation on which the development, use, and application of geographic information is built. There is no doubt that GIS technology has changed how we work with spatial data. GIS technology has revealed and brought to the forefront other allied fields such as remote sensing, photogrammetry, surveying, cartography, spatial statistics, geodesy, and other disciplines that work with spatial data. The use of GIS has made these allied fields accessible to many more people. As a result, the tremendous growth in GIS technology has led to growth in these related fields too, and to approach GIS as a science is a nod to the multidisciplinary nature of geographic information science as emphasized by Goodchild (1992). This is evident by the use of GIS in varied fields ranging from biology to archaeology to criminal justice. Geographic information science involves "research on and research with GIS" (Goodchild 2009).

Many scholars who conducted research on GIS argued the importance of its role in society. This approach emphasized not just hardware, software, and datasets but also the role of people and methods, highlighting a number of issues arising from the use of GIS and information technology on the daily practices of society and organizations, such as ethics of GIS use (Curry 1995), relation between GIS and mass production of maps (McHaffie 1995), geodemographics (Goss 1995), and community involvement and decision-making (Harris et al. 1995). This approach emphasized the importance of social processes on GIS functions. Chrisman stressed the need for broadening the concept of geographic information systems to include social, economic, ethical, and institutional contexts involved in operating a GIS (Chrisman 1997), thus defining GIS as an "organized activity by which people measure and represent geographic phenomena then transform these representations into other forms while interacting with social structures" (Chrisman 1999, 175).

See table 1.1 for different definitions of GIS. The definitions reflect six basic approaches: an information systems approach, an information technology approach, a tool kit approach, a systems approach, a science approach, and a society approach.

Thus, you can think of GIS definitions falling under two main categories, and this also reflects the two main areas of research or focus areas within the field of GIS: one that defines GIS as a system and one that defines GIS as a science and emphasizes the role of social structures on the use of GIS and its social impacts (figure 1.2).

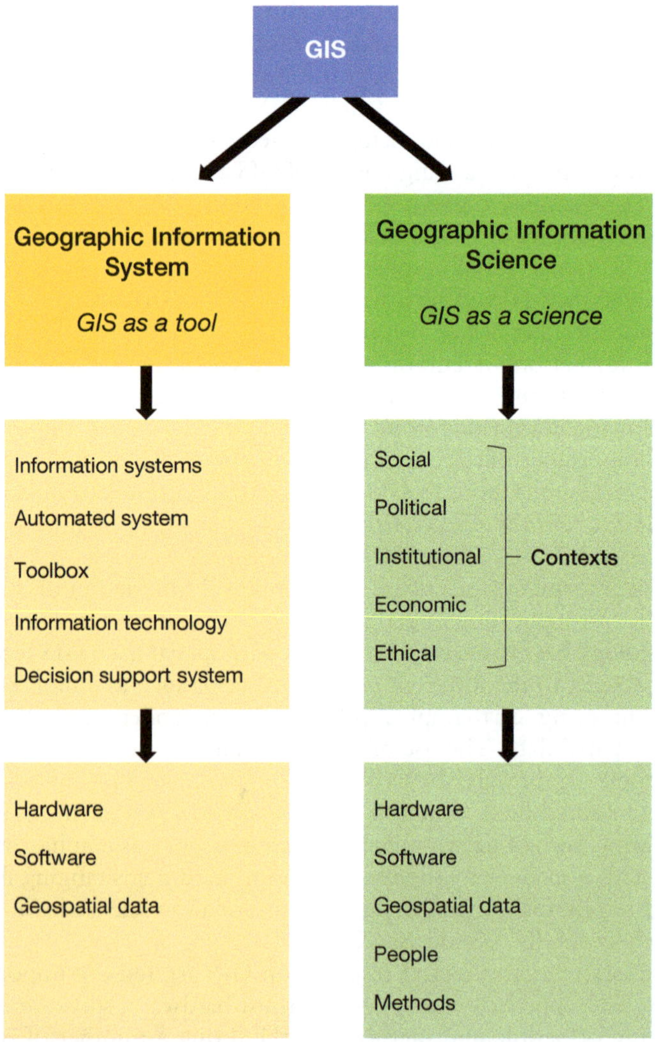

FIGURE 1.2 GIS as a system and GIS as a science.

TABLE 1.1 Definitions of GIS

Approach	Definition
Information Systems	• *a special case of* ***information systems*** *where the database consists of observations on spatially distributed features, activities or events, which are definable in space as points, lines, or areas. A geographic information system manipulates data about these points, lines, and areas to retrieve data for ad hoc queries and analyses.* (Duecker 1979, 106) • *an* ***information system*** *that is designed to work with data referenced by spatial or geographic coordinates. In other words, a GIS is both a database system with specific capabilities for spatially-referenced data, as well as a set of operations for working with the data.* (Star and Estes 1990, 2–3)
Information Technology	• *an* ***automated*** *set of functions that provides professionals with advanced capabilities for the storage, retrieval, manipulation and display of geographically located data* (Ozernoy et al. 1981, 92) • *any manual or* ***computer*** *based set of procedures used to store and manipulate geographically referenced data* (Aronoff 1989, 39) • *an* ***information technology*** *which stores, analyses, and displays both spatial and non-spatial data* (Parker 1988, 1547)
Tool Kit	• *a powerful set of* ***tools*** *for storing and retrieving at will, transforming and displaying spatial data from the real world for a particular set of purposes* (Burrough 1986, 6)
Systems	• *a database* ***system*** *in which most of the data are spatially indexed, and upon which a set of procedures operated in order to answer queries about spatial entities in the database* (Smith et al. 1987, 13) • *a decision support* ***system*** *involving the integration of spatially referenced data in a problem-solving environment* (Cowen 1988, 1554) • *a* ***system*** *with advanced geo-modelling capabilities* (Koshkariov et al. 1989, 259) • *a form of MIS [management information* ***system****] that allows map display of the general information* (Devine and Field 1986, 18) • *A* ***system*** *of hardware, software, data, people, organizations and institutional arrangements for collecting, storing, analyzing and disseminating information about areas of the earth* (Dueker and Kjerne 1989, 7–8) • *a computer* ***system*** *that analyzes and displays geographically referenced information. It uses data that is attached to a unique location* (USGS, n.d.) • *an integrated* ***system*** *of computer hardware, software, and trained personnel linking topographic, demographic, utility, facility, image and other resource data that is geographically referenced* (NASA, n.d.)
Science	• *the set of fundamental* ***issues*** *raised by the technology or the critical issues that arise when the technology is employed. These would include accuracy and uncertainty, scale, and the methods used to capture the infinite complexity of the real world in binary digits* (Goodchild 2009, 3–9)
Society	• *an organized activity by which people measure and represent geographic phenomena then transform these representations into other forms while interacting with* ***social structures*** (Chrisman 1999, 185).

GIS as a system focuses on GIS as a tool that uses a specific software, hardware, and geospatial data to perform a specific task. GIS as a science focuses on the multidisciplinary nature that forms the foundation of GIS and pays attention to the social implications that arise from its use. GIS as a science also emphasizes the role of social, political, economic, institutional, and ethical contexts in shaping GIS use. Thus, in addition to hardware, software, and data, the importance of people who use GIS and make decisions and the methods that are followed to perform specific tasks are emphasized.

So, GIS is multidisciplinary in nature and can be defined as an automated information system that is used for working with geospatial data, and its use is shaped by social contextual factors.

What do these social contextual factors mean? Consider three scenarios: a municipal district in an urban area in India, a municipal district in a community in the United States that places importance on conservation of its natural resources, and a municipal district in a community in the United States that places importance on growth and development in its community. The GIS analysts in all three municipal districts have the same GIS software at their disposal and have similar kinds of datasets on their respective communities. However, how they approach the use of GIS will differ depending on the goals and objectives of the municipal district, which in turn will be shaped by the overall social, political, and institutional environment of the community. The municipal district in India would view GIS as a mechanism to further the state mandate, which is governed by the national political environment, requiring all municipal districts to be modernized and make use of technologies such as GIS. For the municipal district in the United States in a community that would like to further growth and development, GIS would be a means to further real estate sales for commercial development, whereas for the municipal district in a community that values conservation, GIS would be used to further its agenda of natural resource preservation and conservation and map any violations and enforce related laws.

Components of a GIS

When we refer to GIS, we refer to a combination of hardware, software, geospatial data, people, and methods (figure 1.3). Each of these elements has a role to play in the effective use of GIS.

Hardware refers to the actual computer machine on which a GIS operates and is an important part of running GIS. The role of hardware is key for storing, processing, and displaying geospatial data. Today GIS operates on a wide variety of hardware ranging from standalone desktop computers to centralized computer servers to enterprise solutions.

Software enables the functional capabilities for storing, analyzing, and displaying geospatial data. The tools available within the GIS toolset that allow one to store, edit, analyze, or display geospatial data will depend upon the software and the license that one has access to. It includes a database management system, tools to work with geospatial and database data, and a graphical user interface.

Data is a very important component of a GIS. A GIS works with geospatial data, the heart and soul of GIS technology and what sets GIS apart from other information systems. Geospatial data refers to any data that has location information assigned to it. In order to map anything, whether on a paper map or on a digital map, one needs location information. Every time you ask or answer a "where" question you are seeking location information. The locations of hospitals and shopping malls,

FIGURE 1.3 Components of a GIS.

neighborhood boundaries, boundaries of parks, hiking trail routes, and satellite images of neighborhoods are all examples of real-world geospatial data.

Not all data is geospatial. Data about the location of a shopping mall is geospatial data. However, other data about shopping malls such as names of stores within the mall, number of occupants, size of stores, sales, and client zip codes is not geospatial data. The advantage of GIS is that such non-geospatial data can be linked to a geospatial data with location information assigned to it. For instance, you would like to know which zip codes your favorite store services the most. You could create a map showing all the zip code locations that your store services and highlight the top three zip codes. You could then link nonspatial data such as demographic data or income data for each of the zip codes to the zip code locations mapped, creating a detailed database of areas serviced by the store as well as being able to map other characteristics such as socioeconomic or demographic data of the customers serviced by the store.

GIS data consists of two components: a geospatial (spatial) component and an aspatial component, with a file link between the two components making the data complete (figure 1.4). The spatial component of the data enables one to view GIS data on a digital map, which will be displayed on a computer screen. The aspatial component, which contains detailed information about the data, enables a GIS user to analyze the data.

The aspatial component is a tabular data which is often stored in an organization's database management system (DBMS). In a GIS, this tabular data is called an attribute table because it stores all the attribute information of the features you see on a map. A GIS integrates the two components using a file link to associate the spatial component with its respective record in the tabular data. Take the map of the lower 48 US states (figure 1.5). What do you see on this GIS map? You see two datasets on the map: one of the states and one of state capitals.

The state capitals are displayed using a purple-colored point symbology, while the US states are displayed as purple shapes that mimic the actual shape of the states following state boundaries. What you see on the map is the spatial component of the

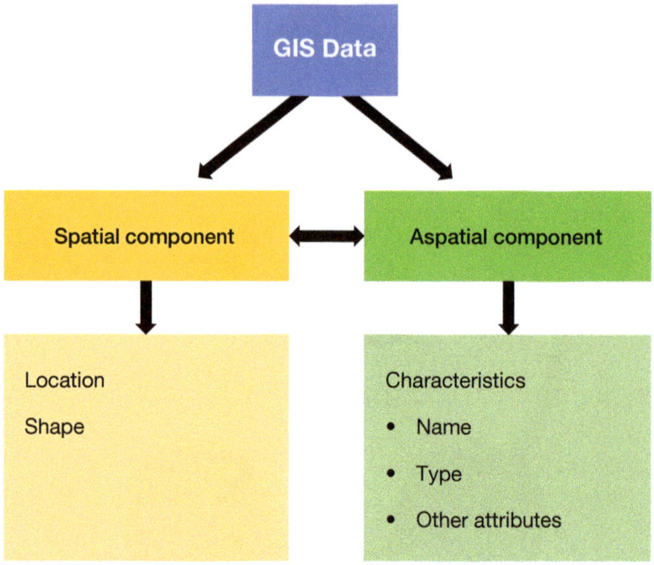

FIGURE 1.4 Components of a GIS data.

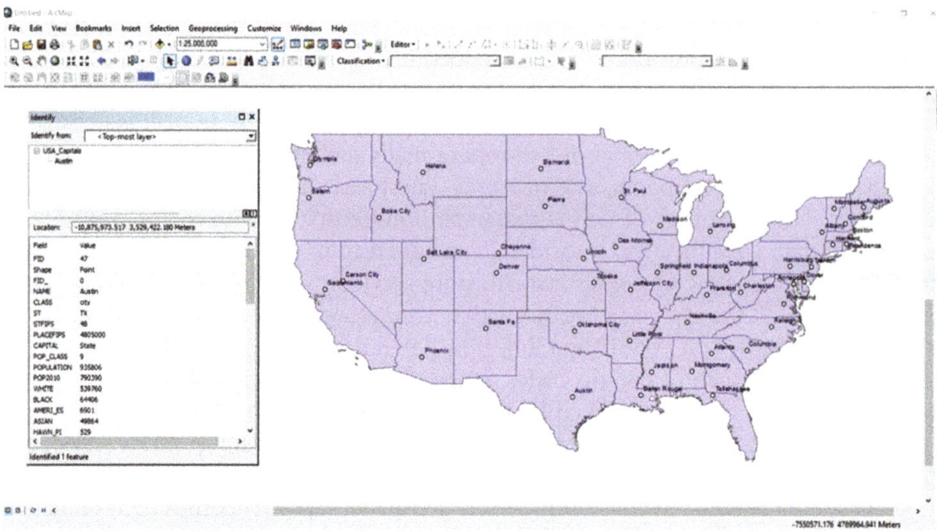

FIGURE 1.5 GIS map of lower 48 US states.

ArcMap interface courtesy of Esri, Inc. USA States layer courtesy of ArcGIS Living Atlas of the World. Data Sources: ESRI, TomTom, Garmin, FAO, NOAA, USGS, EPA, NPS, USFWS

data. When you click on any of the points on the map, a dialogue box opens that displays different types of information about the state capital represented by that particular point, such as name, area, and population. The dialogue box displays the characteristics of the state capital.

How is your map able to display this information? The information is stored as tabular data and is in the background of the spatial component. When called, the GIS software presents the information from the attribute table, which in this case was in a dialogue box that opened when clicking on the point symbol. The spatial and tabular components of the data are linked, and that is how a dialogue window displaying information pertaining to Austin, the state capital of Texas, is displayed when you click on a point symbol that represents Austin on the map and not Madison or any other capital city. Thus a GIS is able to integrate the spatial component and the aspatial/tabular component of a data available in the attribute table, which sets it apart from other information systems.

Let's revisit the earlier example of Thai restaurants. When you search for Thai restaurants on your smartphone's map application, the point markers displayed on your map are the spatial component of the data (figure 1.6). For each Thai restaurant displayed on your map, a detailed tabular data exists in the background. This tabular data carries information that is displayed when you click on any of the point markers on your map. Because the spatial component that is visible on your map and the tabular data that exists in the background are linked, the correct information for each Thai restaurant shows up when you select a point marker.

A GIS cannot be utilized without *people* who operate, design, manage, implement, plan, and make decisions about its use. You need someone to plan how a GIS system would be useful to an organization and champion its cause within an organization, someone to design its use in a real-world scenario within an organizational context, someone to operate the system and implement the project, someone to manage the project, and someone who makes decisions as to whether a GIS is even necessary for an organization, how it will be useful to the overall goals of the organization, and what and how much resources must be allocated to the project. Take a look at the textbox "People in Action" at the end of the chapter. It features a GIS professional and her use of geospatial technology. As you read the interview, you will see the emphasis on people in any GIS project.

And finally, a GIS system will be effective when it is implemented using *methods* that have been thoroughly planned following best business practices and project

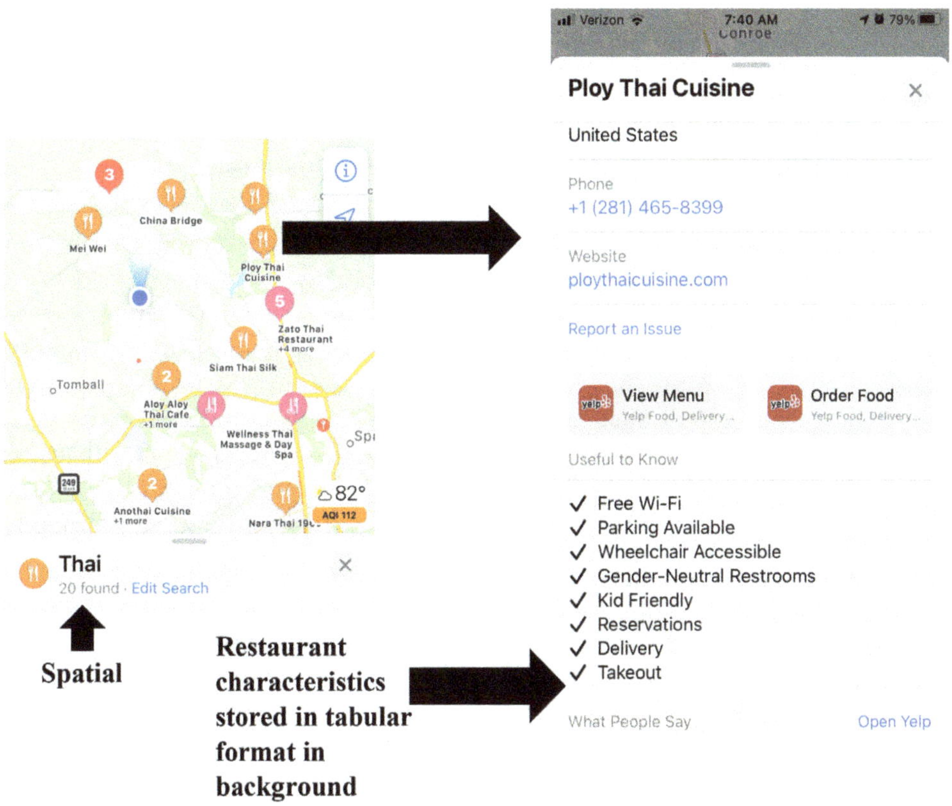

FIGURE 1.6 Spatial and tabular components of GIS data.

implementation procedures. For a GIS system to be utilized effectively by an organization, it must be well integrated with the business process of an organization. A GIS will ultimately be implemented within an organizational context. It does not matter if you have the most advanced hardware and cutting-edge software. Disregarding the organizational environment when making decisions about GIS is a major project implementation blunder and will lead to failure. Again, pay attention to the interview in this chapter's "People in Action." It underscores the importance of thorough planning that follows appropriate project implementation strategies to integrate GIS into an organization.

Location Information

As discussed in the previous section, location information is key for GIS data. Let's go back to the map of the lower 48 US states. How does your map application understand where to place Texas or Louisiana and where to place Texas with respect to Louisiana, or to place Austin in Texas and Baton Rouge in Louisiana? This is because of location information. If the location information was missing, the map application would have no idea how and where to place the states and state capitals. You can think of location as the key to the puzzle that helps to put the different pieces together.

Location information or spatial data is associated with some kind of a geographic reference that aids in identifying where it is located on the earth's surface. Location can be identified in many ways that fall into two broad categories: absolute location with explicit geographic reference and relative location with implicit geographic reference. **Absolute location** is described with respect to a known origin and standard measurement system such as a coordinate system. Examples of absolute location include latitude and longitude and national grid coordinate. We will discuss coordinate systems in chapter 2. **Relative location** is described in reference to another location such as an address, postal code, census tract name, or road name.

A GIS Integrates Information

Have you wondered how when you view a map on a computer screen or mobile device, you are able to visualize all the layers or datasets at the same time although these are separate and distinct datasets? For instance, let's say we want to create a map of a city showing parks, streams, and buildings. If we were working with paper maps, we would have to deal with multiple maps and extract individual information from each paper map and then create a map with all the extracted information. Whereas in the case of a GIS, we are able to view multiple datasets and work with individual characteristic information at the same time. A GIS displays multiple datasets by arranging them in a manner that each dataset, such as parks, streams, and buildings, is displayed at the same time (figure 1.7). It integrates information as a collection of datasets, and each dataset appears in its correct location because of geographic reference.

Many decisions are dependent on such geospatial datasets. For instance, when a new retail chain is proposed to be built in a particular area, the choice of its location is very important. The location of the store should be such that it services the right demographics, is easily accessible, and is not close to a competitor. All of this involves

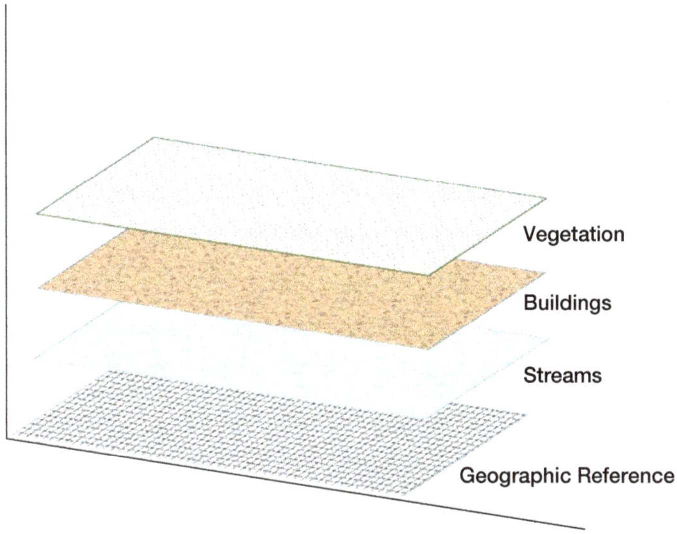

FIGURE 1.7 Multiple datasets in a GIS.

geospatial thinking. Similar thinking would apply to decisions about locating the closest firehouse, opening a new restaurant, determining the fastest route to a school, or evaluating the need for a new street connecting a neighborhood.

Processes Performed by a GIS

A GIS is capable of analyzing data and performing a variety of operations that include query and analysis, data modeling, interpolation, forecasting, and data visualization. However, before any such task can be performed, the following three processes must be performed by a GIS: data input, data manipulation, and data management.

Every GIS task begins with data input. As we already know, in order to use geospatial data in a GIS system, it must be in a digital format. This involves extracting data—spatial and aspatial—from paper maps into digital formats. This process of extracting and converting data from a paper format into a digital format that can be performed by a GIS system is called ***digitizing***. At times this can also be accomplished by digitizing and extracting data from a satellite image or aerial photograph. For instance, let's consider that a new neighborhood is being built in a community and you are tasked with the job of creating geospatial data for the streets, parcels, and other utility lines to be built within the neighborhood. How would you accomplish this since no paper maps of this new neighborhood exist? You can create geospatial data of the parcels, streets, and utility lines by digitizing from a satellite image or aerial photograph. Because of the advancement of technology, many geospatial data already exist in an appropriate digital format and can be used by a GIS system.

Once you have your geospatial data in a digital format, the next step is making sure the data is in a format that is compatible with your GIS system. In this step you are ensuring the digital data that you have access to is usable. Many GIS systems and softwares are available today; we will discuss them in chapter 6. Different types of geospatial data are also readily accessible online today. As a result, geospatial data is available in a variety of different file formats, and not every software will be able to work with every kind of data. This is an important step when a project utilizes data from various sources. Data may be created at different scales, by using different software applications, or by following different technical specifications and hence different file formats, or the data may have unnecessary information not relevant to the project at hand. As a result, the datasets cannot be integrated for immediate use by a GIS system even if they exist in a digital format. So the data must be converted into an appropriate format that is compatible with the GIS system you will be using and the project objectives.

Geospatial data development that involves the two steps of data input and conversion is the most time-consuming and expensive part of any GIS project.

Once you have the geospatial data in an appropriate format, the next step is organizing and managing your datasets. This is an important part of good project management practices. When the project is small in size, one can manage with simple file structures. However, even in this case it is advisable to store and organize data files in a structured manner. For instance, when you work on your GIS class labs and projects, you are advised to download the data and store it in a dedicated folder on your computer, flash drive, or external hard drive. Every time you create a new dataset or have to save an output from a GIS operation, it must go to a dedicated file folder with a meaningful name. If you do not follow this from the start, can you imagine what might happen by the time the semester is about to end? It would be extremely difficult for you to keep track of all your data files and to continue working on a task that requires data files created or stored during a previous activity, and you may end up downloading and storing the same data files multiple times, which is not an efficient practice.

In an environment where the number of users is high and you have different types of users with different geospatial needs and know-how, datasets are voluminous, and multiple departments are involved in using geospatial data and maps, good data management becomes paramount. In such cases a database management system (DBMS) is used to organize and manage geospatial data (we will discuss DBMS in chapter 4). For instance, consider a situation where an organization has been awarded a big project that involves three different departments: A, B, and C. The project must be completed in phases. A person from each department is assigned the task of heading the project. The project is designed and planned in such a manner that department A must first complete a part of the project before department B can begin, and department B must complete its tasks before department C can complete the last phase of the project. In such a situation, what do you think would be the best way to organize the data?

History of GIS

GIS systems are not a recent development. The tasks and operations that GIS perform have been around for many years. In fact, principles of cartography and mapping that are heavily used in GIS date back centuries. For instance, Charles Joseph Minard created thematic maps, a type of map now frequently created using GIS, back in the 1850s (Friendly 2002). (We will discuss thematic maps in chapter 6.) Manual thematic mapping was also used in the planning discipline during the early 1900s. Advances in the field of computer science and engineering and development of spatial theories have shaped the history of GIS.

The origins of GIS as a computerized automated system can be traced back to the late 1950s, when the Canadian government launched the Canada Land Inventory project that laid the foundation of the benefits of computerized mapping and analysis. Multiple projects during the 1960s in the United States and Canada paved the path for the computerized automated system that we know today as GIS. The most influential work in this regard is that of Roger Tomlinson, recognized as the father of GIS. His pioneering work led to the development of the Canada Geographic Information System (CGIS) in 1963. CGIS provided the conceptual and technical understanding of what we recognize as GIS today. Tomlinson developed CGIS that was used by the Canada Land Inventory in their land planning and natural resource management activities. He created a design for a computerized system with the capability to store and expeditiously process large amounts of data. This computerized system adopted a layer approach, the unique characteristic that sets GIS apart, for handling multiple maps. In fact, several design decisions present in current GIS can be traced back to this point in history (Goodchild 2018, 2). Tomlinson coined the term *geographic information system*.

Another influential work during that time included efforts at Northwestern University by Howard Fisher, who created a computerized mapping software known as SYMAP. Fisher subsequently founded the Laboratory for Computer Graphics at the Harvard Graduate School of Design (GSD) in 1965. This marks an important milestone in the development of GIS. The Harvard lab was founded with a grant from the Ford Foundation. The main purpose of the grant was to explore the role of computer mapping in spatializing data. The Harvard lab became the foremost center for research on spatial analysis, and many of the early concepts of GIS were developed and refined there. By the late 1960s, SYMAP was one of the most widely known and utilized computer mapping program (Wilson 2015). Researchers and faculty at the Harvard lab used SYMAP to study a variety of problems.

Students at the Harvard GSD used SYMAP in their projects too. One such student was Jack Dangermond, who joined the lab in 1967 while pursuing a master's degree in landscape architecture. Like other students at GSD, Dangermond used SYMAP in his study of air pollution. He played a pivotal role in refining and further developing SYMAP (Wilson 2015), and drew heavily on those concepts when he founded the Environmental Systems Research Institute (Esri) after graduating from GSD in 1969.

Esri created computer mapping and spatial analytical methods to aid planners in land and resource management. Many of those methods developed during the early years of Esri have become an intrinsic part of current GIS, and today Esri is a leading GIS organization. The roots of vector data design in GIS adopted by Esri early on can be traced back to the Harvard lab. The lab originally started with SYMAP and through the 1970s continued to develop cartographic software packages with interactive computing capabilities such as SYMVU, CALFORM, and POLYVRT. One such program developed at the Harvard lab was ODESSEY, which heavily shaped modern vector GIS (Chrisman 2006, 8). Esri's early GIS software, Arc/Info, adopted the technical framework from ODESSEY.

Many other initiatives were witnessed throughout the United States during the 1960s and 1970s, mainly in the public sector, that further solidified and carved a path forward for GIS. Some of the noteworthy initiatives discussed by Dueker (1979) include the New York State Land Use and Natural Resources (LUNR) system, the Minnesota Land Management Information System (MLMIS), the Polygon Information and Overlay System (PIOS), and the Oak Ridge Regional Modeling and Information System (ORRMIS). One such pivotal development was the Dual

Independent Map Encoding (DIME) system and Geographic Base Files (GBFs) system developed by the US Census Bureau. This was a major milestone in the history of GIS and led to a digital input of the 1970 Census, where the Census Bureau used the DIME-GBF file format to digitize census boundaries, urban areas, and street segments. DIME files were developed using the structures created as part of the POLYVRT program and CGIS, and were widely used for a variety of applications. This was one of the first applications of a GIS file on a large scale, and it fueled development of GIS use in many allied fields.

Throughout the 1960s and 1970s, GIS was practiced using mainframe computers, programming language such as **FORTRAN** was used to write programs to draw maps, and printing and plotting were done using archaic printers and plotters. Visualization was relatively poor, and monochromatic screens did not allow color printing. Maps were designed with black-and-white patterns. Thus GIS was practiced using rudimentary technology and was not yet accessible to the general public. Moreover, it was still an expensive endeavor for many agencies. Advances in the field of computer technology and software engineering in the next decade fueled improvement and growth in the field of GIS.

Two significant changes during the 1980s completely changed the field of GIS. In 1982, IBM introduced the personal computer (PC) to the world. The 1980s also saw the development of more powerful and robust workstations. Growth in the field of electronics engineering and computer science led to more advanced computer hardware compared to previous decades. This enabled GIS softwares such as Arc/Info, IDRISI, and GRASS to transition to the first generation of PCs. GIS was now supported by a platform that was stronger and affordable for many agencies to adopt, leading to a wider adoption of GIS. It also led to the availability of an increased number of GIS softwares.

Throughout the 1980s and 1990s, GIS developed further as a technology as computers became more powerful, affordable, and able to store and process large datasets in a short amount of time. The development of graphical user interface (GUI) dynamically improved GIS usage and helped to make GIS software user-friendly. Moreover, the availability of inexpensive inkjet printers now allowed the use of color in GIS outputs, enabling more features and attributes to be displayed. The 1980s also saw the emergence of GIS infrastructure. The most important among these was the establishment of the National Center for Geographic Information and Analysis (NCGIA) by the National Science Foundation in 1988. The NCGIA created a core GIS curriculum and also established academic research direction for GIS.

The 1990s saw massive growth in the field of GIS. GIS was adopted widely by the public sector as well as private corporations. Its use had become cost-effective with the availability of low-cost desktop GIS programs and the increased use of personal computers, making it easily accessible for public use. GIS softwares were also able to take advantage of developments in the field of software science and engineering, further contributing to its ease of use and public appeal. With the arrival of desktop computers and laptops, GIS use reached new environments outside the confines of research laboratories. The 1990s witnessed the emergence of NCGIA's Research Initiative 19, "GIS and Society: The Social Implications of How People, Space, and Environment Are Represented in GIS" (Mark et al. 1997), that drew attention to the technical, social, and institutional factors that have shaped the development of GIS and in turn have influenced representation and access to GIS usage. The focus was now on including alternative forms of reasoning and understanding of the world into GIS.

The current era of GIS is dominated by internet and mobile technologies that have dramatically changed GIS technology. GPS data and high-resolution satellite images are readily integrated with GIS. Moreover, GIS technology can be easily and readily incorporated into other technologies. GIS datasets are freely available to download from numerous sites, detailed GIS datasets are available at different scales, and GIS maps can be published on the internet. Numerous GIS softwares are available on the market, and users can easily interact with interactive online GIS softwares. Most importantly, the internet has also led to open-source GIS softwares. Growth in geo-analytics and location-based services has widened its use in many fields. GIS is a multibillion-dollar industry today with use in varied fields such as defense, urban planning, natural resource management, land management, utilities, transportation, agriculture, mining, oil and gas, and the government sector.

People in Action Ellen West Nodwell

Ellen West Nodwell, GISP, is the CEO of IntegraShare Dimensions, Inc., a woman-owned business in the United States, based in Texas, that specifically services organizations in areas related to geomatics, geospatial systems, strategy for geospatial adoption, and geospatial data—governance, management, data analysis, and visualization of virtually every geospatial data type.

She is also a founding member and the current chairperson of the APSG Education Foundation, an organization that supports the development of spatial thinkers from K–12 through higher education who later may move into careers in the following areas: geomatics, survey engineering, geospatial technology, and the support of integrity of geospatial data of all types. APSGEF also supports education in these areas for underserved and transitioning students in the same areas of study. The support given by APSGEF is through both pro bono volunteer consulting and teaching as well as through scholarship-type support and program funding.

Ellen has been involved in supporting the use of geographic and geoscientific data for a large portion of her working life—now over 45 years. She has worked in state government as well as private industry, using technology as it has advanced from 10-key adding machines and mainframes to our current technology stacks that utilize "location" in almost every aspect of living—at home, at school, and at work. Ellen is the former Global GIS Manager of Hess Corporation (2006–2013) and was the Spatial Data Coordinator at Anadarko Petroleum (2001–2006). She began using GIS tools ahead of that while employed with Meridian Oil / Burlington Resources as a technical data analyst supporting engineers and geoscientists.

The importance of accurate surveys and maps was impressed upon Ellen at an early age. She has been drawing her own maps since about age five. She is presently teaching her young grandson the importance of maps using a web application that she built for him to navigate around the globe on a computer. She also encourages him to make maps with pencil and paper.

What does your organization/company do?
IntegraShare Dimensions is a service provider/consultancy that provides a full range of services related to GIS (systems), geospatial data-governance, management, and geodetic accuracy assurance, including doing forensic geodesy work. They also make maps and create geospatial databases from antique

(continued)

People in Action *Continued*

maps (or modern imaged or paper maps). The company provides consulting on building geospatial strategy for organizations who seek to use geospatial data and technology across their businesses.

Can you tell the readers a bit about your job? What is your current role and responsibilities?
As a small business owner, I do everything! I am the "admin" as well as the "executive" and I do everything in between. My primary role is as a technologist in both the GIS and the data management disciplines. I have specialty expertise in the petroleum and pipelines industry, as well as having worked in a state government for 10 years, in two mental health hospitals within those, which gives me experience in both government and healthcare settings.

What do you use GIS/geospatial technologies for in your organization?
We use GIS for general mapping and geospatial data visualization—so we "know where things are" and more about these places and their relationships with other places and people, objects, or areas on the earth and on other planets.

I have used GIS for data analysis and intelligence work. As a volunteer with URISA's GISCorps, I have worked in organizations to digitize locations on a map for helping with humanitarian relief—for organizations like World Health, UNESCO, FEMA, and other government agencies (sub-support during storms Maria and Harvey).

In our organization, we use GIS and geospatial technology for business support and humanitarian relief/support.

What brought you into GIS in your field? How did your GIS career begin?
It started with data management—and eventually, I gravitated towards the "layer effect" of GIS! GIS attracted me because it showed "life in layers," which is how my mind is wired. I have been a four-dimensional thinker forever, and GIS was the representation of that in a technology form. I also found relational databases to be attractive, as I think about everything as being connected, even if it is remotely connected! I have found that GIS satisfies both of those concepts very well.

At the time I first met "GIS," I was managing data and supporting engineers and geoscientists in our company, which had been an evolutionary career for me—I began as an engineering group admin and eventually took on more complex tasks that involved analysis, mapping, graphing, and always there was data to manage.

My roles transitioned into more "IT" tasks, and eventually I was promoted into a systems analyst role and moved into corporate IT, where I managed the company's technical databases in a group that had company-wide responsibilities. I was their first corporate "Data Manager." . . .

I guess you would call me a "data wrangler" because I did a lot of that back then, and I still do! Today, it is somewhat easier, as we have great tools like SAFE Software's FME that has yielded huge strides in data interoperability, but we have more data than we did then—datasets are larger—and it takes more computing power, so there is that issue. But I still love GIS—and I love to build interactive visualizations to bring data to life.

How does your company/organization use GIS for business analytics, customer, and employee experience?
We help our customers understand how applying the geospatial perspective can enhance business analysis outcomes, where location is a facet of their business. How customer and employee experience fits in with the people aspect of technology. This can be anything to do with outcomes related to people. We used a quick web application after a flood event to help connect people with resources during a critical time after a disaster—it was a "quick and dirty" application using uploaded Excel spreadsheets from human resources to update information, but it worked, and it saved a lot of time for getting people in need connected to help.

What would be your advice to someone who is interested in pursuing a career in GIS?
It's a great career choice, but pay attention to the basics before you engage in the technology—understand the reasons behind "why" you make the maps, why you build the apps, and why we must manage the data very carefully.

To read the full interview with Ellen West Nodwell, visit this book's companion website at https://bloomsbury.pub/mukherjee-intro-to-gis.

Questions and Activities

1. Discuss to compare and contrast information versus data.
2. Most of us interact with a number of information systems in our daily lives. Identify at least five different information system that you interact with daily and identify the people and processes involved in operating those information systems.
3. What is an information system? Write a one-paragraph description of how a GIS is different from other information systems.
4. The components of a GIS include hardware, software, data, people, and methods. Explain in your own words the significance of each component. Can you think of a real-world example to explain the importance of the last two components?
5. We have covered different definitions of GIS in this chapter. Why do so many definitions of GIS exist? If you were to define GIS, how would you do so? Will you pick one of the definitions from this chapter or create your own definition?
6. Using the information in this chapter, create a timeline diagram for the history of GIS. Highlight the different phases of the history and activities that stand out during that particular phase.
7. What is geospatial data? List at least five geospatial data that you use in your daily life.
8. Search for educational institutions in your region. What GIS courses and degrees are offered at the educational institutions of your choice?
9. Esri is the market leader in GIS today (we will cover Esri in chapter 7). Visit https://www.esri.com/en-us/industries/index and review the different uses of GIS in different fields. Pick a field of your choice and compose a one-paragraph write-up on the use(s) of GIS in that field.
10. Zillow is real estate website where you will find information pertaining to real estate buying, selling, and leasing. Visit Zillow.com and select the "Buy" option. Look for homes for sale in your area or in a zip code of your choice. Do you notice use of geospatial data and mapping? Compose a one-paragraph write-up explaining how this website is making using of GIS.

Key Terms

absolute location (p. 12)
data (p. 3)
FORTRAN (p. 16)
geographic information system (GIS) (p. 4)
geospatial data (p. 8)
information (p. 4)
information system (p. 4)
relative location (p. 12)

References

Aronoff, Stanley. 1989. *Geographic Information Systems: A Management Perspective.* Ottawa, ON: WDL Publications.

Bourgeois, David. 2014. "What Is an Information System?" In *Information Systems for Business and Beyond*, 5–74. Open Textbook Challenge by the Saylor Academy.

Burrough, Peter A. 1986. "Principles of Geographic Information Systems for Land Resources Assessment." *Geocarto International* 1, no. 3: 54. https://doi.org/10.1080/10106048609354060.

Chrisman, Nicholas. 1982. "The Role of Quality Information in the Long-Term Functioning of a Geographic Information System." *Cartographica: The International Journal of Geographic Information and Geovisualization* 21, no. 2 & 3: 303–12.

———. 1997. *Exploring Geographic Information Systems.* New York: John Wiley & Sons.

———. 1999. "What Does 'GIS' Mean?" *Transactions in GIS* 3, no. 2: 175–86.

———. 2006. *Charting the Unknown: How Computer Mapping at Harvard Became GIS.* 1st ed. Redlands, CA: Esri Press.

Clarke, Keith. 2011. *Getting Started with Geographic Information Systems.* 5th ed. Upper Saddle River, NJ: Pearson Prentice Hall.

Cowen, D. J. 1988. "GIS versus CAD versus DBMS: What Are the Differences?" *Photogrammetric Engineering and Remote Sensing* 54: 1551–54.

Curry, Michael. 1995. "Rethinking Rights and Responsibilities in Geographic Information Systems: Beyond the Power of the Image." *Cartography and Geographic Information Systems* 22, no. 1: 58–69.

Devine, H. A., and R. C. Field. 1986. "The Gist of GIS." *Journal of Forestry* 84, no. 8: 17–22.

Dueker, Kenneth. 1979. "Land Resource Information Systems: A Review of Fifteen Years Experience." *Geoprocessing* 1: 105–28.

Dueker, Kenneth, and Daniel Kjerne. 1989. *Multipurpose Cadastre: Terms and Definitions.* Falls Church, VA: American Society for Photogrammetry and Remote Sensing.

Friendly, Michael. 2002. "Visions and Re-visions of Charles Joseph Minard." *Journal of Educational and Behavioral Statistics* 27, no. 1: 31–51.

Goodchild, Michael. 1992. "Geographical Information Science." *International Journal of Geographical Information Systems* 6, no. 1: 31–45.

———. 2009. "Geographic Information Systems and Science: Today and Tomorrow." *Annals of GIS* 15, no. 1: 3–9.

———. 2018. "Reimagining the History of GIS." *Annals of GIS* 24, no. 1: 1–8.

Goss, Jon. 1995. "Marketing the New Marketing: The Strategic Discourse of Geodemographic Information Systems." In *Ground Truth*, edited by John Pickles, 130–70. New York: Guilford Press.

Harris, Trevor M., Daniel Weiner, Timothy A. Warner, and Richard Levin. 1995. "Pursuing Social Goals through Participatory GIS: Redressing South Africa's Historical Political Ecology." In *Ground Truth*, edited by John Pickles, 113–29. New York: Guilford Press.

Koshkariov, A. V., V. S. Tikunov, and A. M. Trofimov. 1989. "The Current State and the Main Trends in the Development of Geographical Information Systems in the USSR." *International Journal of Geographical Information Systems* 3, no. 3: 257–72.

Maguire, David. 1991. "An Overview and Definition of GIS." *Geographical Information Systems: Principles and Application* 1: 9–20.

Mark, David M., Nicholas Chrisman, Andrew U. Frank, Patrick H. McHaffie, and John Pickles. 1997. "The GIS History Project." http://www.ncgia.buffalo.edu/ncgia/gishist/bar_harbor.html.

McHaffie, Patrick. 1995. "Manufacturing Metaphors: Public Cartography, the Market and Democracy." In *Ground Truth*, edited by John Pickles, 113–29. New York: Guilford Press.

NASA. n.d. "Geographic Information Systems (GIS)." EarthData: Open Access for Open Science. Accessed February 8, 2022. https://www.earthdata.nasa.gov/learn/backgrounders/gis.

Ozernoy, Vladimir M., Dennis R. Smith, and Alan Sicherman. 1981. "Evaluating Computerized Geographic Information Systems using Decision Analysis." *INFORMS Journal on Applied Analytics* 11, no. 5: 92–98.

Parker, H. D. 1988. "The Unique Qualities of a Geographic Information System: A Commentary." *Photogrammetric Engineering and Remote Sensing* 54, no. 11: 1547–49.

Smith T. R., S. Menon, J. L. Starr, and J. E. Estes. 1987. "Requirements and Principles for the Implementation and Construction of Large-Scale Geographic Information Systems." *International Journal of Geographical Information Systems* 1: 13–31.

Star, Jeffrey, and J. E. Estes. 1990. *Geographic Information Systems: An Introduction*, 2–3. Englewood Cliffs, NJ: Prentice Hall.

USGS. n.d. "What Is a Geographic Information System (GIS)?" Accessed February 8, 2022. https://www.usgs.gov/faqs/what-geographic-information-system-gis.

Wilson, Matthew. 2015. "Celebrating the Advent of Digital Mapping." *ArcNews*, Winter 2015. https://www.esri.com/about/newsroom/arcnews/celebrating-the-advent-of-digital-mapping.

Wright, Dawn J., Michael F. Goodchild, and James D. Proctor. 1997. "Demystifying the Persistent Ambiguity of GIS as 'Tool' versus 'Science.'" *Annals of the American Association of Geographers* 87, no. 2: 346–62.

Projections and Coordinate Systems 2

Learning Objectives

- Articulate the importance of location information and explain how the information is recorded and stored in a GIS
- Understand the relation between the field of cartography and GIS
- Understand what map scale is and how it is represented, and articulate the difference between large-scale and small-scale maps
- Compare different models used to approximate the shape and size of the earth and the importance of these models in GIS
- Understand and explain map projection surfaces, distortions, and their use in GIS
- Understand coordinate systems and identify the difference between projected coordinate systems and geographic coordinate systems
- Recognize different coordinate systems used within the US and worldwide

Location Information

As discussed in chapter 1, a GIS is capable of doing two things simultaneously: display real-world features on a map and present detailed characteristics of the feature when called through point and click. The characteristics or attributes of the feature are stored in a table called an *attribute table*. A GIS stores the information, linking appropriate information in the attribute table to the appropriate feature, and presents it as a unified data on the map. A GIS thus stores both the map information and the attribute information of feature characteristics. However, the fundamental way a GIS communicates with its user is through a map. The map can be on your computer screen or smartphone or any other device.

Consider a scenario where you have decided to meet a friend for lunch at a new Thai restaurant located in a large shopping area. How will you communicate to your friend where to meet? You must specify the location of the restaurant. You can do that in a number of ways. You can specify the name of the restaurant, which can give your friend a basic idea of its location provided your friend is familiar with the restaurant and the area. If your friend is new in town and is not familiar with the area, she does not have all the requisite information to know your whereabouts. Now if you tell her that you are in the Landings Bay area, you have given her another type of location information. However, Landings Bay is a large area and she still lacks the requisite information to locate you. You can give her the name of the shopping area or the street where you have parked or the freeway she must take to get to the area or the exit she must take from the freeway. These are all different forms of location information that will provide your friend some sort of a reference of where you are. It may help your friend to get to the shopping area, but she will still not know exactly where you or the restaurant is located.

In order for your friend to know your or the restaurant's precise location or for her to be able to get to the general area, the different types of information that you shared with your friend to describe your whereabouts must make sense. And for that, we would need a system that acts as a reference for the information that you provided. This would be a map on the friend's smartphone map application or a map available through the car's navigation system or a map of the shopping area available outside the parking structure. We all face such situations in our daily life where we are searching for a location. It could be the location of a shop, community, school, hospital, or park. In any of these cases the information can be described by using relative location, which is location information with implicit geographic reference described relative to another location.

However, many applications such as emergency management, 911, and disaster management cannot function with relative locations because relative locations describe a general area, not precise points. Here absolute location information is required, described using coordinates. Moreover, relative locations do not cover every point on the surface of the earth. Think of how you would describe a location in the middle of an ocean or a river or a forest. There are no zip codes or street names and numbers. Similarly, there is no uniformity in the way relative locations are described. Every region and country follows its own standards. So if locations are described using relative location, a location from one region may not be understood by someone from a different region.

You may have tabular data with detailed information about the characteristics of a feature. But without location information you cannot map anything to display or visualize patterns or distribution, nor can you conduct any spatial analysis. For instance, if we go back to the example of the Thai restaurant, you want to go to a restaurant that is close to where you live, has at least a four-star ratings from customers, and offers dine-in facilities. All of this information about all the Thai restaurants in your city is stored in a table. You can review the information one by one in the table. However, if you have over 100 Thai restaurants in your city, it would be an arduous task to narrow down all restaurants in your area and then further narrow down the information based on your criteria. You will still not be able to visualize the information to assess any other information such as accessibility. Viewing this information on a map makes it simpler and far more convenient. However, in order for this information to be displayed on a map and for you to be able to narrow down your search based on your criteria, there needs to be a system that would enable the information to be displayed on a map.

Location information is thus the essence of geospatial data and the key characteristic of geospatial technology and GIS. Geospatial technology cannot function without location information. Therefore, there has to be a way to measure location at every point on the earth so that it can be referenced and identified on a map. Moreover, this system used to measure and describe location must be created using some sort of a uniform framework or reference so that it can be utilized and understood by everyone, no matter where they are. A system created using a uniform framework will also make it possible to compare location at one point with location at another point on the earth.

Finally, when we talk about features on the surface of the earth, we are referring to a three-dimensional space. Representing and examining features on a map, whether it is a paper map or a digital map, refers to a two-dimensional space. Thus, we also need a methodology or a system that allows us to transform earth features on a surface that we can visualize and examine. This system, framework, or methodology is what we will be covering in this chapter. And to discuss these topics we will visit cartographic fundamentals and methods.

Cartography and Maps

Cartography is the science of mapmaking. It is the theory behind map construction and use. The field of cartography was particularly influenced by the advent of geospatial information and technology. Nonetheless, key concepts of cartography are very important in a GIS, such as an understanding of the shape and size of the earth, datum, map projections, coordinate systems, and scale. These concepts are the building blocks of mapping. In today's digital world we are all accustomed to interacting with digital maps, and principles of cartography have evolved to adapt to the modern world. However, theories of cartography have a long tradition going back to the Greek civilization when important cartographic principles were developed by Greek scholars such as Ptolemy, Herodotus, Anaximander, and Eratosthenes. Even today we use some of those principles.

A map is a graphical representation of the earth used to represent features, patterns, activities, and phenomena on the earth's surface using different symbols. A map is an abstraction that does not show every element of the earth. Rather, it provides us a generalized view of the earth and the environment. If you look around at your surrounding, you will notice that they consist of too many features, patterns, activities, events, and phenomena generating an infinite amount of data. It is practically impossible to measure, store, process, and map an infinite amount of data. Therefore, maps present a generalized view of the earth representing selective features and phenomena.

For instance, take the example of the three maps in figure 2.1. All of the maps show the town of Conroe in Texas. However, you will notice that the town looks different in each of the maps. The amount of features and their shape change in each map and so does the boundary of the town. Each map has also captured a different amount of detail. As you go from the top to the bottom, the features captured and displayed on the map increase. So does the amount of detail shown on the maps. Thus, as you go from the top to the bottom, you see more features and details on the maps. However, do you think that the map at the bottom displays every feature that exists in the town of Conroe? That is not possible. The map presents a generalized view of selective features only.

The increase in the amount of detail captured as you go from the top to the bottom map can be explained using an important cartographic concept called *scale*. If you look at any map, whether it is a paper map or a digital map, it will be smaller in size than the surface it represents. All maps represent a systematic reduction of reality, and the amount of reduction in size shown on a map is known as map scale. Map scale is defined as the ratio of a distance on the map to the same distance on the ground. A map scale tells you the relationship between these two distances. This relationship can be described in three ways: representative fraction, verbal scale, or a graphic scale bar.

One of the most common ways to express scale is by using representative fraction (RF). RF is the ratio between map and ground distance. It is written as 1/x or 1:x. The unit of measurement is same for the numerator and the denominator. 1 is the distance on the map and x is the distance on the ground. So a map scale expressed as 1:5000 or 1/5000 means 1 inch on the map represents 5,000 inches on the ground or 1 centimeter on the map represents 5,000 centimeters on the ground.

A *verbal scale* is a sentence describing the relationship between map and ground distance. Verbal scale is expressed as "1 inch is equal to 5,000 inches" or "1 inch is

FIGURE 2.1 Generalized maps of Conroe, Texas.

Sources: Esri, HERE, Garmin, Intermap, Increment P Corp., GEBCO, USGS, FAO, NPS, NRCAN, GeoBase, IGN, Kadaster NL, Ordnance Survey, Esri Japan, METI, Esri China (Hong Kong), © OpenStreetMap contributors and the GIS User Community

equal to 50 miles." Verbal scale is descriptive in nature. However, it is not necessary that the unit of measurement be the same on both sides. Often a mix of units is employed to use familiar units on both sides such as "1 inch is equal to 50 miles." Therefore, verbal scales can be confusing for some and can make it difficult to compare two

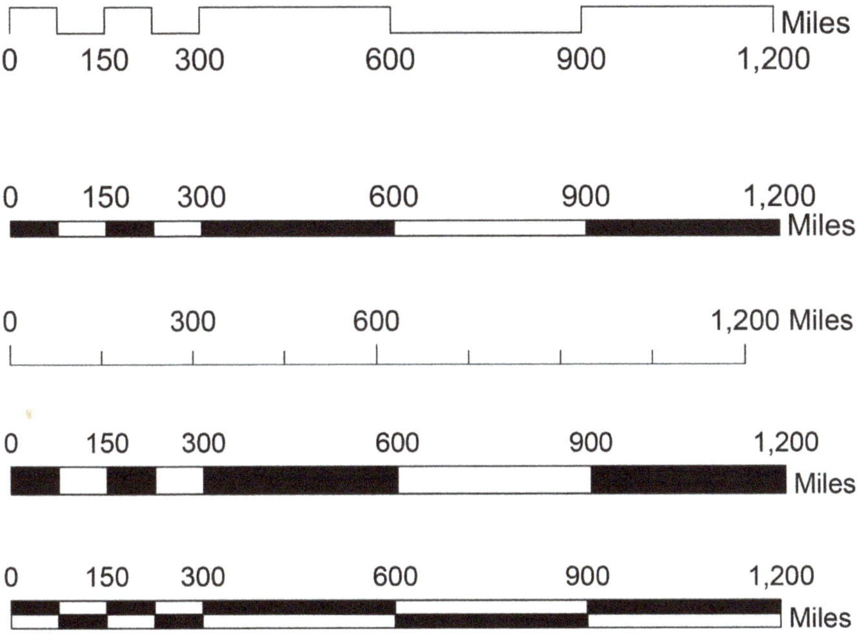

FIGURE 2.2 Graphic scale bar.

Credit: Scale bar created using Esri ArcGIS

maps. If the units of measurement are different on both sides in a verbal scale, the units must first be converted to a RF, discussed above, before comparing maps.

The third way to show a map scale is by using a *graphic scale bar*, shown in figure 2.2. It looks like a small ruler printed on a map. If a map is enlarged or reduced, something that happens a lot when working with geospatial data in a GIS, the graphic scale bar changes size proportional to the size of the map. An RF or verbal scale will not do that. A graphic scale bar is therefore especially useful in web maps.

Let's revisit the three maps in figure 2.1. Each map is drawn at a different scale: the scale of the map at the top is 1:200,000; the map in the middle is 1:50,000; and the map at the bottom is 1:5000. When you compare the three maps, the map at the top is a small-scale map compared to the other two maps. The map at the bottom is a large-scale map compared to the two maps at the top. So we are going from a small-scale to a large-scale map from top to bottom (figure 2.3). In other words, the closer the ratio is to 1:1, the larger the scale. It is all in the details of the map and the area covered. A small-scale map covers a large area with less detail, whereas a large-scale map covers a small area with great detail.

Map scale plays an important role in how much detail and the features a map will display. For instance, on a small-scale map that shows the entire United States, only state capitals will be displayed as dots, whereas on a large-scale map the focus will be on the state of Texas and more cities in Texas will be displayed along with other features such as roads and freeways. On a larger-scale map, the focus will be on the city of Houston and dots will represent key features in the city along with roads, parks, and streams. If the map scale further increases, one can focus on a neighborhood in Houston, and buildings and parks in the neighborhood will now be displayed as polygon shapes instead of dots along with inner neighborhood streets. As the scale keeps getting larger, different types of information can be shown on the map. Thus the choice of map scale is key in determining what and how much information a map will display and how the information will be represented on a map.

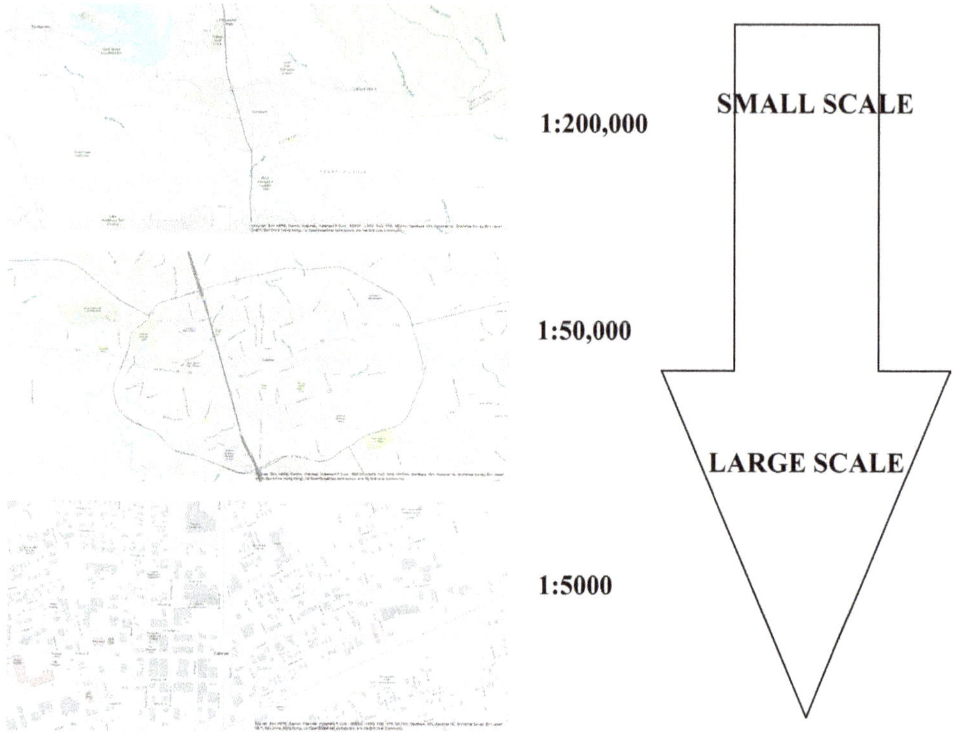

FIGURE 2.3 Small-scale versus large-scale map.

As you go from the bottom map to the top map in figure 2.3, there is a systematic reduction in scale and proportional reduction in the number of features and details that you see on the map. In the bottom map, buildings and parks are represented as polygon shapes, and one can clearly visualize the inner streets and street network. In the map in the middle, one can no longer visualize the building polygon shapes other than the large school buildings, and instead of the inner streets one can now only visualize the main roads in Conroe. And finally in the map at the top, none of these details can be visualized other than the boundary of Conroe and the main freeway. This decline in scale and corresponding reduction in detail as we move from a large-scale to a small-scale map is known as **map generalization**. This is applicable to both paper maps and geospatial data in digital formats.

When working with geospatial data in a GIS, a user can zoom in and zoom out and accordingly the map can be enlarged or reduced. A GIS user will not necessarily work with the data at the same scale at which it was created. A GIS provides its user the capability to focus on a specific area at a different scale than the original data. Digital maps on your smartphone's mapping application allow you to visualize a series of maps that systematically become large or small in scale. These are known as *multiscale maps* because the maps are at different or multiple scales with a varying amount of details and features represented by the individual scales. What you see is a seamless map, allowing you to zoom in and zoom out of the map. As you zoom in on the map, more detail and more features tied to the individual scale are shown. As you zoom in further, there is an increase in detail displayed on the map. Multiscale maps have been designed in such a way that the degree of generalization changes with each map scale. For instance, in the three maps of Conroe in different scales, the large-scale map displays the main freeway. As one zooms into the map, one can visualize the highways, arterial streets, inner streets, and the street network.

Shape of the Earth

Location can be measured, recorded, and shown on a map in many ways. However, before any of these tasks can be accomplished, we must first ascertain the shape and size of the earth, which is done using mathematical calculations. In order to do so, we must define a geometrical shape that approximates the shape of the earth. The quest to determine the shape of the earth goes all the way back to the Greek era. Humankind has known for a very long time that the earth is spherical in shape. This knowledge is attributable to the Greek philosopher Aristotle, whose observations during the fourth century BC convinced philosophers and scholars in subsequent centuries of the spherical shape of the earth.

Calculations leading to an estimate of the size of the earth can also be attributed to Greek philosophers. The first scientific calculation of the circumference of the earth was conducted by the Greek scholar Eratosthenes in 250 BC. The earth was considered to be a sphere until the mid-1600s. Crucial work conducted by prominent scholars and scientists during the late 1600s through the 1800s in the UK and France led to the development of key scientific theories that established the theories of gravity and the earth's rotation. These theories were highly relevant in subsequent work establishing that the shape of the earth closely resembled an oblate ellipsoid, or spheroid, whose equatorial radius is 13 miles longer than its polar radius. Thus, it was established that the earth resembled a shape that was larger at its center than at its poles. The equatorial radius is called the *semi-major axis* and the polar radius is called the *semi-minor axis* (figure 2.4). You can think of an ellipsoid as a three-dimensional shape created by rotating a two-dimensional ellipse.

However, the shape of the earth is neither a sphere nor a precise ellipsoid. The surface of the earth is uneven and irregular, full of varied features such as mountains, trenches, oceans, and valleys. The varied landmass causes differential gravitational pull on the earth's surface. For instance, mountains and hills have more mass than

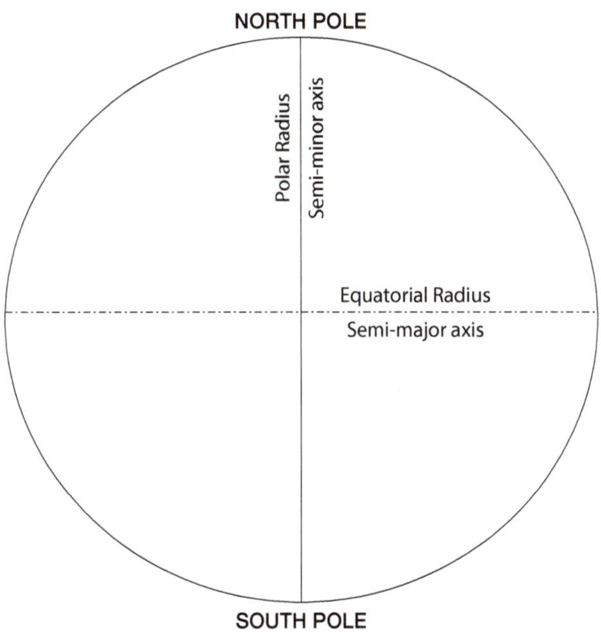

FIGURE 2.4 Earth as an oblate ellipsoid/spheroid. Equatorial radius is 13 miles (21 kilometers) longer than the polar radius.

valleys, causing stronger gravitational pulls in mountainous regions. This variation in size, shape, and mass distribution of features on the earth's surface causes differential gravitational pull on the earth. Due to the differential gravitational forces, some parts of the earth do not align exactly like an ellipsoid. Therefore, another model to approximate the shape of the earth was proposed called the geoid, which places the earth's surface at mean sea level in an open ocean without any disturbances, waves, or swells. A geoid is an irregularly shaped surface which defines zero elevation. The United States Geological Survey (USGS) defines a geoid as "an imaginary sea level surface that undulates or has a wavy surface extending through the land masses over all of the earth including oceanic areas." Figure 2.5, taken from the National Oceanic and Atmospheric Association (NOAA) website, shows the United States geoid where areas in yellow and orange have a slightly stronger gravitational field because of the Rocky Mountains.

There is a significant difference between an ellipsoid and a geoid, as the ellipsoid assumes that the earth's surface is homogeneous and smooth, whereas the geoid accounts for the varied densities in the earth and landmass on the surface that causes different gravitational pull on the earth's surface. A geoid model approximates the earth more accurately. Figure 2.6 shows a comparison between the geoid model and an ellipsoid.

However, the geoid model involves very complex mathematical calculations. In order to simplify the model, various ellipsoids/spheroids are calculated to approximate the geoid model. Thus an ellipsoidal model has been used as a general representation of the earth's shape.

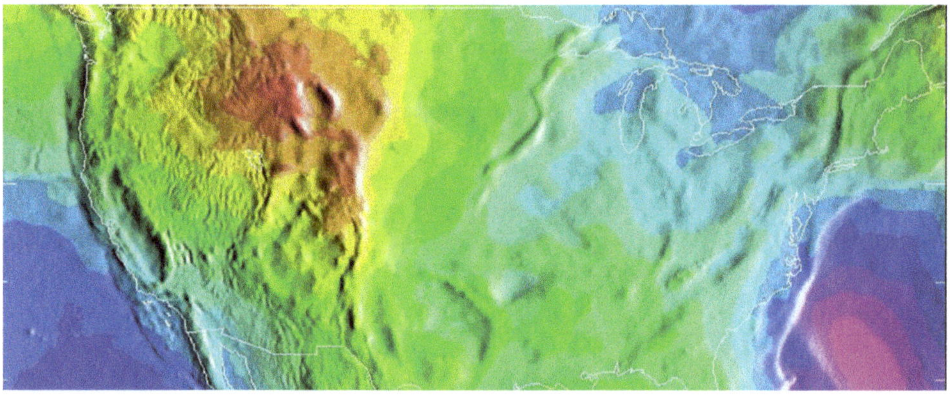

FIGURE 2.5 United States geoid.

Credit: NOAA; public domain

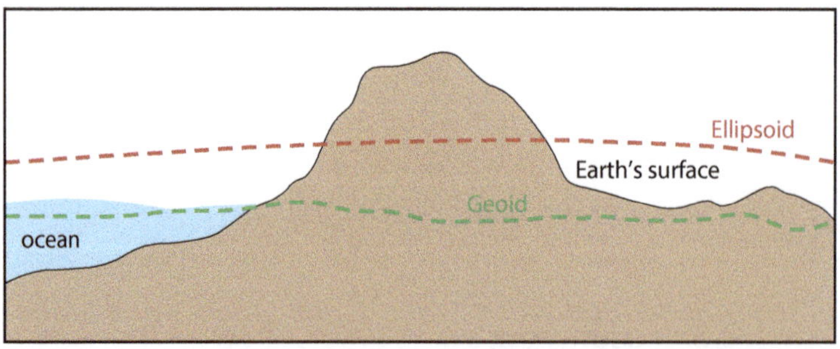

FIGURE 2.6 Comparison between the geoid model and an ellipsoid.

Credit: USGS; public domain

Several efforts have been made since the 19th century to measure the shape and size and determine ellipsoids representing the earth. Advances in the field of geodesy, surveying, space technology, and global positioning system (GPS) have made it possible for scientists and researchers to more accurately make these measurements and also account for the most minor changes in the earth's surface in the measurements. Many ellipsoids have been established and are used around the world. Table 2.1 is a short list of ellipsoids developed and adopted in different parts of the world. Ellipsoids are differentiated from each other by the lengths of their semi-major and semi-minor axes. A particular ellipsoid works well for a specific region or geographic area because that ellipsoid will approximate a geoid model most accurately for that area.

As shown in table 2.1, new measurement efforts have created new ellipsoids and have been adopted as countries move from an older model to a new ellipsoidal model. For instance, the Clarke 1866 ellipsoid was a good fit for North America during the late 1800s and was used as a reference for further computations to calculate latitude and longitude for mapping purposes in North America. Later in 1972, the United States adopted the WGS72 ellipsoid. This was later replaced by a new ellipsoid, the Geodetic Reference System of 1980 (GRS80), that was adopted at the XVII General Assembly of the IUGG in Canberra, Australia, in December 1979. Unlike the Clarke 1866 ellipsoid, the GRS80 ellipsoid was adopted globally and is hence a global ellipsoidal system. The GRS80 was a geocentric ellipsoidal system

TABLE 2.1 Ellipsoid Measurements

Name	Date	Semi-major Axis (m)	Semi-minor Axis (m)	Area of Use
Plessis	1817	6,376,523	6,355,862.9333	France
Everest	1830	6,377,299.365	6,356,098.359	India
Airy	1830	6,377,563.396	6,356,256.909	Great Britain, Ireland
Bessel	1841	6,377,397.155	6,356,078.963	Central Europe, Chile, Indonesia, Japan
Clarke	1866	6,378,206.4	6,356,583.8	North America
Clarke	1878	6,378,190	6,356,456	North America
Clarke	1880	6,378,249.145	6,356,514.870	Africa, France
Helmert	1906	6,378,200	6,356,818.17	Egypt
Hayford	1910	6,378,388	6,356,911.946	USA
International	1924	6,378,388	6,356,911.946	Global
Krasovsky	1940	6,378,245	6,356,863.019	Soviet Union
WGS66	1966	6,378,145	6,356,759.769	USA
Australian	1966	6,378,160	6,356,774.719	Australia
South American	1969	6,378,160	6,356,774.719	South America
WGS72	1972	6,378,135	6,356,750.52	USA
GRS80	1979	6,378,137	6,356,752.3141	Global ITRS
WGS84	1984	6,378,137	6,356,752.3142	Global GPS

based on the earth's center of mass. These measurements were further refined in 1984 making use of technological advancements and new methods of calculation to develop the more recent World Geodetic System of 1984 (WGS84). WGS84 is a geocentric or earth-centered system adopted worldwide and was developed specifically as a reference system for global positioning systems.

Calculation of an ellipsoidal model is very important because it allows further mathematical calculations to develop a model with which to measure location and elevation on the earth's surface. This mathematical model is known as datum. A **datum** is an abstract system that provides a reference framework to measure locations on the surface of the earth relative to the surface of an ellipsoidal. This is important information for surveyors, geodesists, and cartographers. For instance, when giving someone directions to get to your location, you would provide some sort of a starting point that they are familiar with, like a landmark or a cross street, to reach you. You can think of a datum as that starting point that acts as a reference. Surveyors and geodesists use datums to create starting points for maps and surveys that require accurate and consistent location information.

Three main datums are used in the United States: **horizontal datum**, **geometric datum**, and **vertical datum**. A horizontal datum is a system created using a collection of points on the earth whose latitude and longitude have been accurately determined by NOAA's National Geodetic Survey (NOAA 2023). Horizontal datums are used to collect and measure positions on the surface of the earth. A geometric datum is a system for collecting and measuring positions relative to an ellipsoidal model of the earth. A vertical datum is a system created using a collection of points on the earth whose height has been accurately determined with respect to a nationally defined reference surface such as mean sea level. Vertical datums are used to measure and describe land elevations.

Hundreds of datums have been developed and adopted and are used by countries the world over. Some datums are local and apply to a specific region or continent, while others are global and provide measurements for the entire world. Datums may be geocentric, where they are earth-centered, or they can be local depending on an ellipsoidal model. In a local datum the ellipsoid aligns to closely match a specific point on the earth's surface. In this case, a point on the surface of the ellipsoid is matched to a particular position on the surface of the earth and serves as the starting point of the datum. Table 2.2 shows some of the datums used in different parts of the world and the ellipsoids they are based on.

Three datums are commonly used in the United States: NAD27, NAD83, and WGS84. The North American Datum of 1927 (NAD27) was developed using the Clarke 1866 ellipsoid for measurements of North America and was used for mapping and surveying purposes until the early 1980s. The datum measurements were centered at Meades Ranch in Kansas and involved several thousand control points across the United States. These control points were later corrected in the 1980s for distortions that had accumulated over time, when the United States adopted a new ellipsoid, the GRS80. With the arrival of satellite technology and advanced measuring devices and calculation methods, there was a need to update NAD27. Consequently, the National Geodetic Survey developed the North American Datum of 1983 (NAD83) for the North American continent. The new datum was developed using a dense mix of terrestrial and Doppler satellite data along with several hundred thousand control points. NAD83 is the most current datum used in North America today. However, since 2007 the National Geodetic Survey has been working on a modernization program that will replace NAD83 with four new terrestrial reference frames. The new reference frames will be accessed via reference stations that continuously receive Global Navigation Satellite System (GNSS) data such as the GPS. The

TABLE 2.2 Datums Used in Different Parts of the World

Datum	Ellipsoid	Origin and Use
OSGB36	Airy 1830	Great Britain, Ireland
WGS72	WGS72	Global
WGS84	WGS84	Global
ED50	International 1924	Potsdam
PSAD56	International 1924	La Canoa, Venezuela
NAD27	Clarke 1886	Meades Ranch, Kansas
NAD83	GRS80	Global
OSSN70	Airy	Herstmonceux, UK
MINNA	Clarke 1880 RGS	Minna, Nigeria
ETRS89	GRS80	Europe
GDA94	GRS80	Australia
PZ-90	Krasovsky 1940	Russia
Indian Geodetic Datum	Everest spheroid	Kalyanpur, Central India

modern spatial reference system is scheduled to be available in 2026. The WGS84 datum was developed by the US Department of Defense in 1987 for use by global positioning systems using the WGS84 ellipsoid. Unlike NAD83 that is predominantly used in the United States and Canada, WGS84 datum is used all over the world.

As a GIS user working with maps or datasets from different sources or time periods, knowledge about datum is very important because data created on one datum may not align with data created using another datum. Figure 2.7 displays the shift between NAD27 and NAD83 shown with a dashed cross at the corner of the map. Let's say your class has been assigned to work on a GIS project focusing on a specific area. You acquire all the datasets that you need for the project. One of the datasets is using NAD27 datum and the other is using NAD83 datum. When you try to display

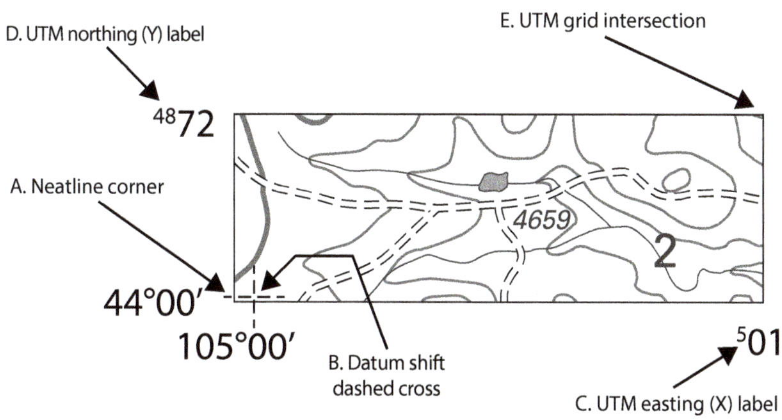

FIGURE 2.7 Southwest corner of a USGS quadrangle in northeastern Wyoming. Shift between NAD27 and NAD83 shown with a dashed cross at the corner of the map.

Credit: USGS; public domain

the two datasets on any GIS application, they will not necessarily align. You will notice that the dataset using NAD27 datum is off by a few hundred meters from the data using NAD83. This is because both datasets are using different reference information for measurements. Therefore, when working with geospatial data, the best approach is to keep all your geospatial datasets in the same datum.

The process of changing measurements from one datum to another is known as datum transformation. This involves changing ellipsoids and a complex mathematical computation using transformation algorithms. GIS softwares and preprocessing software packages that accompany GPS units come preloaded with datum transformation algorithms.

Map Projections

Now that we have established that the shape of the earth can be estimated to roughly resemble an ellipsoid, or spheroid, there is no doubt that real-world features exist in a three-dimensional space. Nonetheless, methodological approaches to measure and calculate distance, direction, and area on a spherical surface are complex and difficult. A globe is the only true representation of the earth's surface on a three-dimensional space. However, unlike paper maps or maps on your smartphone, globes are not easy to carry. They also cannot show the kind of detail that paper maps can display. Think of the kind of details you might find on a road map or a reference map or a topographic map. A globe cannot provide that level of detail. Therefore, maps are created to understand the world in a two-dimensional space.

Whether you work with paper maps or digital maps on a computer screen or your smartphone, you are interacting with the map in a two-dimensional space. Thus, features and their locations are represented from a three-dimensional space on a two-dimensional space. This transformation of real-world features and their locations from their places on the earth's surface onto a flat surface is called map projection. This transformation from a three-dimensional surface to a two-dimensional surface involves complex geometrical calculations and is a must in order to be able to work with geospatial data.

So what does it mean to transform to a two-dimensional surface? Does it mean directly transforming onto a sheet of paper or a computer or mobile device screen? The answer is no, because an ellipsoid or a spheroid that approximates the shape of the earth is such that it cannot be flattened without geometrical distortions. Mathematicians call such a surface an *undevelopable surface*. In mathematics a developable surface is a smooth surface with zero curvature that can be rolled or unrolled into a flat sheet. Cylinders, cones, and planes are considered developable surfaces. So the whole purpose of map projection is to move real-world features and their locations from a mathematically undevelopable surface to a developable surface involving a series of geometrical transformations.

Map Projection Surfaces and Parameters

All of us at some point have played a game where we cast hand shadows on a wall. You can think of a projection surface as similar to the surface of the wall on which the shadow is cast. The size and shape of the shadow will be influenced by the light source and the wall surface. How close or far away the hand is from the light source will determine the type of shape your hand shadow will display. Similarly, whether the shadow is cast on a corner wall or a curved wall or on a flat wall surface will have an impact on the shape and size of the shadow. You can think of the earth as your hand, and the shadow that is cast on the wall is the map projection.

Imagine a transparent or see-through globe with a light bulb inside it. If you turn the light bulb on, it will cast a shadow that will show the continents and lines of latitude and longitude. If you wrap a piece of paper on this see-through globe and somehow extract the lines of latitudes and longitudes and the continents from the shadow on the paper and then unroll the paper, what you would essentially have is a globe projected onto a two-dimensional surface. The flat surface on which the earth's features and its locations are projected is called a developable surface. Developable surfaces that are used for this purpose are a cylinder, a cone, and a flat plane. Accordingly, there are three projection families based on the developable surface: cylindrical projection, conic projection, and azimuthal/planar projection, shown in figure 2.8.

The appearance of the map projection cast by the globe will depend on the type of projection surface (cylindrical, conic, or azimuthal/planar projection), how the surface is placed with respect to the globe, and the location of the light source. In a cylindrical projection the globe is projected on a cylinder, where the cylinder touches the globe along a small circle or slices through the globe. Similarly, in a conic projection the globe is projected on a cone, where the cone either touches the globe along a small line or slices through the globe. In a planar projection the globe is projected on a flat plane that either touches the globe at a point or slices through the globe.

Depending on how the projection surface and the globe are arranged, we can have a secant projection or a tangent projection, as shown in figure 2.9. In a secant projection the projection surface will intersect or cut or slice through the globe, whereas in a tangent projection the projection surface will touch the globe either at a point or along a line. The place where the globe touches the projection surface is called the point or line of tangency. The light source can be positioned at infinity or at the center of the globe, or opposite the point of tangency or parallel along the axis. Accordingly, we have orthographic, stereographic, or gnomonic planar projections, and in the case of cylindrical or conic projections we have central cylindrical or cylindrical equal area or central conic projections.

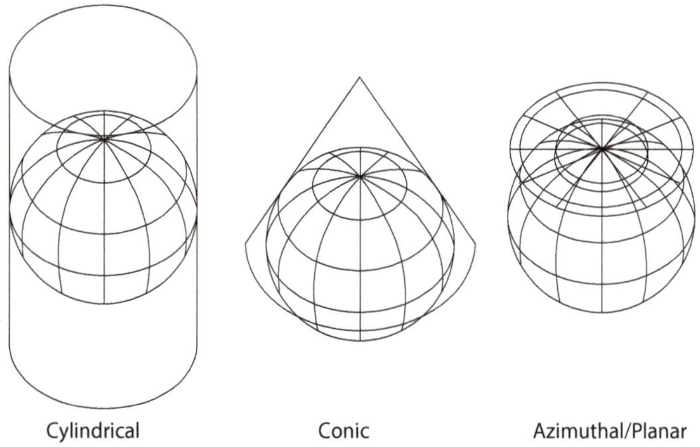

FIGURE 2.8 Map projections based on developable surface.

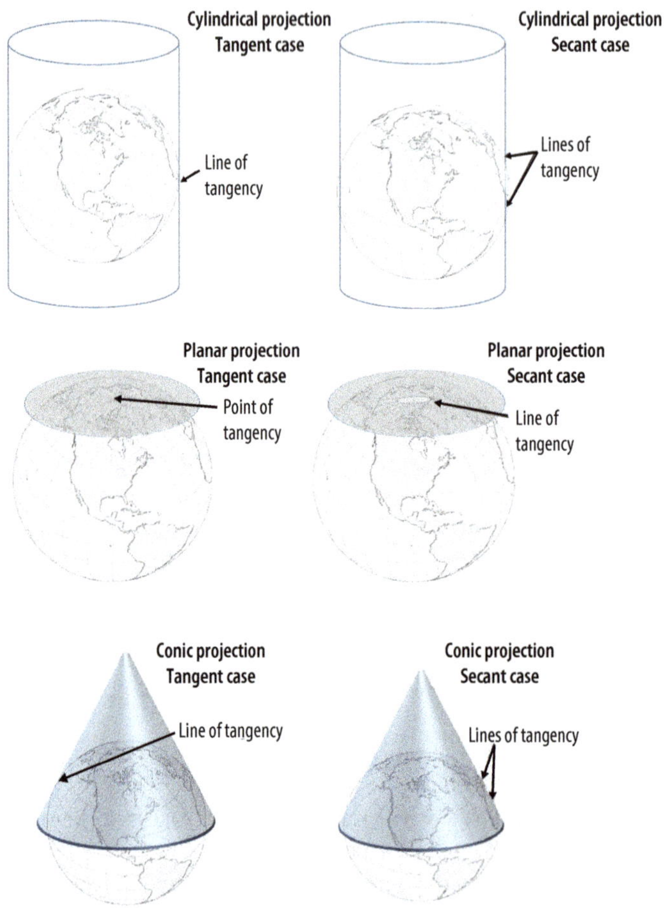

FIGURE 2.9 Tangent and secant projections.

Another parameter used to distinguish projection systems is the aspect. An aspect refers to the location of the point or line of tangency on the projection surface. In other words, it refers to the location where the projection surface is placed on the globe. Projection surfaces can be placed at or along the equator, at poles, or at any angle (figure 2.10). Accordingly we have equatorial, polar/transverse, or oblique projections. Table 2.3 summarizes the different projections based on their parameters.

Map Projections and Distortions

All map projections distort features on the earth's surface during the transformation process. When cartographers construct projections, this distortion is assigned to one of four categories based on the properties that are preserved during the transformation from a three-dimensional surface to a two-dimensional surface. These properties are area, shape, distance, and direction.

When the relative size of regions on the earth is preserved everywhere on the map, the projection is called an **equivalent,** or **equal-area, projection**. Equivalent projections preserve areal relationship. Albers equal area and cylindrical equal area are examples of equal area/equivalent projection. Many GIS application require area calculations in different analyses. In such cases it is important that the geospatial data

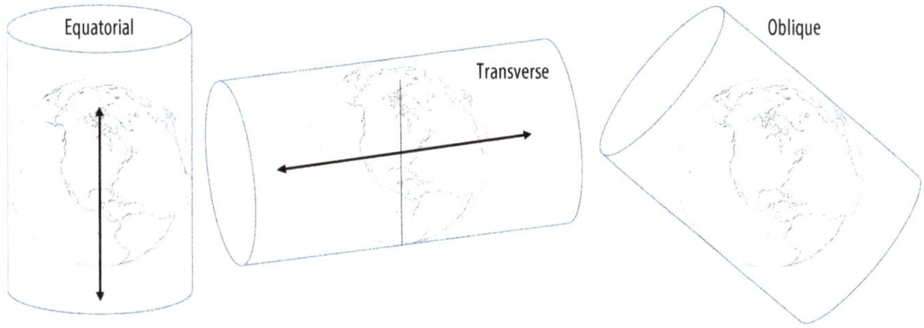

FIGURE 2.10 Projections based on aspect.

TABLE 2.3 Summary of Different Projection Parameters

Parameters	**Names**		
Surface	**Cylinder**	**Cone**	**Plane**
Surface Relation	*Tangent*: projection surface will touch the globe *Secant*: projection surface intersects/cuts/slices through the globe		
Light Source	*Central cylindrical*: at center of globe *Cylindrical equal area*: light with parallel rays along the polar axis	*Central conic*: at center of globe	*Orthographic*: at infinity *Stereographic*: on surface of globe opposite the point of tangency *Gnomonic*: at center of globe
Aspect	*Equatorial*: point/line of tangency at/along the equator *Transverse*: point/ line of tangency at or encircles the poles *Oblique*: point/line of tangency at any other angle	*Equatorial*: point/line of tangency at/along the equator *Transverse*: point/ line of tangency at or encircles the poles *Oblique*: point/line of tangency at any other angle	*Equatorial*: point/line of tangency at/along the equator *Polar*: point/line of tangency at or encircles the poles *Oblique*: point/line of tangency at any other angle

being used is projected using an equal area/equivalent projection where the area is mapped equally across the surface. Here all regions on the map will be shown with their correct area as on the ellipsoid. Figure 2.11 shows a map of the world created using equal area projection. As you can see in the top and bottom portions of the map, the shape of the regions is distorted. This is because on a globe, lines of longitude converge at the poles. However, when mapping using equivalent projection the convergence points are mapped as lines. The poles are thus displayed as lines like the equator, stretching all the regions on the map in an east–west direction. In order to maintain the same area throughout the map, shape is distorted.

Now compare figure 2.11 with figure 2.12. The map in figure 2.12 has been created using Mercator projection, which is a conformal projection. A **conformal projection** preserves the property of shape. Lambert conformal conic is another example of conformal projection. In conformal projection, angles on the globe are preserved on the map, thereby preserving shape. A point feature on the globe will also appear as a point on the map. Conformal projections preserve shape in small areas

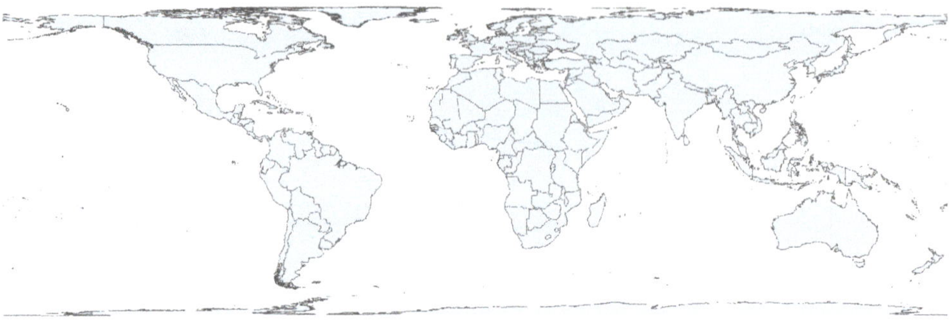

FIGURE 2.11 Equivalent projection: cylindrical equal area.

ArcMap interface courtesy of Esri, Inc. Data Sources: Esri, Garmin International, Inc.; U.S. Central Intelligence Agency (The World Factbook); National Geographic Society

FIGURE 2.12 Conformal projection: Mercator.

ArcMap interface courtesy of Esri, Inc. Data Sources: Esri, Garmin International, Inc.; U.S. Central Intelligence Agency (The World Factbook); National Geographic Society

by maintaining the same map scale in all directions. However, in order to maintain shape, size is distorted and so is area. This distortion in size will vary in amount throughout the map.

As you can see in the map in figure 2.12, created using Mercator projection, there is significant distortion in area, particularly in regions farther away from the equator. In this map, Greenland appears to be the same size as Africa, and Antarctica appears to be a massive landmass. However, in reality we all know that Greenland is much smaller in size than shown on the map, about the size of Mexico, and Antarctica is slightly larger than Canada in size. Nonetheless, because Mercator projection preserve angles on the globe in order to preserve shape, it is very useful for navigation purposes or any activity that involves guiding and analyzing motion of objects on the surface of the earth, such as nautical and aeronautical charts and maps. If you have ever used Google Maps, MapQuest, or OpenStreetMap, you have in fact utilized a map using Mercator projection, a conformal projection.

A projection that preserves direction from the center of the map to any other location is called a **true direction,** or **azimuthal**, **projection**. It is not possible to accurately display directions from all points on the earth on a flat map. In azimuthal projection, certain types of direction lines that are drawn directly from the center of the map are displayed as straight lines. However, it only preserves direction for one point at a time on which the projection is centered, for example, meridians emanating from the poles in the case of the North Pole gnomonic projection shown in figure 2.13. A straight line between any other two points will not preserve

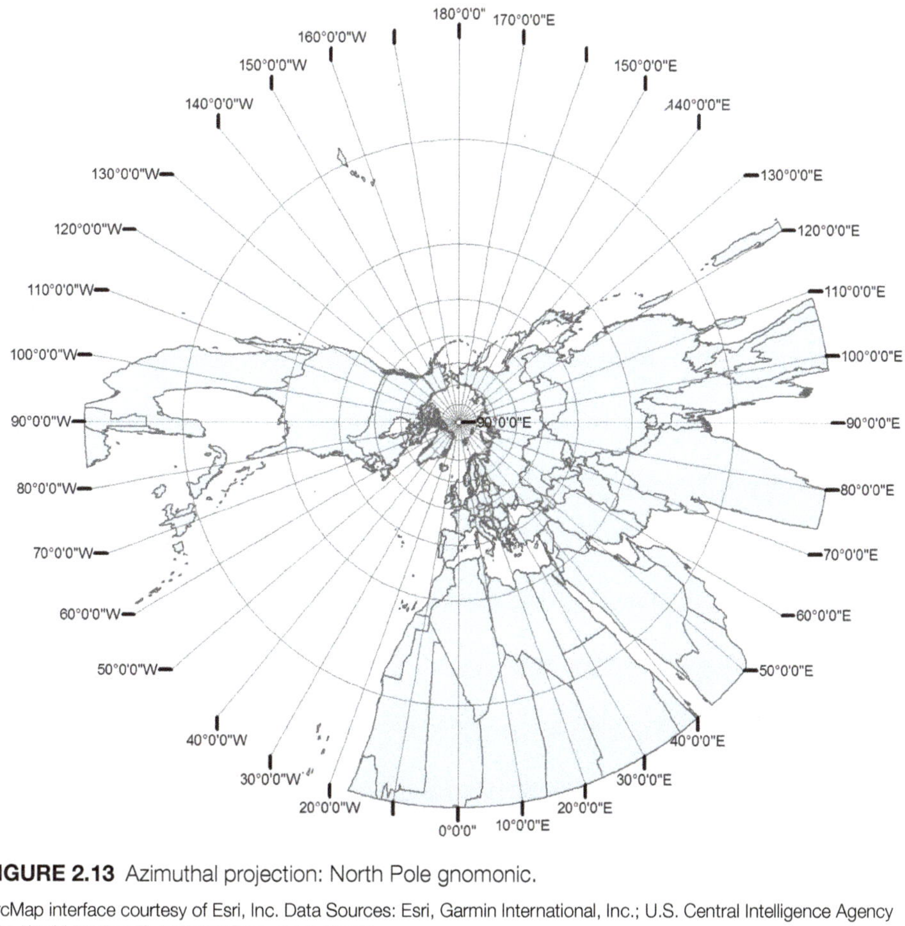

FIGURE 2.13 Azimuthal projection: North Pole gnomonic.

ArcMap interface courtesy of Esri, Inc. Data Sources: Esri, Garmin International, Inc.; U.S. Central Intelligence Agency (The World Factbook); National Geographic Society

true direction. Gnomonic, stereographic, Lambert azimuthal equal area, and orthographic are all examples of true direction/azimuthal projection. Azimuthal/true direction projections are used to create maps that show the shortest path between two points in a three-dimensional space, that is, the earth's surface, known as the great circle route. Such maps are very useful for air and water navigation.

A projection that preserves distance is called a **equidistance projection**. It is practically impossible to maintain true distance across a flat map. Equidistance projections preserve the spherical great circle distance and display true distance between one or two points and another point on a map along a line connecting those points. Azimuthal equidistant, equidistant conic, and equidistant cylindrical are examples of equidistance projection. The map shown in figure 2.14 was created with azimuthal equidistant projection. If you look closely at the map, you will notice that Australia and New Zealand cannot be recognized at all because their shape and size are distorted. However, true distance is preserved in this map between the center of the map, shown by a black dot, and any other point on the map. At the same time, distance between all other points will be distorted. Equidistant projections are very useful for long-distance route and travel planning and mapping purposes.

As discussed above, all map projections preserve either shape, area, direction, or distance. However, in the process of preserving one property, other properties

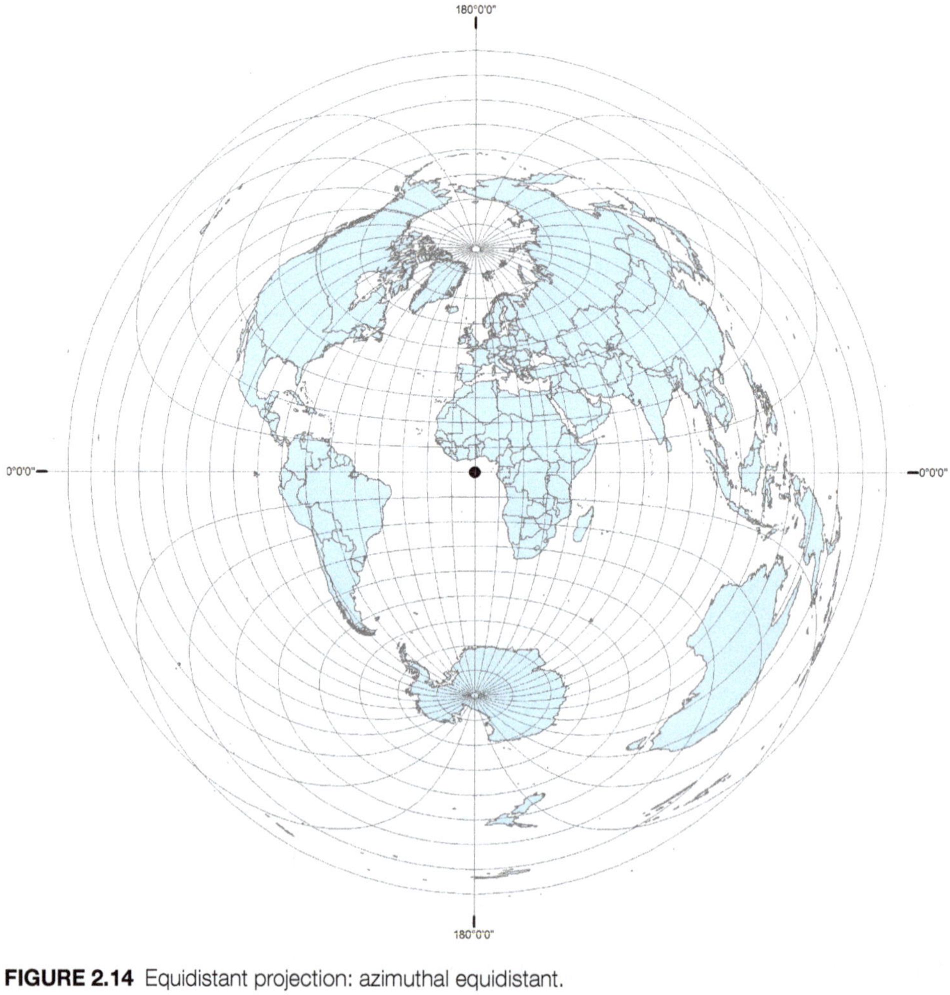

FIGURE 2.14 Equidistant projection: azimuthal equidistant.

ArcMap interface courtesy of Esri, Inc. Data Sources: Esri, Garmin International, Inc.; U.S. Central Intelligence Agency (The World Factbook); National Geographic Society

are greatly distorted, as you saw in the case of Mercator and azimuthal equidistant projections. No projection preserves all four properties. Distortions are most pronounced near the edges of the map, particularly when the map displays large areas and you move away from the center of the map. In order to solve this problem, cartographers have devised a new category of projection: *compromise projection*. These projections maintain a balance by not completely preserving any one property. Instead, they moderately distort all properties.

A popular example of such a projection would be a Robinson projection, shown in figure 2.15. Compromise projections spread the distortion evenly across the map. This way the distortion is not pronounced in any one map area. However, you now will not have a specific area on a map with zero distortion. If you look at the map in figure 2.15 that was created using Robinson projection, all of the continents appear similar in shape and size to what you would observe on a globe. Compare the map in figure 2.15 with the same map created using either Mercator projection that preserves shape (figure 2.12) or azimuthal equidistant projection that preserves distance (figure 2.14). World maps created by the National Geographic Society and many wall maps of the world are created using Robinson projection.

Another example of a compromise projection is the Winkel tripel shown in figure 2.16, created by the German cartographer Oswald Winkel. This projection minimizes distortion of shape, area, and distance. Hence, it is neither conformal nor equal area nor equidistant. Rather it is a compromise projection that minimizes all three geometric distortions.

Another category of projection is *interrupted projection*. In this type of projection, the globe is divided in order to minimize distortion within a section of the projection, and better parts of the projection are repeated within each section. The Goode interrupted homolosine projection shown in figure 2.17 is an example of an interrupted projection. It is a combination of Mollweide and sinusoidal projections creating a composite of pseudocylindrical equal area projections. It consists of six interrupted sections and uses Mollweide projection from 40°N or 40°S to the pole and sinusoidal projection from the equator to 40°N or 40°S, and it shows discontinuity where both of the projections meet. In creating maps using these projections, the continuity of the earth's surface is lost by interrupting the projection.

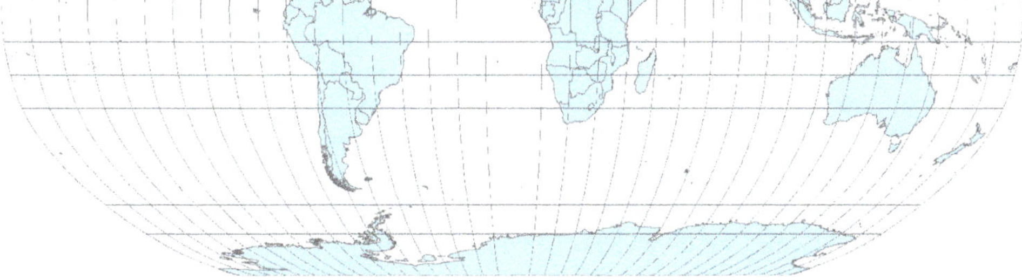

FIGURE 2.15 Compromise projection: Robinson.

ArcMap interface courtesy of Esri, Inc. Data Sources: Esri, Garmin International, Inc.; U.S. Central Intelligence Agency (The World Factbook); National Geographic Society

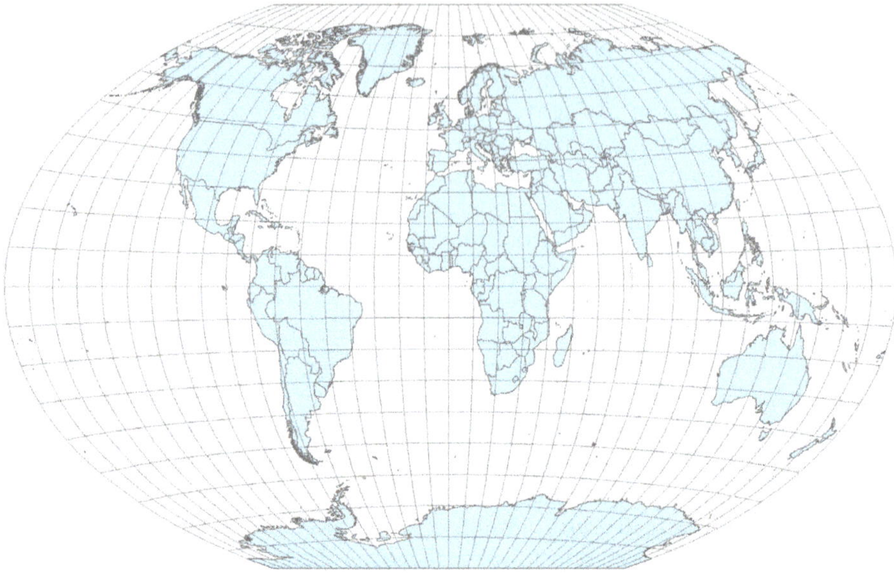

FIGURE 2.16 Compromise projection: Winkel tripel.

ArcMap interface courtesy of Esri, Inc. Data Sources: Esri, Garmin International, Inc.; U.S. Central Intelligence Agency (The World Factbook); National Geographic Society

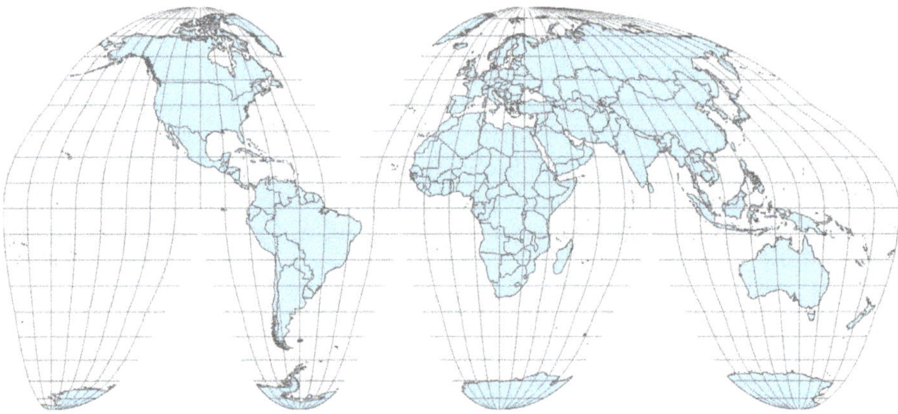

FIGURE 2.17 Goode interrupted homolosine projection.

ArcMap interface courtesy of Esri, Inc. Data Sources: Esri, Garmin International, Inc.; U.S. Central Intelligence Agency (The World Factbook); National Geographic Society

Cartographers have devised several projections with different parameters, as discussed in the previous section. GIS applications provide users a variety of different projections in ArcGIS and other GIS softwares. In fact, the current version of ArcGIS software provides users an option to select from 104 different map projections.

So how do you determine the right map projection? The most important consideration in making this decision must be the location and the purpose of the map. If the goal is to calculate area or analyze a parameter that is dependent on area, you must select a projection that preserves that particular property, that is, equivalent projection. On the other hand, if distance between two points is important, an equidistant projection must be used. In other words, the projection must match the GIS operation. The aim is to minimize distortion in your map. This is particularly important when the area involved is very large because as an area gets larger, errors due to map projections become more significant.

Location is also an important factor. Certain countries work with specific map projections, with a specific ellipsoid and a datum, and some projections are more suitable for specific geographic areas. For instance, Fredrick Pearson recommends a guideline to select map projections based on the latitude of the area mapped. Per this recommendation, equatorial regions must be mapped using cylindrical projection, mid-latitude regions using conic projection, and polar regions with planar projections (Pearson 1984).

However, there are many projections of each type, and one must decide which one from the specific type to select. Even at the local level, counties and cities may adopt a specific map projection with a particular ellipsoid and datum to minimize distortion. This is very important when a GIS user is working with geospatial data derived from different sources, which is usually the case. The GIS user must ensure that all the datasets and maps use the same projection. If the GIS user makes the most basic error of comparing or putting together geospatial data or maps that were created using different projections based on different ellipsoids and datum, significant errors may arise in subsequent mapping and analysis. And, finally, selection of a map projection also influences the overall map design and must therefore be given serious consideration in addition to the location and objectives of the map.

Coordinate Systems

Have you ever described your location to someone or described how someone can reach you? When we do so, we describe our location or path to our location with reference to something else. For instance, you would describe your location by providing the path to get to your location by saying, "Once you reach Elm Street, go to the second stop sign, take a left, and keep going for about one and half miles until you see a Starbucks to your left, then take a right at the next traffic light and you will see my car parked in front of the store." Instead, if you had to describe the location of the store, you would simply provide the street address, "53 West Palm Street." This is also a location description with reference to something else: Go to West Palm Street and locate building number 53.

Location information can be described in numerous ways. As discussed earlier, location descriptions fall into two main categories: absolute location and relative location. Absolute location is the location of a feature at a fixed point on the earth's surface, while relative location refers to the location of a feature with reference to another location. In either case, in order to describe location, it is imperative to have a system or framework for defining locations on the earth's surface. When you describe absolute location, it is the location of a feature at a fixed point. Why is it a fixed point? Because it is fixed with respect to an origin that starts at zero point. It is possible to have this origin with "zero" point because of the system or framework for defining locations on earth.

A system to define locations on earth can be devised once the shape and size of the earth is known. In fact, Greek scholars were the first to develop a methodical approach of placing reference lines on the earth's surface. The Greek astronomer, mathematician, and geographer Hipparchus proposed the concept of parallels and meridians and a numbering system for them. The system of parallels and meridians is called graticule, and the numbering system is called *latitudes* and *longitudes*. A system to represent and define locations on the earth's surface is called a coordinate system. Coordinate systems are reference systems for defining real-world locations of features within a standardized common framework. Any point on the surface of the earth can be located by using this standardized system.

Absolute locations are described using latitudinal and longitudinal coordinates. Lines that run in an east–west direction around the globe are called **latitudes** or **parallels** (figure 2.18). The equator serves as the starting point with 0 degrees for measuring latitudes. Measurements to the north of the equator are called *north latitudes* and measurements to the south of the equator are called *south latitudes.* Measurements of south latitude are sometimes described using negative values. Lines of latitude are measured in angular units from the equator and range from 0 to 90 degrees north and south of the equator.

Similarly, lines that run in a north–south direction from the North Pole to the South Pole are called **longitudes** or **meridians**. The prime meridian serves as the starting point for measuring longitudes. Measurements to the east of the prime meridian are *east longitudes* and measurements to the west of the prime meridian are *west longitudes.* Measurements of west longitude are sometimes described using negative values. Lines of longitude are measured in angular units from the prime meridian and range from 0 to 180 degrees east and west of the prime meridian. For instance, absolute location of Wupatki National Monument in Flagstaff, Arizona, can be described as 35 degrees, 32 minutes, 52.5552 seconds north of the equator (35° 32' 52.5552" N) and 111 degrees, 23 minutes, 31.9920 seconds west of the prime meridian (111° 23' 31.9920" W or –111°23'31.9920" longitude). An important point to note is that the lines of longitude are not spaced equally throughout the globe because lines of longitude are farthest apart at the equator and get closer as they converge at the poles.

Latitudes and longitudes can also be described in decimal degrees by using the formula *decimal degrees = degrees + minutes/60 + seconds/3600.* So the location of Wupatki National Monument in decimal degrees would be 35.547932, -111.392220.

As we have discussed, latitude and longitude coordinates are used to pinpoint absolute locations. However, the system of latitudes and longitudes identifies and locates features on a round or curved surface of the earth and therefore is not suitable for measuring area, length, and direction on maps depicted on a two-dimensional surface such as a sheet of paper, a computer screen, or a smartphone. As discussed earlier in this chapter, paper maps or maps on a computer screen or

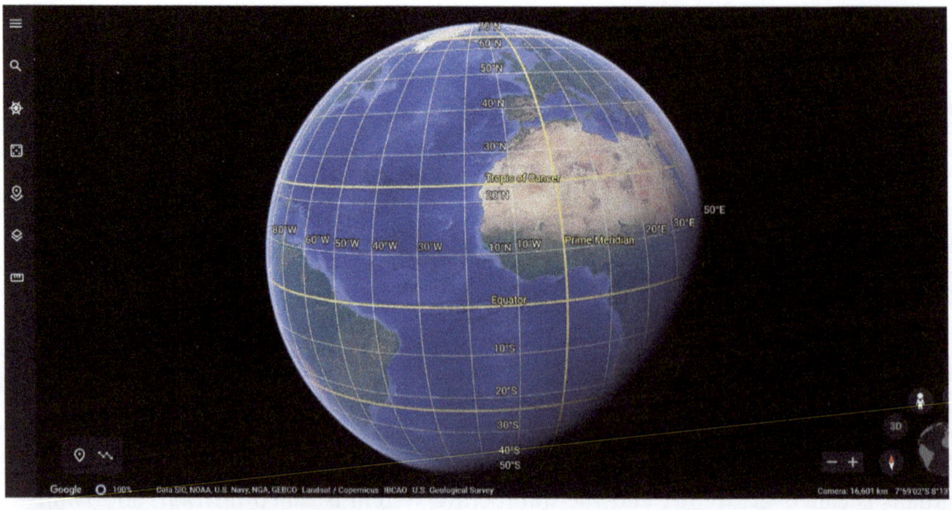

FIGURE 2.18 Latitudes and longitudes.

Credit: Google Earth

your smartphone's mapping application are created by projecting the curved three-dimensional surface of the earth onto a flat two-dimensional surface. Accordingly, in a GIS two types of coordinate systems are used, geographic coordinate systems and projected coordinate systems.

A geographic coordinate system (GCS) defines where the feature or the data is located on the earth using a three-dimensional spherical or an ellipsoidal surface. In a GCS, locations are defined using latitudes and longitudes. A GCS consists of a datum that includes an ellipsoid, a prime meridian, and units in degrees.

A projected coordinate system (PCS) is defined on a flat, two-dimensional surface, and location on the flat surface is defined using Cartesian coordinates. Location is described by using x, y coordinates. A PCS is always based on a GCS. Additionally, a PCS consists of a map projection, projection parameters specific to a particular location, and linear units of measurement. In a PCS, location is measured with respect to horizontal and vertical axes and one does not have to deal with converging meridian lines at the poles. A PCS is useful for determining distance, direction, and area.

Next we will discuss few projected coordinate systems used in the United States.

Universal Transverse Mercator Coordinate System

The Universal Transverse Mercator (UTM) coordinate system is a planar coordinate grid system based on a secant transverse Mercator projection, hence the name. The UTM system divides the earth into 60 north–south zones and location is determined by x and y coordinates in those zones. Since transverse Mercator projection greatly distorts polar regions, UTM covers the earth from 84° N to 80° S and so is not used for mapping polar regions. Each UTM zone is 6° longitude in width. The zones are numbered individually starting from zone 1 at 180 degree West longitude and goes from west to east. So zone 1 covers 180° to 174° W, zone 2 covers 174° to 168° W, zone 3 covers 168° to 162° W, and so on. The United States is covered by zones 10 through 19 (figure 2.19). For instance, the state of Texas falls into UTM zones 13, 14, and 15; Harris County in Texas falls into UTM zone 15; and Chicago, Illinois, is in UTM zone 16. New Delhi, India, is in UTM zone 43, and Toronto, Canada, is in UTM zone 17 (figure 2.20).

Each zone has its own central meridian and individual parameters based on its bounding longitudes. Moreover, each zone has separate origins for the northern and southern hemispheres. The unit of measurement to describe location is meters, and x

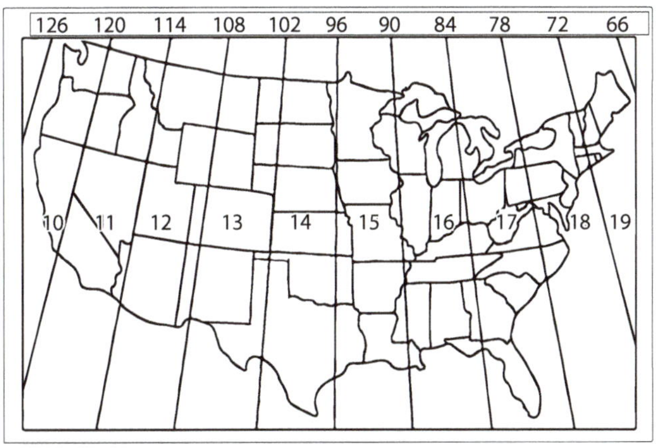

FIGURE 2.19 UTM grid for the United States.

Credit: USGS

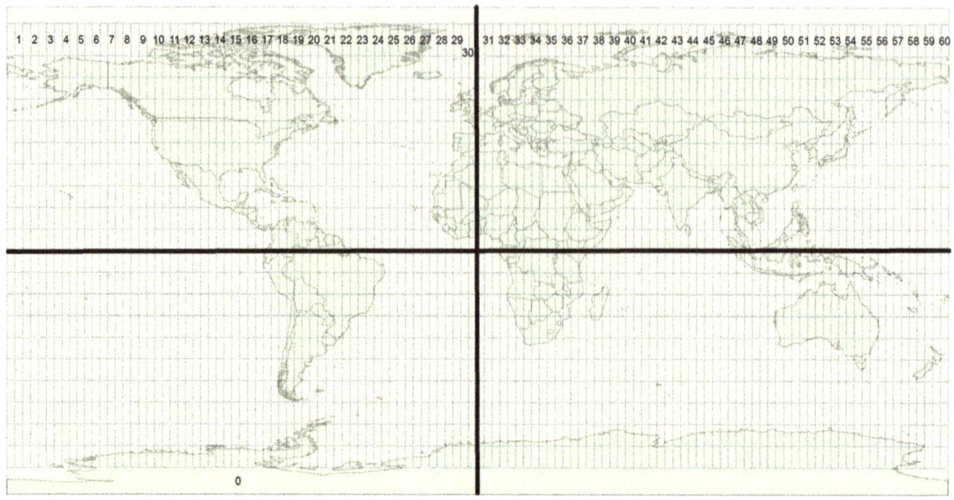

FIGURE 2.20 World UTM grid.

ArcMap interface courtesy of Esri, Inc. Data Sources: Esri, Garmin International, Inc.; U.S. Central Intelligence Agency (The World Factbook); National Geographic Society

and y coordinates are referred to as easting and northing coordinates. In the northern hemisphere, UTM northing is calculated by measuring the number of meters a feature is north of the equator. The northing measurement for Wupatki National Monument is 3,933,880.14 meters, indicating that it is 3,933,880.14 meters to the north of the equator. In the southern hemisphere, the measurement is a bit different, with distance measured in meters from the equator and represented as a negative value. Northing is calculated by adding 10,000,000 meters to the negative value, and this is referred to as *false northing*. This value was determined by system designers in order to avoid a negative value for distance measurements in the southern hemisphere.

For easting calculation, there is no one prime meridian. Each UTM zone has its own central meridian, which runs through the center of the zone. The central meridian is assigned a value of 500,000 meters, and measurements are taken to the east by adding meters or to the west by subtracting meters from 500,000 meters. Let's take the example of Wupatki National Monument. It is in UTM zone 12 and has an easting value of 464,449.41 meters. This indicates that this location is approximately 35,550.59 meters to the west of zone 12's central meridian. This value for easting measurements is known as *false easting* because it is based on an imaginary central meridian in the zone. When the UTM coordinate system is used, it is important that the zone, false northing, and false easting are all specified.

Since UTM is based on transverse Mercator projection, a conformal projection, shape and local angles are preserved without distortion for small areas. Area and distance are distorted, however, like in other conformal projections. The distortion is minimized by using secant projection and small zones. Since the UTM zones cover the entire world, it is a useful referencing system for mapping applications to be shared worldwide. It is also used in military applications. Modern GPS receivers are programmed to work with UTM parameters.

UTM zones do not go beyond 84° N and 80° S. The 60 UTM zones start converging at the poles, and as a result every few miles a new zone would be detected. In order to cover the polar regions, a rectangular coordinate system called the **Universal Polar Stereographic (UPS)** system was created. It consists of a north zone and a south zone. The north zone covers the area from 84° N to the North Pole (figure 2.21). Similarly, the south zone extends from the pole to 80° 30' S. It is based on the stereographic projection customized for each pole. All maps of the polar regions are based on the UPS system.

Military Grid Coordinate System

The Military Grid Reference System (MGRS), or the military grid, was adopted for use by the US Army in 1947 and is used by many organizations and countries worldwide. It is used in conjunction with the UTM and UPS systems. The purpose of creating the military grid was to avoid using long numerical coordinates to pinpoint locations by using a lettering system. In this system, each of the 60 UTM zones are divided into strips of 8° latitude, except for one strip in the northernmost part that is 12° latitude. The 60 UTM zones are thus divided into quadrilaterals that measure 6° longitude by 8° latitude. Each of the quadrilaterals are assigned a letter from C to X starting at 80° S latitude. The letters A, B, Y, and Z are reserved for use with the UPS system in the polar regions. The letters

FIGURE 2.21 Universal Polar Stereographic (UPS) system.

ArcMap interface courtesy of Esri, Inc. Data Sources: Esri, Garmin International, Inc.; U.S. Central Intelligence Agency (The World Factbook); National Geographic Society

I and O are not utilized to avoid confusion with similar-looking digits. Each 6° × 8° quadrilateral represents approximately 1,000 square kilometers on the ground. Each quadrilateral is described using alphanumeric codes, where you first refer to the zone number and then the letter designated to the row.

For instance, Wupatki National Monument is located in a quadrilateral assigned with code 12S, where 12 refers to UTM zone 12 and S refers to the quadrilateral designated with the letter S. Each quadrilateral is further divided into 100,000-meter-square cells. Each cell is assigned a two-lettered alphabetical code, the first letter for the row and the second letter for the column. Wupatki National Monument is located within a 100,000-meter-square cell assigned the code 12SVE. Within this cell successively accurate locations can be isolated by dividing and subdividing each subsequent cell and assigning numeric codes. For instance, each 100,000-meter cell is further divided into 10 cells, each measuring 10,000 meters, that are assigned numeric codes. In this 10,000-meter cell, Wupatki National Monument is located at 12SVE65. The 10,000-meter cell is further subdivided into 10 cells yielding 1,000-meter individual cells that are assigned numeric codes. In this 1,000-meter cell, the location of Wupatki National Monument can be further narrowed down to 12SVE6598. The process is continued until the desired accuracy is achieved. Locations can be pinpointed within 1-meter-square cells. In our example, 12SVE6598935027 represents the location of Wupatki National Monument within a 1-meter-square cell. Here, 65989 is the easting and 35027 is the northing, representing the number of meters east and north of the southwest corner of the square, respectively.

United States National Grid

UTM values are used as the basis for another coordinate system, the **United States National Grid (USNG)**. The US Department of Homeland Security proposed the use of the USNG for all referencing and mapping applications in the United States, so that users of various devices that utilize geospatial data and require accurate ground coordinate information have a uniform reference system nationwide. This system is used widely by the military and by organizations such as the USGS, Federal Emergency Management Agency (FEMA), and the National Guard.

The USNG system is based on the UTM coordinate system and the alphanumeric referencing system used in the MGRS. It is similar to the MGRS with the only difference being that it uses NAD83 US datum. For every location it uses a single identifier for the northing, easting, and zone information. One of the principal uses of the USNG is in the USGS National Map Viewer. The map viewer displays USNG coordinates, as shown in figure 2.22.

State Plane Coordinate System

The **State Plane Coordinate System (SPCS)** was created in the 1930s to define parcel boundaries in the United States with the goal of simplifying land parcel perimeter and area calculations and achieving accuracy within a foot of the true measurement. In order to achieve this level of accuracy, each state was divided into zones with an individual grid for each zone. Today SPCS is widely used by city and county governments as well as surveying, engineering, and utility companies that require highly accurate land surveying information. It is a projected coordinate system like UTM and covers the entire United States. Each state has its own map in the SPCS system.

The SPCS is based on two projections: transverse Mercator and Lambert conformal conic. States that are elongated north–south or have an east–west extent are

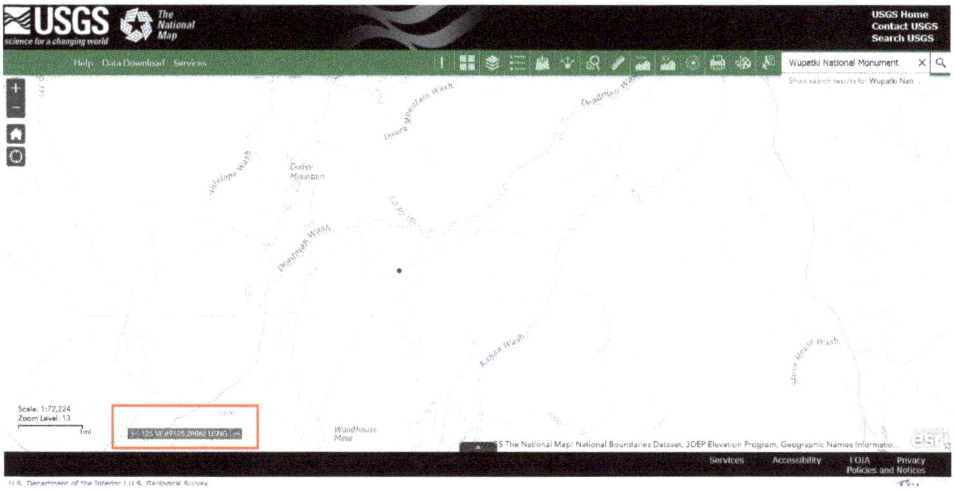

FIGURE 2.22 USGS National Map Viewer view of Wupatki National Monument showing USNG coordinates.

Credit: National Boundaries Dataset, 3DEP Elevation Program, Geographic Names Information System, National Hydrography Dataset, National Land Cover Dataset, National Structures Dataset and National Transportation Dataset, USGS Global Ecosystems, US Census Bureau TIGER/Line data, USFS road data, Natural Earth Data, US Department of State HIU, NOAA National Centers for Environmental Information

drawn using Lambert conformal conic projection, while states that are elongated east–west or predominantly have greater north–south extent are drawn using transverse Mercator projection. For instance, states like California, Texas, and Wisconsin are drawn using Lambert conformal conic projection, while states like New York, New Mexico, and Arizona are drawn using transverse Mercator projection. Each state is divided into zones following state or county boundaries (figure 2.23). The number of zones varies from one to six. Rhode Island has one zone, Louisiana has two zones, Wisconsin has three zones, Texas has five zones, and California has six zones. The zones were revised when the SPC system was modernized in 1983 and switched from NAD27 to NAD83 datum, GRS80 ellipsoid, and metric units of measurement.

Each zone in the SPCS has its own central meridian and baseline. Position of the central meridian and baseline will vary from state to state depending on the projection used to map the state. For each zone, its origin is measured to the west and south of the zone. For states mapped using Lambert conformal conic projection, the origin is approximately 2,000,000 feet west of the central meridian, and for states mapped using transverse Mercator projection, the origin is 500,000 feet west of the central meridian. This approach for measuring origin was adopted to ensure that x and y coordinates are positive numbers. Coordinates are measured by calculating the distance from these false origin points, giving the user false northing and false easting. SPCS coordinates are described by referencing the false easting in feet, false northing in feet, the state, and the zone.

Other Systems

Several other coordinate systems are used within the United States and worldwide. Many countries use UTM or MGRS or alternatively have their own reference system. European countries like Belgium, Finland, Ireland, Italy, the Netherlands, Sweden, Switzerland, and the United Kingdom have a national grid system that covers their region. Similarly, Australia and New Zealand have their own reference system.

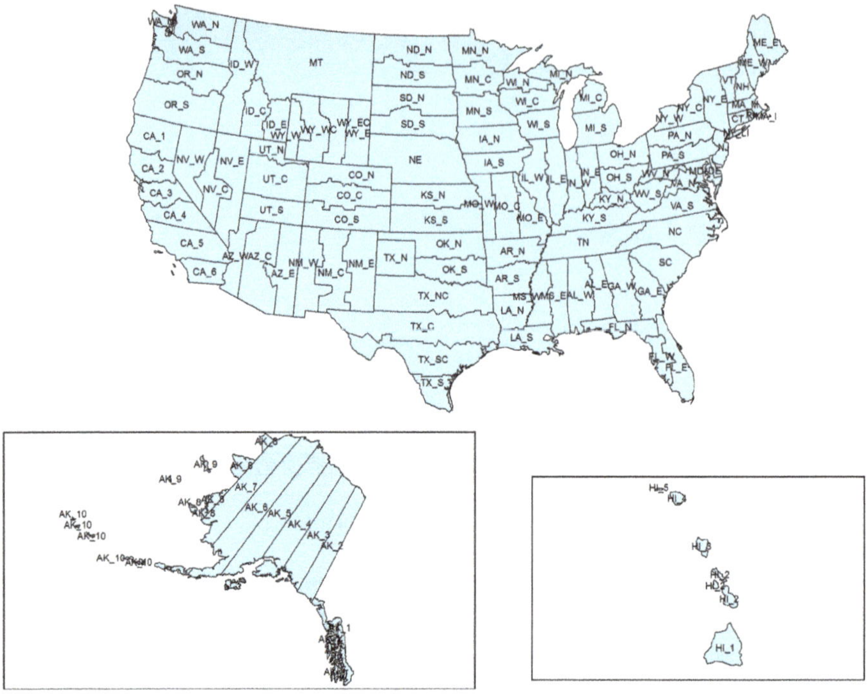

FIGURE 2.23 State Plane Coordinate System zones.

ArcMap interface courtesy of Esri Inc. Data Sources: Esri, TomTom, U.S. Department of Commerce, U.S. Census Bureau

Within the United States, use of GIS within state and local government agencies has led to many states developing special coordinate systems customized to their specific requirements or region. For instance, state and local government agencies in Wisconsin frequently use the Wisconsin Transverse Mercator reference system that was created by shifting their UTM zone to the center of the state. Similarly, other states like Georgia, Illinois, Kentucky, Oregon, and Pennsylvania have created such special state coordinate systems by shifting the central meridian of their UTM or SPCS zone to the center of the state in order to consolidate multiple zones into one. Many commercial vendors of mapping applications and geospatial data have also developed proprietary coordinate systems customized to be used with their specific products.

Ultimately a GIS application must have the capability to move between map projections, coordinate systems, datums, and ellipsoids and as a GIS user it is your responsibility to ensure that the appropriate parameters and specifications are used when working with multiple maps or when acquiring geospatial data from multiple sources. As GIS users we have many options to select from because there are many datums, map projections, and coordinate systems. It is imperative that a GIS user knows which datum, projection, and coordinate system to use and how that will impact mapping, measurement, and analysis.

Questions and Activities

1. Refer to Esri's StoryMaps on map projections created by Bojan Šavrič and Melita Kennedy, available at https://storymaps.arcgis.com/stories/ea0519db9c184d7e84387924c84b703f. Review all of the map projections, pick any five map projections from different projection families, and write a short paragraph describing each.

2. Another useful guide for map projections and their uses has been developed by Bill Rankin and is available at http://www.radicalcartography.net/index.html?projectionref. Review the table of different projection systems and applications. Can you think of at least five applications from your daily life where any of the listed projection systems may have been used?
3. You can explore map projections interactively on other websites too. Visit the following web pages and explore different projection systems and how they change your map display.
   a. Map Projections Transitions by Jason Davies at https://www.jasondavies.com/maps/transition.
   b. Projections Explorer available at https://robertsgeospatial.com.au/products/projection-explorer.
4. In this chapter, we discussed different coordinate systems. Quite often you may have to convert from one coordinate system to another; for instance, convert from GCS to UTM or SPCS. A few utilities are freely available online to do these conversions. Some convert GCS to UTM, while others covert GCS to either UTM or SPCS. The utilities are available at:
   a. http://rcn.montana.edu/Resources/Converter.aspx
   b. https://www.latlong.net/lat-long-utm.html
   c. https://www.earthpoint.us/Convert.aspx

   In order to utilize any of the above utilities, you will need GCS coordinates; a good resource is available at LatLong.net (https://www.latlong.net). You can use this website to find GCS coordinates by inputting a place name such as Texas State Capitol and you will get the coordinates in decimal degrees as well as in degrees, minute, and seconds.

   Let's use the example of the Sears Tower in Chicago, Illinois. You must have two tabs open in your browser. On one tab open any of the GCS coordinates locator websites. On the second tab open one of the GCS to UTM converter websites (we will use http://rcn.montana.edu/Resources/Converter.aspx). Find the latitude and longitude coordinates of Sears Tower. Now copy the coordinates into the appropriate boxes in the GCS to UTM converter webpage. Click on "Convert Decimal Degrees" (if you copy and paste the latitude and longitude in degrees, minutes, and seconds, make sure to paste in the boxes that require DMS and click on "Convert Degrees, Minutes, Seconds") to determine the UTM zone and easting and northing of Sears Tower.

   Try few other examples on your own. LatLong.net allows you to search for latitude and longitude coordinates by address. Alternatively, you can input place names to search for coordinates. Use the address of your university building or dorm or your favorite restaurant and use the website to locate its GCS coordinates and then convert those into UTM coordinates.
5. Next, we will use Earth Point (https://www.earthpoint.us/StatePlane.aspx) to convert GCS coordinates to State Plane. The web page allows you to convert State Plane to latitudes and longitudes or convert latitudes and longitudes to State Plane. We will convert latitudes and longitudes to State Plane. Use the example of Sears Tower again. Find the latitude and longitude coordinates of Sears Tower, then copy the coordinates into the appropriate boxes and click the "Calc" button. Make sure to input the SPCS zone. Use figure 2.23 to determine the SPCS zone. The web page also gives you the option to display the coordinates on Google Earth (you must have Google Earth installed to utilize this option) by clicking on the "Fly To On Google Earth" button. Determine State Plane coordinates of other locations in your area and display them on Google Earth.

Key Terms

cartography (p. 23)
conformal projection (p. 35)
conic projection (p. 33)
coordinate system (p. 41)
cylindrical projection (p. 33)
datum (p. 30)
datum transformation (p. 32)
developable surface (p. 33)
ellipsoid, or spheroid (p. 27)
equidistance projection (p. 38)
equivalent, or equal-area, projection (p. 34)
geographic coordinate system (GCS) (p. 43)
geoid (p. 28)
geometric datum (p. 30)
graticule (p. 41)
horizontal datum (p. 30)
large-scale map (p. 25)
latitudes, or parallels (p. 42)
line of tangency (p. 33)
longitudes, or meridians (p. 42)
map (p. 23)
map generalization (p. 26)
map projection (p. 32)
map scale (p. 23)
Military Grid Reference System (MGRS) (p. 45)
planar projection (p. 33)
projected coordinate system (PCS) (p. 43)
representative fraction (RF) (p. 23)
secant projection (p. 33)
small-scale map (p. 25)
State Plane Coordinate System (SPCS) (p. 46)
tangent projection (p. 33)
true direction, or azimuthal, projection (p. 37)
United States National Grid (USNG) (p. 46)
Universal Polar Stereographic (UPS) system (p. 44)
Universal Transverse Mercator (UTM) coordinate system (p. 43)
vertical datum (p. 30)

References

American Congress on Surveying and Mapping. 1983. *The North American Datum of 1983: A Collection of Papers Describing the Planning and Implementation of the Readjustment of the North American Horizontal Network.* American Association for Geodetic Surveying Monograph No. 2. https://www.ngs.noaa.gov/PUBS_LIB/NAD1983_Collection_of_Papers.pdf.

Kimberling, Jon A., Aileen R. Buckley, Phillip C. Muehrcke, and Juliana Muehrcke. 2009. *Map Use.* Redlands, CA: Esri Press.

Moore, Larry. 2001. "Datum Shifts and Digital Map Coordinate Displays." US Geological Survey. https://www.wvgis.wvu.edu/resources/standardsGuidelines/datum_shifts_v2.pdf.

NOAA (National Oceanic and Atmospheric Administration. 2023. National Geodetic Survey, "Datums and Reference Frames." Last modified December 7, 2023. https://www.ngs.noaa.gov/datums/index.shtml.

Pearson, Frederick. 1984. *Map Projection Methods.* Blacksburg, VA: Sigma Scientific.

Shellito, Bradley. 2018. *Introduction to Geospatial Technologies.* New York: W. H. Freeman–Macmillan Learning.

Snyder, John. 1987. *Map Projections: A Working Manual.* US Geological Professional Paper 1395. Washington, DC: US Government Printing Office.

Snyder, John, and Philip M. Voxland. 1989. *An Album of Map Projections.* US Geological Survey Professional Paper 1453. Washington, DC: US Government Printing Office.

3 GIS Spatial Data Models

Learning Objectives

- Understand how a GIS models real-world features and feature relationships
- Explain the difference between a discrete object and a continuous object
- Explain what a data model is and its importance in GIS
- Discuss the two spatial data models in GIS, vector and raster
- Describe the key characteristics of a vector data model
- Understand and articulate the different structures of a vector data model
- Explain the key characteristics of vector data model structures
- Describe the key characteristics of a raster data model
- Compare vector data model versus a raster data model
- Describe the different file formats for vector and raster data files
- Use GIS data and conduct some basic operations

Representing Real-World Features

When we talk about GIS or refer to geospatial data or maps, we are referring to digital data and maps. A GIS works with digital data and maps. If you look at your smartphone's map application, what do you see? How is your map displayed? How are datasets on your map displayed? If you view your map using the "Default" option, highways, streets, roads, and streams are displayed as straight or curved lines. Water bodies such as lakes or ponds and green spaces are displayed as blocks of varied shapes. A convenience store, grocery store, or restaurant is displayed as a point with the appropriate symbology. If you select the "Satellite" option for your maps, you will be able to view a map that resembles a picture or photograph of the environment, where you see actual houses, cars, trees, roads, and highways displayed on your map. You will also be able to see swimming pools in backyards of houses and cars parked in parking lots or in front of houses. When you select the two options for your map (figure 3.1), the map displays two different types of geospatial data. Accordingly, the features displayed on your map look different.

In the previous chapter we discussed how to measure location of features on the earth's surface and transform from a three-dimensional surface to represent features on a two-dimensional map surface. The question now is, how do we represent real-world features on a map? As we already know, a GIS can be used to perform a variety of tasks such as determining the most appropriate and optimal location for a grocery or retail chain, a library, or a convenience store based on key criteria; finding the shortest route to navigate from point A to point B; calculating rate of change over a time period as deforestation continues; investigating the incidence of a disease in a geographic area and examining underlying factors; investigating the rate of crimes in a geographic area; examining changes in land use–land cover patterns and their impact on the environment; locating the closest restaurant, grocery store, gas station, or theme park and calculating the fastest route to get there; and investigating

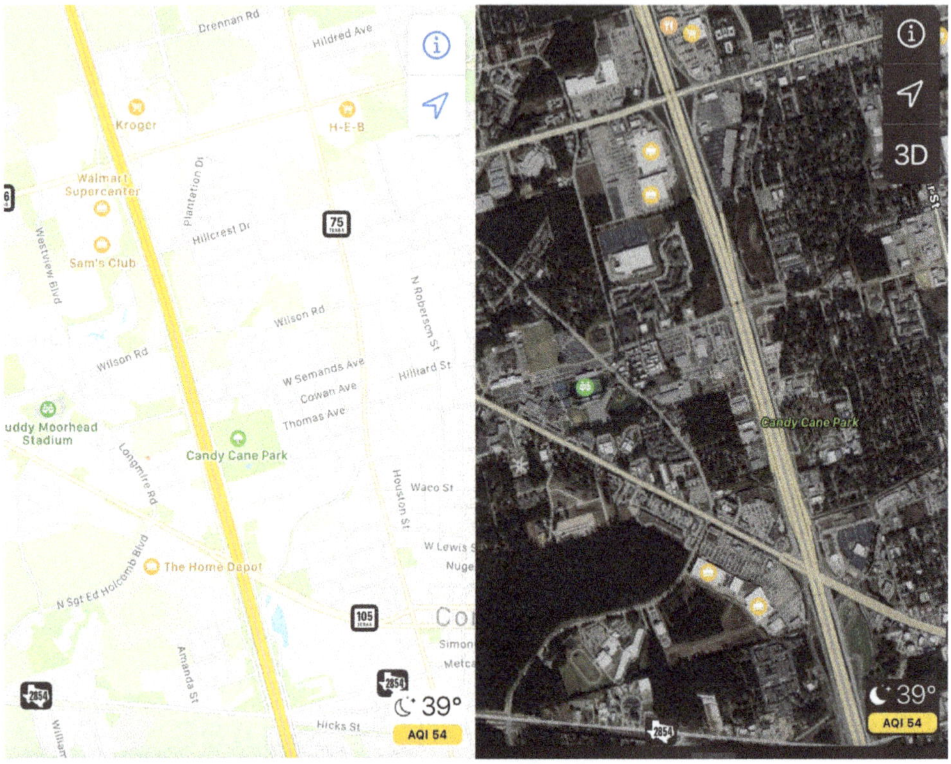

FIGURE 3.1 Map settings on a smartphone's mapping application: Default option versus Satellite option.

the best neighborhood to purchase a house based on specified criteria. A GIS can perform such tasks and much more.

What sets GIS apart from other software applications is its ability to analyze and map geospatial data, data that is tied to a geographic location. For instance, a software may calculate the shortest path from point A to point B, but if it does not have the capability to map the route or provide directions to go from point A to point B, the application will have very restrictive utility. Similarly, a software application may calculate the total amount of area lost to deforestation, land use–land cover change for a specified time period, or the total number of cases or incidence rate of a certain disease. But if the application is unable to calculate where the change has occurred or where the incidence of disease is higher or growing, it will be of little use. The ability to map real-world locations of features, events, or activities and examine spatial patterns and relationships that mimic real-world scenarios is the defining characteristic of a GIS.

A GIS performs such tasks in a computerized environment. Remember, in addition to other components, a working GIS consists of a hardware and software. In other words, a GIS is a computer-based, automated system that is either installed on your desktop, laptop, or mobile device or is available on the web via the cloud. In any case, no matter what platform is used, a GIS provides a digital environment to represent, analyze, and investigate real-world features, events, and activities. And for this to happen, real-world features must be represented in a digital environment.

Once real-world features have been transformed from a three-dimensional environment of the earth's surface so that they can be represented and displayed on a two-dimensional surface, they must now be depicted somehow on this two-dimensional surface, whether on a computer, mobile device, or any other digital environment. If the data is depicted on the two-dimensional surface of a paper map, it must be converted into a digital format before it can be used in a GIS. A GIS therefore represents real-world features and phenomena in a digital environment such that they closely represent real-world scenarios. It does so by viewing the world in two ways.

One way views the world as a collection of discrete objects. This is called a discrete object view, where objects or features have a definite location, a fixed starting point and end point, and clear definable boundaries. Take the example of your university campus. Every street on your campus starts and ends at a fixed point, every building has a clear and definable boundary separating it from the neighboring building. You will find several other examples of discrete objects around your university campus such as street signs, trees, benches, computers and desks in classrooms, parking lots, and cars parked in the parking lots. Thus, in a discrete object view, real-world features and phenomena are conceptualized as separate objects.

The other way of viewing the world is a continuous surface view, where objects or features are not separate entities with fixed location and definable boundaries. Instead, the world is viewed as made of infinite points and each point has a value representing a constantly varying surface. In this world view, values are available at every point. Thus, the world is conceptualized as a surface with values at every point that vary. Take the example of temperature. Temperature does not have a fixed location nor does it have a fixed, definable boundary. You will have temperature readings at every point in an area, even if you choose not to measure at every location, and the values vary gradually as you move from one area to another. Similarly, elevation varies gradually across a landscape, transitioning from high elevation to low elevation, and so do snowfall and rainfall. These are all examples of a continuous surface that exists throughout with no space between observations.

In order to create this abstraction of the real world, a GIS uses a mathematical construct or set of rules to represent the real world in a computerized environment. This construct is called a geospatial data model. A data model is an abstract method for representing real-world features and characteristics in a digital environment for a GIS. A data model will determine how the computer represents and manages features, phenomena, and characteristics. It will determine how the data is organized for use by a GIS software, how the software will access the data to work with it and analyze it, and how the data will be displayed for viewing. It will decide whether you will view highways, roads, and streams as lines and green spaces and buildings as blocks of varied shapes or whether you will view highways, roads, streams, green spaces, and buildings exactly as they appear in the real world, as if you were looking at a picture of your surroundings.

As discussed in chapter 1, GIS data consists of two components: a spatial component and an aspatial component. Accordingly, a GIS will have two data models, one for the spatial component and one for the aspatial component. In this chapter, we will discuss data models for the spatial component of a GIS data. There are two data models for representing the spatial component of a GIS data: a vector data model and a raster data model (figure 3.2). All GIS datasets are represented as either a vector data model or a raster data model.

Vector Data Model

A vector data model represents real-world features and phenomena as points, lines, or polygons. These are referred to as vector objects and form the foundation of a vector data model. Here, complex objects are formed by connecting simple

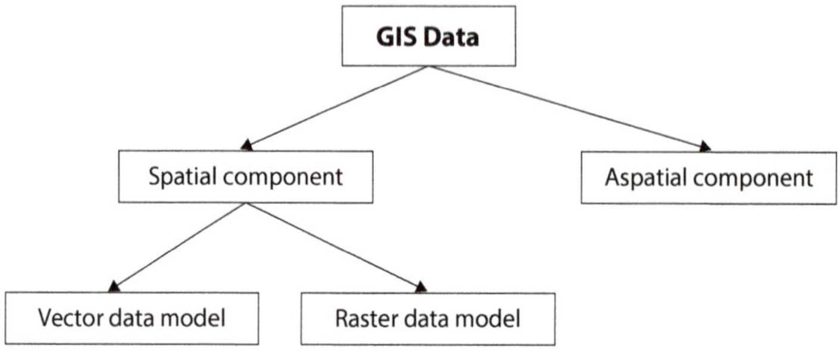

FIGURE 3.2 Geospatial data models.

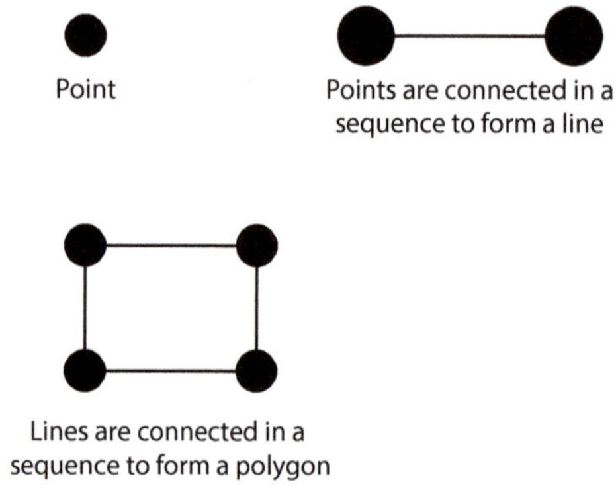

FIGURE 3.3 Point, line, and polygon.

objects. Points represent the most basic and simplest vector object and are the building blocks of a vector data model. Points are connected in a sequence to form a line, and lines are connected in a sequence to form an enclosed area or a polygon (figure 3.3).

Vector objects are characterized by their spatial dimension. Points are zero-dimensional, with no length and no width, and are associated with a set of x, y coordinate points. A **point** represents a singular, distinct feature location on the ground. A feature represented as a point on a map can be a true point feature on the ground such as a ground control point or it can be a feature that is conceived of as a point for mapping purposes at a specific scale. For instance, on a world map, cities would be represented as points, and on a large-scale map of a city, buildings would be represented as points. Other examples of point features include hospitals, oil wells, major cities, state capitals, benchmarks, survey points, manholes, police stations, fire stations, and weather stations. The x, y coordinates of point features are recorded and stored in latitudinal and longitudinal units or in feet or meters depending on the coordinate system used. Points in figure 3.4 represent cities in the US state of Texas.

A **line** is a one-dimensional vector object that has length and no width and is formed by connecting a sequence of points with a set of x and y coordinates. Lines represent vector objects that have length and location coordinates. They may be straight or curved. All lines have a starting point and an end point, referred to as *nodes*, shown in figure 3.5. Nodes also refer to points at the intersection or junction of multiple lines. Lines may consist of multiple points between the starting point and end point to give shape to the line. These points are referred to as *vertices* or *vertex*. Lines that connect two nodes are referred to as *arcs*. Examples of line features

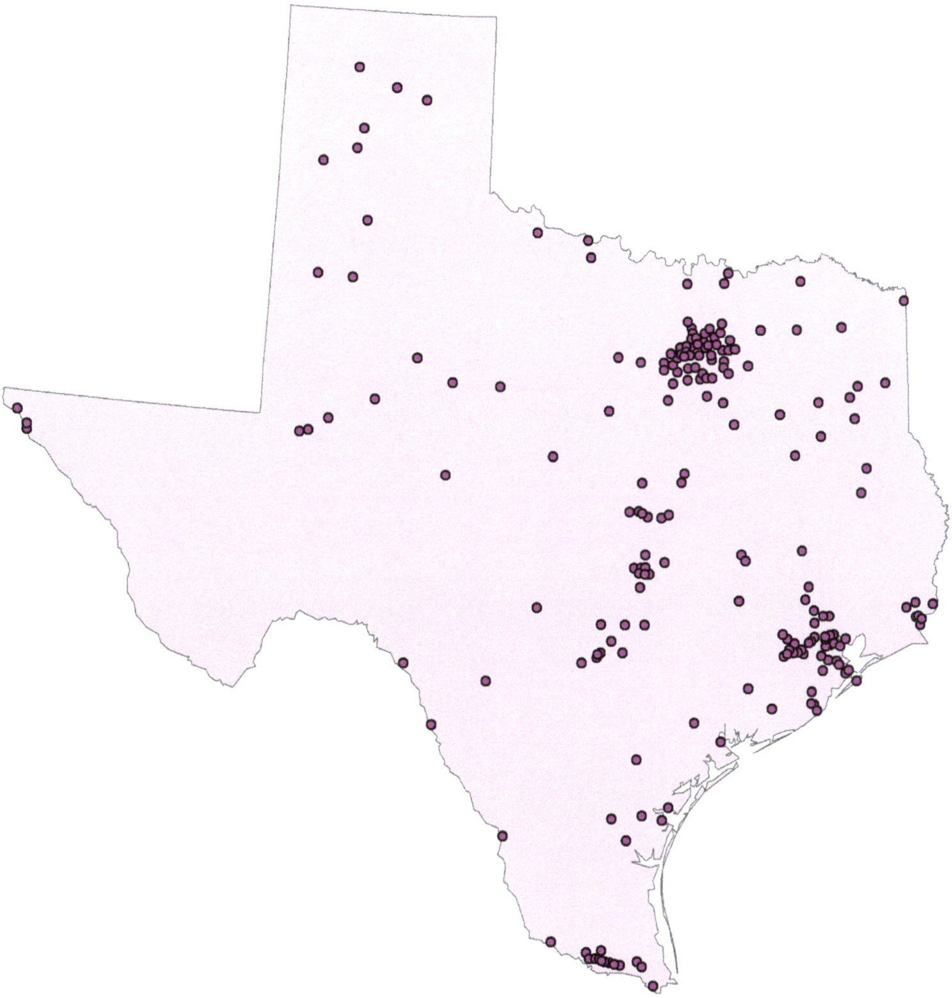

FIGURE 3.4 The points in this map represent cities in the state of Texas. Each point has a unique x and y location.

ArcMap interface courtesy of Esri, Inc. Data Sources: Esri, TomTom, U.S. Department of Commerce, U.S. Census Bureau, National Oceanic and Atmospheric Administration (NOAA), National Ocean Service (NOS), National Geodetic Survey (NGS)

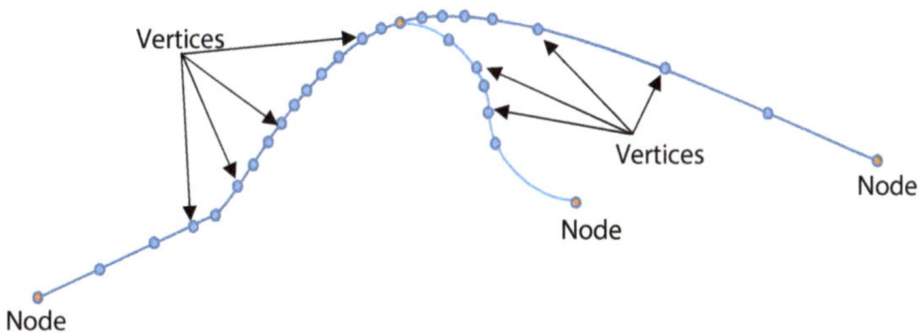

FIGURE 3.5 Line vector object formed by connecting a sequence of points.

include political boundaries, parcel boundaries, roads, highways, streams, and fault lines. Lines in figure 3.6 represent interstate highways in Texas.

A **polygon** is a two-dimensional vector object that has length and width. Polygons are also referred to as areas. They are closed features that are formed by connecting a series of lines in a sequence going in a specific direction, shown in

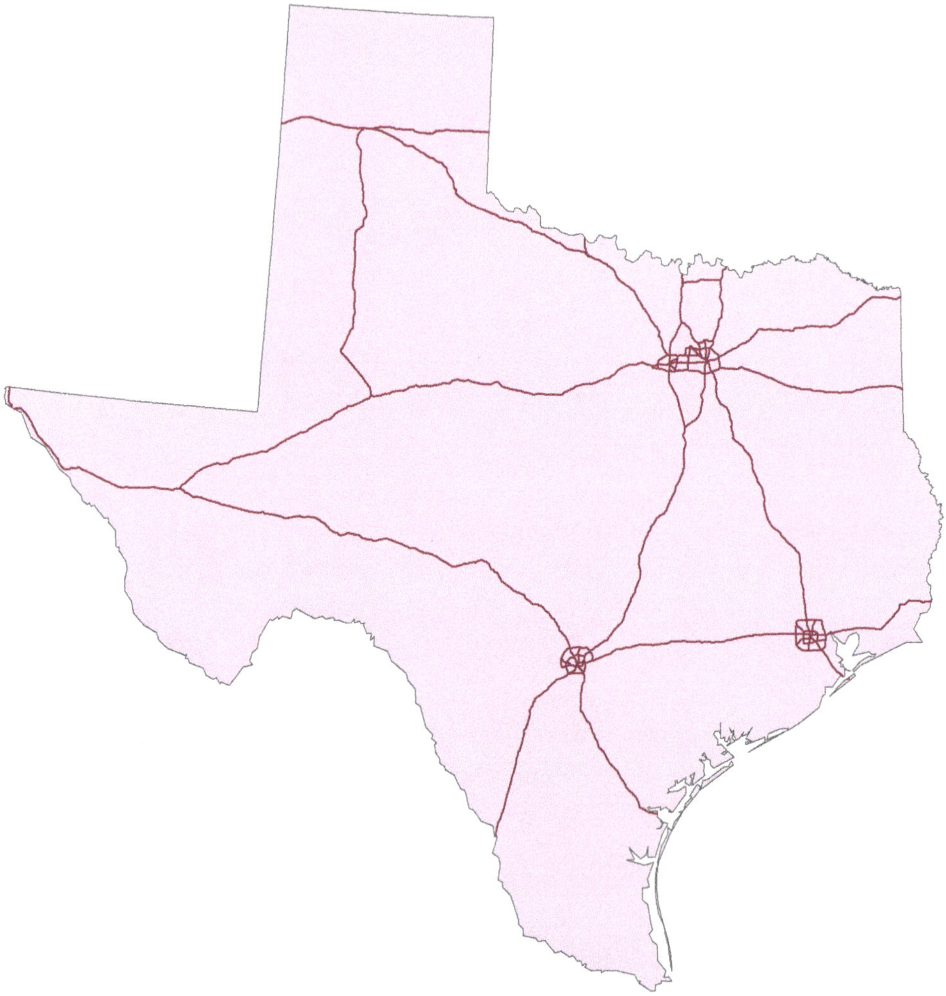

FIGURE 3.6 The lines in this map represent interstate highways in the state of Texas.

ArcMap interface courtesy of Esri, Inc. Data Sources: Esri, TomTom, U.S. Department of Commerce, U.S. Census Bureau

figure 3.7. In the case of polygons, the starting node and the ending node are the same, so the starting node of the first line and the end node of the last line that form a polygon are the same. Polygons that share a common boundary share common nodes. Polygon features have size, area, and perimeter properties. Polygons are used to represent features like land parcels, building footprints, states, counties, census tracts, and land use. The 254 counties in the state of Texas are represented as polygons in figure 3.8.

Scale plays an important role and is a key consideration when determining if a vector object will be displayed as a point or as a polygon. On a small-scale map of a country, cities will be displayed as point features with no area or size measurement, whereas on a large-scale map of a state, cities will be displayed as a polygon or area feature allowing a GIS user area and perimeter calculations. Similarly, on a small-scale map, rivers and creeks will be displayed as lines, whereas on a large-scale map, rivers and creeks may be displayed as polygons allowing one to calculate the area occupied by a creek or river.

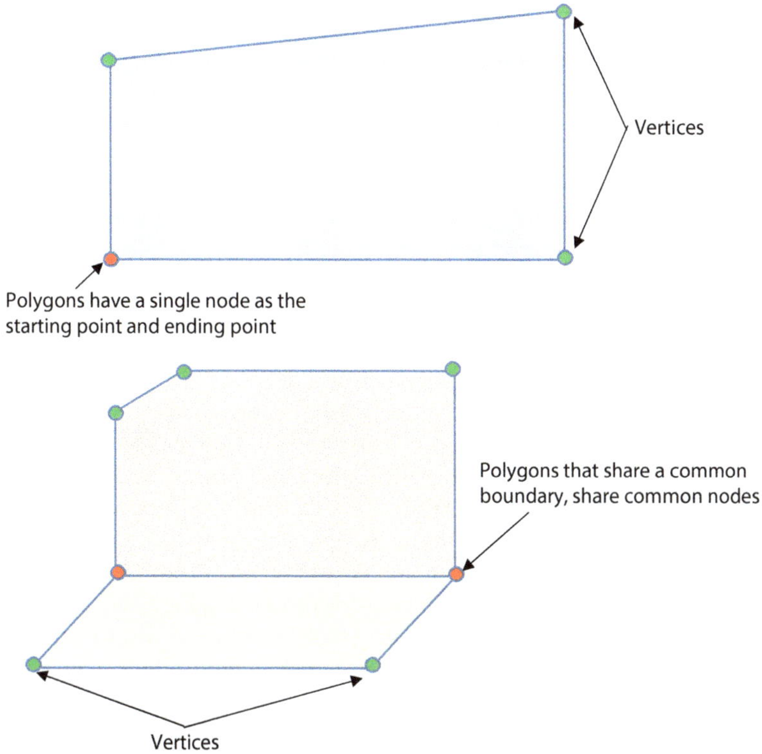

FIGURE 3.7 Polygon vector object formed by lines.

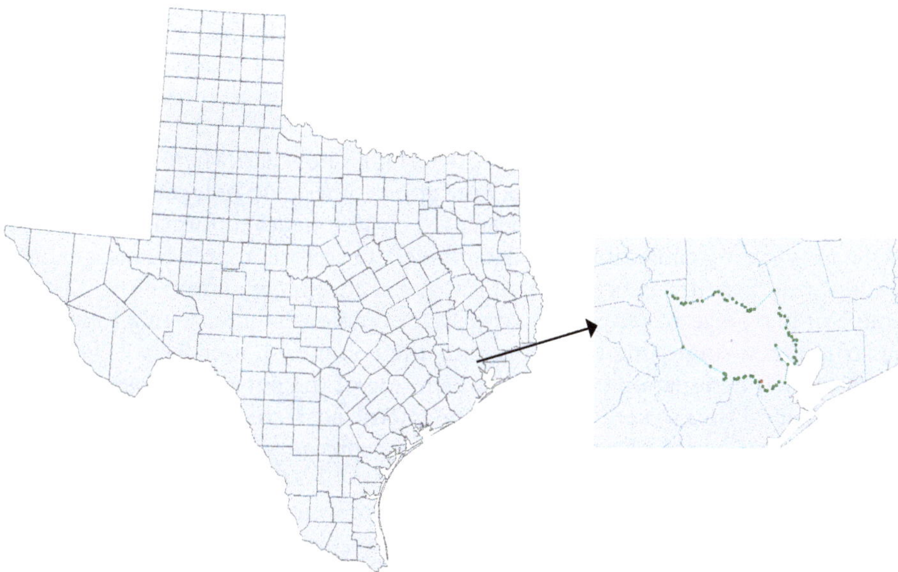

FIGURE 3.8 The polygons in this map represent the 254 counties in the state of Texas. Each county polygon has a unique shape and area. A polygon representing Harris County is shown, formed by a sequence of lines, which are formed of nodes and vertices.

ArcMap interface courtesy of Esri, Inc. Data Sources: Esri; U.S. Department of Commerce, Census Bureau; U.S. Department of Commerce (DOC), National Oceanic and Atmospheric Administration (NOAA), National Ocean Service (NOS), National Geodetic Survey (NGS)

Vector Data Model Structures

There are different structures of vector data models, and one way to differentiate them is to think in terms of how the structure models feature behavior. The first generation of vector objects had a simple structure consisting of lines with arbitrary starting points and end points. In this vector data model, points, lines, and polygons lacked any structure and are represented by a string of x and y coordinates. Here lines resemble spaghetti, hence the structure of the vector data model is called a spaghetti data model. Complex shapes are formed by adding more lines, or strands of spaghetti, to the structure. The spaghetti data model does not store feature relations. Features are stored independent of each other as geometrical objects. As a result, if two polygons share a common boundary, each polygon will be made of individual lines with its own set of x and y coordinates. Thus, the line representing the boundary that is shared by the two polygons will be stored twice. This gives rise to redundancies in the dataset and minimizes the efficiency of the data model. Spaghetti data models are therefore not suited for applications that involve advanced analytical computation. Due to their simple structure, spaghetti vector data models are suitable for simple mapping and basic graphics production.

In contrast to the spaghetti data model is another structure of vector data models, a topological data model. There is no doubt that geospatial datasets model real-world features and phenomena. In order to do so, a GIS needs to know how features connect with other features and understand the relation between the features. For instance, if two streets cross, they must meet at an intersection or a cross street. Similarly, if two parcels or zip codes are next to each other and share a common boundary, a GIS must know that a single identical boundary exits between the two polygons and understand that knowledge to be able to model this real-world relationship on a map. If a street has an overpass or a bridge across it, a GIS must know and understand that the two lines will not intersect. This ability to understand the environment and how features connect with other features in their surroundings and behave, and understand feature relations is called topology. The concept of topology is derived from geometrical mathematics and is used in GIS to record, understand, and describe logical spatial relationship between features.

A topological data structure logically determines how and where nodes connect points and lines on a map; determines how points, lines, and polygons share geometry; and models vector object relationships. This topological information accompanies datasets in a topological vector data model. Therefore, topological spatial data has additional data files associated with it that store topology information. Take the example of two adjacent polygons that share a common boundary. In a topological data structure, a single line that represents the common boundary will be stored with an explicit designation explaining which side of the polygon belongs with the line. Topological data structures preserve geometric characteristics and spatial properties of features, enabling efficient projection, reprojection, or transformation of geospatial datasets.

Topological data structures consist of three important topological rules that are useful in understanding spatial relationships in GIS: adjacency or contiguity, connectivity, and containment. Adjacency describes how one polygon is related to another polygon in its neighborhood and if they share a common boundary. The adjacency rule requires that arcs or lines connecting to form a polygon must have direction, describing its starting point and end point via "from node" and "to node." This determines adjacency information for polygons. Adjacent or contiguous polygons share a line or an arc, and this information is key in defining the left and right side of each line or arc, shown in figure 3.9 and table 3.1. In a topological data model, this information is stored with the attribute information.

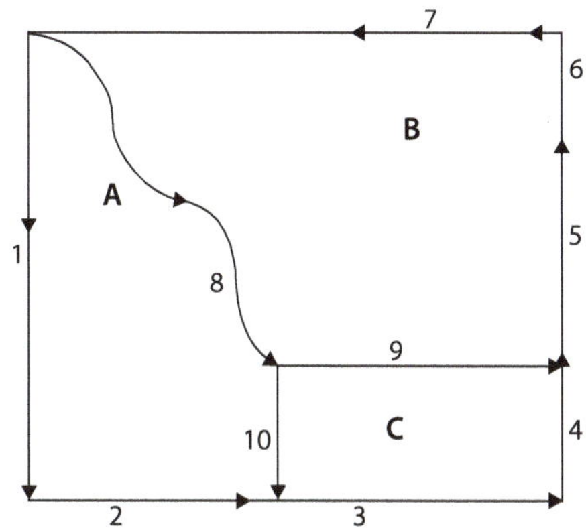

FIGURE 3.9 Topological rule of adjacency.

TABLE 3.1 Polygon Adjacency Topology

Adjacency

Arc	Left Polygon	Right Polygon
1	Universe	A
2	A	Universe
3	C	Universe
4	C	Universe
5	B	Universe
6	B	Universe
7	B	Universe
8	B	A
9	B	C
10	C	A

The topological rule of connectivity establishes how lines are formed by connecting points at the starting point and at the end point describing arc-node topology. Per this rule, each line has a "from node," which is its starting point, and a "to node," which is its end point, shown in figure 3.10 and the arc-node topology shown in table 3.2. The rule helps determine how lines intersect with one another. Lines or arcs that share a common node are connected. In figure 3.10, the computer program will able to determine that it can move from arc 1 to arc 2 and then to arc 3 because they share a common node, but it cannot move from arc 1 to arc 5 or arc 9 because they do not share a common node.

The third topological rule of containment describes how spatial objects are enclosed within a polygon. Arcs connect to surround an area and form a polygon. This defines a polygon-arc topology, making a note of arcs that form a polygon, shown in figure 3.11, and the polygon-arc topology, shown in table 3.3. Each arc is stored once. Thus contiguous polygons with a shared boundary do not overlap, minimizing data redundancy and amount of data stored.

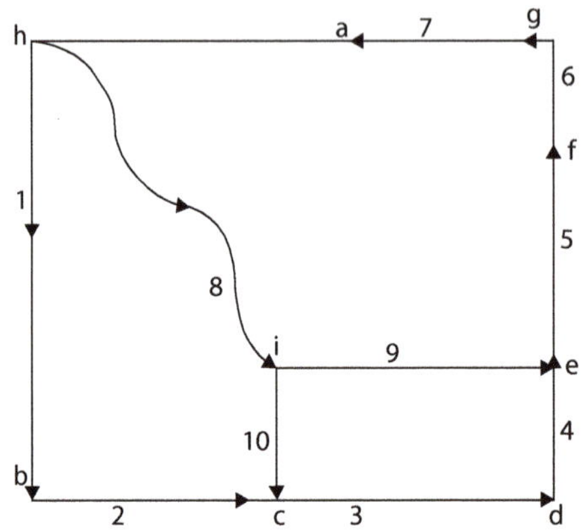

FIGURE 3.10 Topological rule of connectivity.

TABLE 3.2 Arc-node Topology

Connectivity

Arc	From Node	To Node
1	a	b
2	b	c
3	c	d
4	d	e
5	e	f
6	f	g
7	g	a
8	h	i
9	i	e
10	i	c

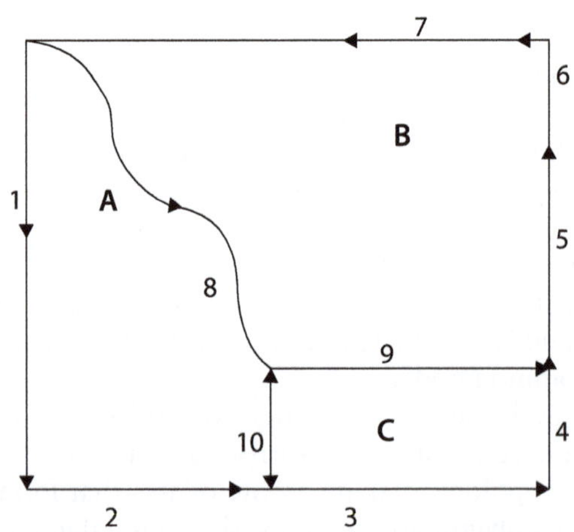

FIGURE 3.11 Topological rule of containment.

TABLE 3.3 Polygon-arc Topology

Containment

Polygon	Arc List
A	1, 2, 10, 8
B	8, 9, 5, 6, 7
C	9, 10, 3, 4

These three topological data rules are central to the functioning of vector data. They allow software programs to analyze and determine spatial relationship between features in order to logically model real-world phenomena. For instance, when using a GPS or a GPS-enabled map on your phone to navigate, let's say you must get on a bridge through an overpass on a road to go to a specific location. How does the mapping application know that the line shown on the map is actually an overpass bridge that passes over the road and it is not another road or street that intersects the main road? The map uses the topological rule of connectivity of road network to determine whether the line on the map is an intersection or a bridge or an overpass. If topological information was missing, the map would have no knowledge of the bridge on the road and would not be able to navigate you. Without topological information it would appear as though there are two roads that intersect at a junction.

Topology also enables automated detection and elimination of errors in vector data. Topological data is very useful for many operations requiring spatial analysis; for instance, finding the shortest path, length of a road segment, or section of a river. Such types of analysis heavily rely on the rule of "from node" to "to node" and use that information along with the attribute information to calculate length, distance, volume, and more.

Another way to differentiate structures of vector data models is to think in terms of how the vector data structure stores spatial and attribute information. In a georelational data model, the spatial and attribute information are stored in separate files. The spatial data is stored in a graphics file and the attribute information is stored in a relational database, and the two are linked by a common ID. The spatial features are linked to their respective record in the database by means of an ObjectID. An example of a georelational data model is shown in figure 3.12. Spatial information for the 254 counties in Texas is stored in one file, and the attribute information is stored separately in the attribute table. Each record in the attribute table is associated with its respective spatial file via the ObjectID. In this case, Harris County with ObjectID 2818 is selected in the table and is highlighted in the map too. In a georelational data model, the software must go back and forth between the spatial data and its corresponding attribute information.

Georelational data models have been used for several years now and are the most popular vector data structures. Esri's ArcGIS shapefile is a popular georelational vector data model structure. Shapefiles do not contain topological information and are therefore also spaghetti data models. They contain features as points, lines, or polygon vector objects. Feature geometries are stored as shapes that consist of x and y coordinates. Shapefiles take less disk space than other GIS data files. They consist of multiple data files that together comprise an actual shapefile.

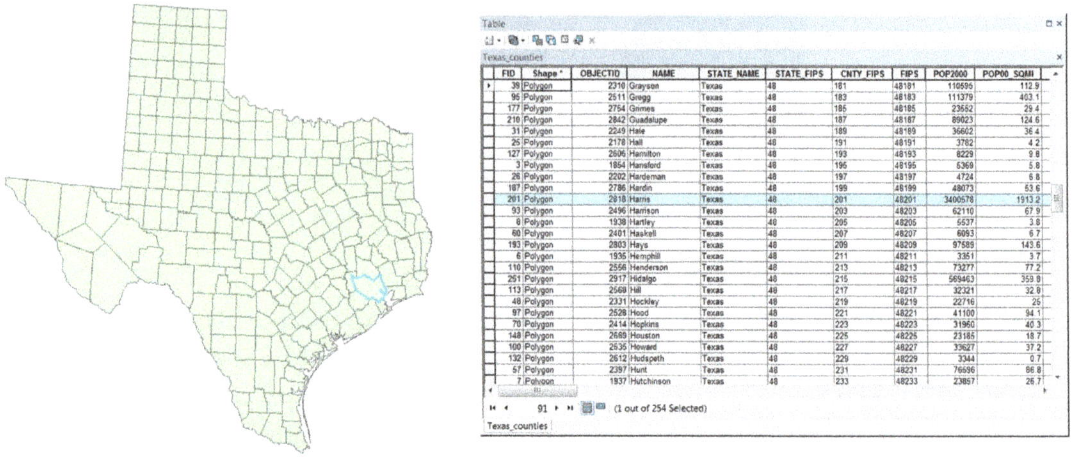

FID	Shape *	OBJECTID	NAME	STATE_NAME	STATE_FIPS	CNTY_FIPS	FIPS	POP2000	POP00_SQMI
39	Polygon	2310	Grayson	Texas	48	181	48181	110595	112.9
96	Polygon	2511	Gregg	Texas	48	183	48183	111379	403.1
177	Polygon	2754	Grimes	Texas	48	185	48185	23552	29.4
210	Polygon	2842	Guadalupe	Texas	48	187	48187	89023	124.6
31	Polygon	2249	Hale	Texas	48	189	48189	36602	36.4
25	Polygon	2178	Hall	Texas	48	191	48191	3782	4.2
127	Polygon	2606	Hamilton	Texas	48	193	48193	8229	9.8
3	Polygon	1854	Hansford	Texas	48	195	48195	5369	5.8
26	Polygon	2202	Hardeman	Texas	48	197	48197	4724	6.8
187	Polygon	2786	Hardin	Texas	48	199	48199	48073	53.6
201	Polygon	2818	Harris	Texas	48	201	48201	3400578	1913.2
93	Polygon	2496	Harrison	Texas	48	203	48203	62110	67.9
8	Polygon	1938	Hartley	Texas	48	205	48205	5537	3.8
60	Polygon	2401	Haskell	Texas	48	207	48207	6093	6.7
193	Polygon	2803	Hays	Texas	48	209	48209	97589	143.6
6	Polygon	1935	Hemphill	Texas	48	211	48211	3351	3.7
110	Polygon	2556	Henderson	Texas	48	213	48213	73277	77.2
251	Polygon	2917	Hidalgo	Texas	48	215	48215	569463	359.8
113	Polygon	2568	Hill	Texas	48	217	48217	32321	32.8
48	Polygon	2331	Hockley	Texas	48	219	48219	22716	25
97	Polygon	2528	Hood	Texas	48	221	48221	41100	94.1
70	Polygon	2414	Hopkins	Texas	48	223	48223	31960	40.3
148	Polygon	2669	Houston	Texas	48	225	48225	23185	18.7
100	Polygon	2535	Howard	Texas	48	227	48227	33627	37.2
132	Polygon	2612	Hudspeth	Texas	48	229	48229	3344	0.7
57	Polygon	2397	Hunt	Texas	48	231	48231	76596	86.8
7	Polygon	1937	Hutchinson	Texas	48	233	48233	23857	26.7

FIGURE 3.12 Georelational data model.

ArcMap interface courtesy of Esri, Inc. Data Sources: Esri; Census Bureau; U.S. Department of Commerce (DOC), National Oceanic and Atmospheric Administration (NOAA), National Ocean Service (NOS), National Geodetic Survey (NGS)

Every shapefile consists of a *.shp file that contains the geometric shape, a *.shx file that is an index file linking the spatial file with the database file, and a *.dbf file that stores the attribute information. These are the three key files that any shapefile requires to function. In addition, a shapefile may contain other data files too. When working with shapefiles, if you want to move the shapefile from one location to another, care must be taken to ensure that all the associated data files are relocated or the shapefile will not function. Chances are that in your introductory class you will mostly make use of shapefiles when working on labs.

Another structure for vector data models based on how spatial and attribute information is stored is the **object-based vector model**. This is a newer structure and is replacing georelational data models. In an object-based vector model, geospatial data are treated as objects, and features and attributes are stored as a single unit. This minimizes the overhead in matching features to attributes. Thus, an object-based vector model is more efficient than a georelational data model. A geodatabase is an example of an object-based vector data model. Geodatabases are very versatile and maintain and store topological information. They contain a variety of geographic datasets such as multiple-feature datasets, tables, rasters, geometric networks, and topology. Vector data are organized as feature datasets and feature classes in a geodatabase. Geodatabases support a variety of data management functions and allow users to define and apply rules and relationships to datasets that follow sophisticated geospatial relations.

The vector data model and its different structures are best suited for representing discrete objects. Figure 3.13 summarizes the structure of a vector data model.

Raster Data Model

GIS datasets are also represented as raster data models. If you have worked with digital photographs, you are already familiar with raster datasets. Every time you have taken a photograph and saved it in your device in a JPEG, TIFF, GIFF, PNG, or other file format, or clicked a selfie or photograph of your surroundings with your smartphone, or edited an image or a picture in JPEG, GIFF, TIFF, or PNG

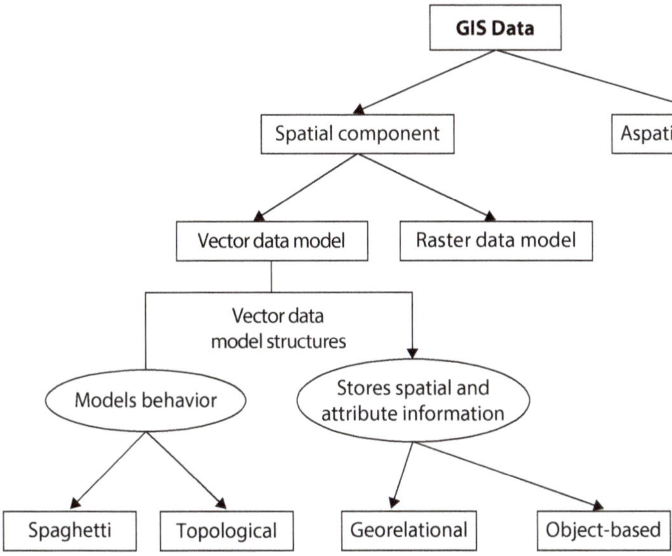

FIGURE 3.13 Vector data model and its structure.

FIGURE 3.14 Zoomed-in raster data.

format using a commercially available software such as Photoshop or Adobe Illustrator, you have worked with a raster dataset. Take the example of any photograph. What do you notice when you zoom into the image? Depending on the resolution, the image will start appearing pixelated at some point, as shown in figure 3.14.

You will notice that when you have zoomed enough into an image, it consists of several tiny square pixels, or "picture elements," and together these pixels create a coherent image. The difference between the images that you generate by clicking a photograph or downloading an image from the web and a raster data that you use in a GIS is that the ones used in GIS contain spatial reference or location information.

A **raster data model** represents the world as a collection of evenly spaced tiny squares of pixels called *grid cells*. Grid cells can also be rectangular, hexagonal, octagonal, or any other geometric shape. However, since square shapes present a simple geometric form to work with, the majority of GIS datasets are built on square pixels. A raster data consists of grid cells arranged in rows and columns. Cell size in a raster data is determined by measuring one side of a square cell and represents the spatial resolution of the raster data (figure 3.15).

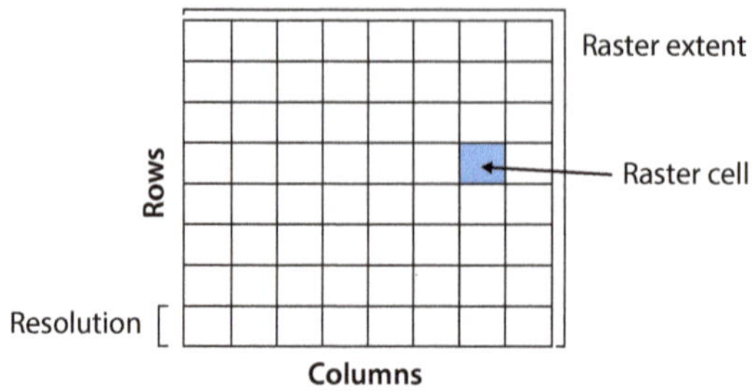

FIGURE 3.15 Grid cells of a raster data model.

The area of each cell (length of the grid cell × width of the square cell) represents the area covered by each grid cell on the ground. A raster data with a spatial resolution of 10 meters will cover an area of 10 m × 10 m = 100 square meters on the ground. Similarly, a raster data with a cell size of 30 meters will cover an area of 30 m × 30 m = 900 square meters on the ground. Thus, a raster data with a cell size of 10 × 10 meters is of a higher spatial resolution compared to a raster data with a spatial resolution of 30 meters. As spatial resolution decreases, the geographic area or the area on the ground covered by each grid cell increases and quality of raster data diminishes, whereas as spatial resolution increases, the geographic area covered by each grid cell decreases and quality of the raster data increases. Raster data with high spatial resolution will result in significantly large file sizes and computational needs for analysis and display. It is thus a trade-off between spatial resolution and file size and computing needs.

Each grid cell in a raster data contains a single value that represents the attribute or the characteristic of a feature or a phenomenon modeled by the raster data at that location (figure 3.16). This value is numeric in nature, stored in either integer or floating point format. The individual cell value of a raster data is called a **digital number** or **pixel value**. In the absence of data, a cell is assigned an arbitrary value of NoData represented by such values as –9999 or 256. This denotes the absence of data at that specific geographic location represented by the cell. Raster datasets represent either continuous data such as elevation, precipitation, and temperature or categorical data that can be grouped into different categories such as land use–land cover and soil type.

1	1	0	0	0	1	1	1	1	5
1	1	1	4	4	4	4	2	2	2
0	0	0	2	2	2	2	2	1	1
1	1	1	3	3	3	3	3	3	3
4	4	4	4	4	2	2	2	2	2
1	1	1	1	1	0	0	0	0	2
0	0	0	0	0	0	0	0	5	5
1	1	1	2	2	2	2	2	1	1
3	3	3	3	3	5	5	5	5	5
3	3	3	3	2	2	2	1	1	1

FIGURE 3.16 Each grid cell in a raster data contains a single value.

In a raster data, the digital number or pixel value represents different types of data. In other words, each pixel value can represent values for a phenomenon represented by continuous data such as temperature, precipitation, or elevation; magnitude values of a phenomenon; land use–land cover class; or spectral values (figure 3.17). For instance, in the case of image rasters such as a satellite image or aerial photograph (figure 3.17a), each pixel value corresponds to the wavelength of light reflected by the feature on the ground and captured by the camera sensor. The values are recorded as pixel values representing different colors, where each color corresponds to a specific pixel value. If you scan a paper map, the output will be an image file, which is a raster data. In this case, the pixel value of each grid cell will correspond to an index value representing a specific color captured by the scanning device, such as USGS's digital raster graphic (DRG) data shown in figure 3.17b. In the case of a raster data representing continuous data such as elevation, shown in the digital elevation model (DEM) in figure 3.17c, each pixel value represents an elevation value at that location on the ground. In the land use–land cover map shown in figure 3.17d, each pixel value represents a specific land use–land cover category at that point on the ground such as vegetation, built-up area, and water.

Cell size is an important factor in a raster data model. As we already know, each cell or pixel in a raster data is assigned only one value. Take the case of a raster data with a spatial resolution of 5 kilometers. Each cell will represent an area of 5 km × 5 km = 25 square kilometers. Let's say this is a raster data model representing categorical data, such as land use–land cover. It is highly possible that an area covering 25 square kilometers consists of more than one type of a land use–land cover class represented by multiple features at that location on the ground, and each feature will have a specific pixel value. Similarly, a raster data model representing continuous data such as relative humidity, depth, or height may have different values for an area covering 25 square kilometers. However, a cell or a pixel cannot be assigned more than one value. A raster data model will either average the pixel values to assign a single value to the cell in the case of a continuous data or assign the value of the dominant

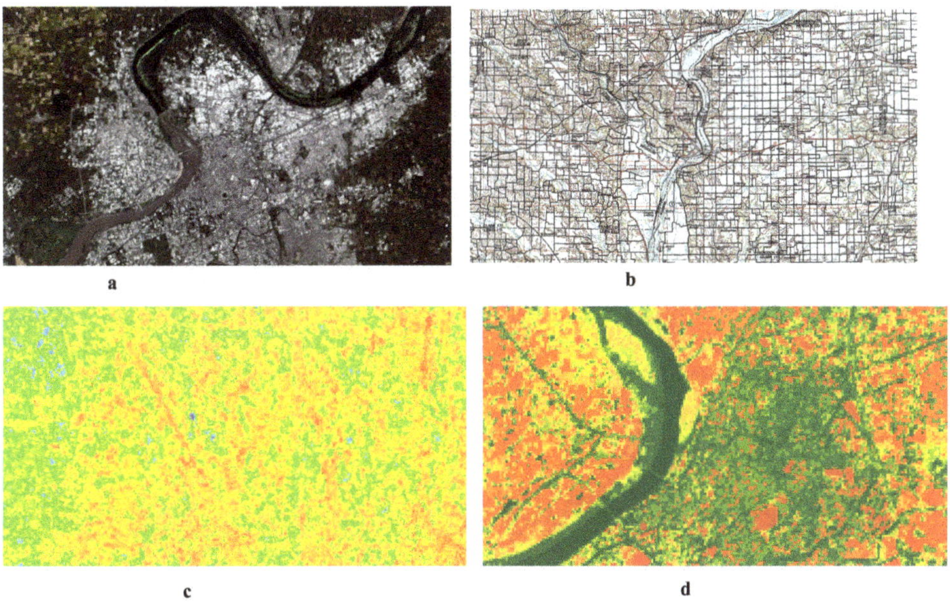

FIGURE 3.17 Examples of raster data: (a) satellite image, (b) digital raster graphic (USGS), (c) digital elevation model, (d) land use–land cover map.

Data source: (a, c, d) Mukherjee and Singh (2020); (b) USGS; public domain

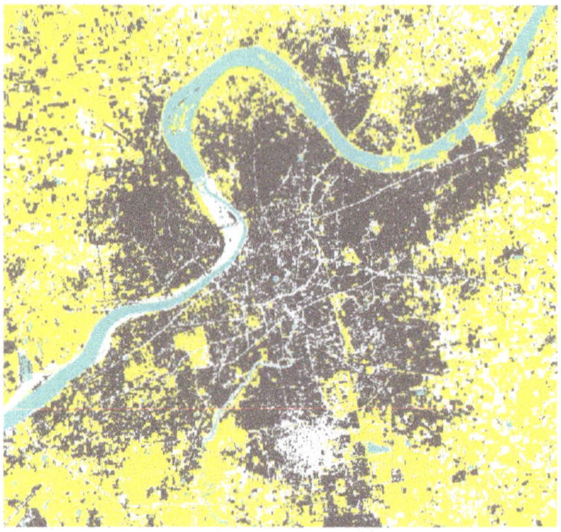

OID	VALUE	COUNT	CLASS
0	1	16646	Water body
1	2	59025	Bare soil
2	3	101740	Vegetation
3	4	84214	Built up

FIGURE 3.18 Attribute table of a raster data model representing categorical data.

ArcMap interface courtesy of Esri, Inc.

Credit: Mukherjee and Singh (2020)

feature to the cell in the case of a categorical data. Therefore, in a raster data with lower spatial resolution where each cell covers a large geographic area, the values associated with each cell will not be as precise as a raster data with higher spatial resolution where each cell covers a smaller geographic area.

Raster data models that represent categorical data will have an attribute table (figure 3.18). In the attribute table, pixel values are assigned to each category. The attribute table here consists of four columns: *OID*, *Value*, *Count*, and *Class*. The first three columns—*OID*, *Value*, and *Count*—are default columns generated by the system. *OID* is a unique identification number for each row in the table. The *Value* column holds a list of unique cell values in the raster data. The *Count* column provides a list of the total number of cells, with the cell value in the *Value* column. The *Class* column provides a list of categories assigned to each cell value. Examining the attribute table will give you the total number of pixels or cells in the raster data that belongs to each category. For instance, as shown in figure 3.18, there are 16,646 cells/pixels in the raster data with cell value 1 that belong to land cover class water body, and 59,025 cells/pixels with cell value 2 that belong to land cover class bare soil. Raster data models representing continuous data do not have an attribute table.

Comparing the Two Spatial Data Models

Vector data models are more precise and accurate than raster data models. The building blocks of a vector data model are point vector objects with zero spatial dimension, described by a set of x, y coordinates that represent the object's actual precise geographic location. Complex vector objects are formed from these simple vector objects, with precise location information providing vector data models with accurate location information. Vector data models can therefore be used to measure, examine, and view precise location information. Changing the scale of visualization does not impact the manner in which a vector data object is displayed. In other words, you can zoom into a vector data object and the way a point, line, or a polygon is displayed will not change, whereas in a raster data model, as you zoom in, the data will start appearing pixelated.

A vector data model has a compact structure compared to a raster data. Therefore, vector data files are smaller in size. Raster data tend to be larger in size, with high storage and computational requirements. Because of its data structure, vector data models are better suited for certain types of analysis, particularly those that require precision or accurate location information or where spatial relations are key for analysis purposes such as error detection, network analysis, shortest path analysis, cluster analysis, and linear referencing. Vector data models, however, use complex algorithms in the background to analyze data and can lead to higher computational needs for large datasets.

Raster data can be generated easily, using inexpensive devices. Today, everyone has access to a camera, and through the ubiquitous smartphone, anyone can click a picture anywhere, generating raster data. Raster data models of the earth are also readily available, and we have access to several different types of satellite data with varying spatial resolutions. It is a lot easier to visualize and understand raster data models of the earth because they provide a worldview that looks exactly like it is in reality.

Compare the vector view and the raster view of Austin, Texas, shown in figure 3.19. In the vector view, the city boundary is shown as a light yellow polygon and the streets are shown as fuchsia pink lines. Compare this with the raster view of the city derived from Google Earth. In the raster view, once you search the location and display it on your map, you can actually see all the features present on the ground depending on the scale. You can visualize the streets, streams, lakes, and buildings. Notice that Lake Travis, which is clearly visible in the raster view, is not visible in the vector view because the vector view will only display those features or vector objects that have been added to the map. In the vector map shown in figure 3.19, only the city boundary and roads have been added to the map, whereas in the raster view, you get a picture or snapshot of everything or almost everything that is present in the environment. This snapshot is displayed as a group of cells or pixels that gives you a coherent picture.

Raster data files are large in size, particularly when the area covered is large and spatial resolution is high. This in turn leads to high computation requirements for data analysis and visualization. Raster data also poses problems if there is a need to reproject the data. As discussed in chapter 2, map projections distort the shape, size, angle, or area of features. Geometric transformations that occur during the map reprojection process alter the output layer. As a result, reprojection of raster datasets that consist of an array of pixels poses significant challenges. Raster data is not well suited for certain kinds of spatial analysis, particularly those that require working with discrete data or where precise feature or location information is critical for the analysis.

File Formats

Both vector and raster data files are available in different formats. Esri's shapefile and geodatabase file formats are two of the most well-known vector data file formats that model earth features and phenomena as points, lines, and polygons.

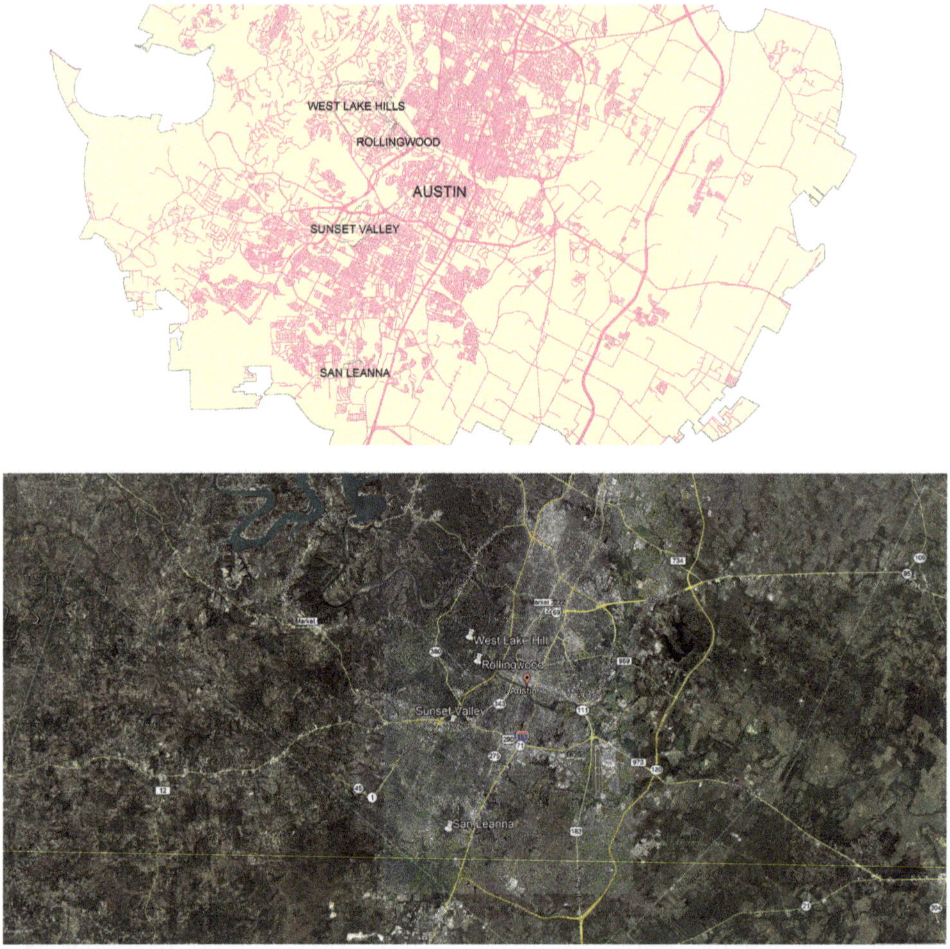

FIGURE 3.19 Vector versus raster view.

Credit: Raster view from Google Earth. Data Sources: Esri; U.S. Department of Commerce, U.S. Census Bureau

Other vector file formats are also widely used, such as Google's Keyhole Markup Language (KML) files. If you have used Google Earth, you are familiar with KML files. KML files are used by earth browsers such as Google Earth, Google Earth Pro, and Google Maps to display geographic data. GPS data files such as GPS Exchange Format, or GPX, derived from GPS devices are another widely used vector file format. Digital Line Graph, or DLG, files are vector format data created by the US Geological Survey. DLGs are vector data extracted from topographic maps in DLG format. The files can be used in most GIS applications. Additionally, other application packages such as IDRISI, MapInfo, and AutoCAD have their own proprietary vector data formats such as IDRISI's VCT files, MapInfo's TAB files, and AutoCAD's DXF files.

Similarly, many raster data file formats are available to use in GIS for spatial analysis. Some of them are proprietary formats such as Esri's GRID file format. Another proprietary raster data file format is ERDAS Imagine's IMG files. Single and multiband satellite image data are stored as IMG files and analyzed in ERDAS Imagine. Another widely used raster file is GeoTIFF. GeoTIFF files are extensively used for GIS and remote sensing applications. Software applications such as ENVI and IDRISI also have their own proprietary raster data files. The digital elevation model, or DEM, format produced and distributed by the USGS is another popular raster

FIGURE 3.20 Vector data overlaid on raster data.

Credit: Raster image derived from ArcGIS Online

data file format that is used in GIS. Other raster data files that we all are familiar with are the common formats used to store images such as JPEG (Joint Photographic Experts Group), PNG (Portable Network Graphics), GIF (Graphics Interchange Format), BMP (Bitmap), and MrSID (Multiresolution Seamless Image Database). However, JPEG, GIF, PNG, and BMP file formats are not used in GIS for spatial analysis.

It is common to use both vector and raster data in GIS projects and applications where raster data such as a satellite image or aerial photograph is used as a basemap to overlay vector data of roads, streams, and buildings shown as points, lines, and polygons on top of the raster data. As shown in figure 3.20, zip code boundaries represented by red polygons, an airport represented by a green point, and hospitals represented by blue points are overlaid on a basemap of a world topographic image derived from ArcGIS Online. Such usage of vector and raster data in conjunction with one another is a fairly common practice.

Raster data can be converted to vector data and vice versa. GIS software packages provide conversion tools or options to convert from one format to another. In vector to raster conversion, points, lines, and polygons are converted into an array of grid cells. The raster data will be based on any one attribute in the vector data. So if you want to create a raster layer of different attributes from the vector data, you would have to create multiple raster data. Raster to vector conversion can be used to convert raster data composed of an array of pixels into vector data represented by points, lines, and polygons. The accuracy of the vector data will depend on the spatial resolution of the raster data. The higher the spatial resolution of the raster data used for conversion, the more accurate the vector data will be.

Questions and Activities

1. Compare discrete objects with continuous objects. For each of the following types of data, indicate which ones are discrete and which ones are continuous objects:
   a. benches in your classroom
   b. computers in your classroom
   c. buildings on your college campus

d. the oxygen level in your classroom
e. sales tax rate
f. relative humidity in your area of residence
g. average daily temperature
h. soybeans harvested in Illinois

2. Summarize the salient characteristics of a vector data model. What are the advantages and disadvantages of a vector data model?
3. Why is topology important for a vector data model? Can you think of a few real-world applications of topology that you encounter on a daily basis?
4. Table 3.4 describes the topology of a map with two polygons, A and B. (Note: A dash [–] indicates that the left or right polygon is outside the map.) Draw the map so that it has the correct topology. Label all points, lines, and polygons with their correct identifiers on your map. Make sure to show arrows to indicate the direction and the *Begin Point* and *End Point*. Draw the map with the left and right polygons oriented as though you were standing on the *Begin Point* and looking towards the *End Point*.

TABLE 3.4

Line ID	Begin Point	End Point	Left Polygon	Right Polygon
5311	5	2	B	A
5312	2	3	B	–
5313	4	5	–	A
5314	5	6	–	B
5315	3	6	B	–
5316	4	1	A	–
5317	1	2	A	–

5. Draw a diagram from the topological information in table 3.5. Note that all arcs are not necessarily straight lines. They can be curved lines too.

TABLE 3.5

Arc	From Node	To Node
A	1	2
B	3	2
C	3	1
D	1	4
E	3	4
F	5	5
G	4	2

6. Use the topological information given in table 3.6 to create polygons. Use arc and arc direction to combine polygons. Note that all arcs are not necessarily straight lines. They can be curved lines too.

TABLE 3.6

Polygon	Number of Arcs	Arc List
A	3	A, D, G
B	3	C, D, E
C	1	F
D	4	B, E, G, -F

7. Summarize the key characteristics of a raster data model. What are the advantages and disadvantages of a raster data model?
8. What type of a vector object would you use to represent the following features: shopping mall, political boundary, fence around a parcel, park, subway route, arterial roads, streets in a neighborhood, stream, land parcel, college campus?
9. In the question above, do you think you can represent some of the features using multiple vector objects? Why? Name those features and what other vector object you would use to represent them.
10. Many online resources allow members of the public to view GIS maps and data and perform basic GIS tasks. GIS data for many city and county governments are available online for the public to view and analyze. These maps display both vector and raster data.

    One such example is the city of Huntsville, Texas. Open your web browser and go to https://www.arcgis.com/apps/webappviewer/index.html?id=de44957e55ac4ffc98fa10d26f4a88e7. In the search option available on the page, type 4650-001-0-00100. This is the GEO ID of H-E-B, a grocery supermarket chain located in Huntsville. Hit enter on your keyboard or click on the search tool (looks like a magnifying glass). The map will zoom to the parcel of the grocery store and a pop-up window will open with all the attribute information.

    a. What can you tell about the type of data displayed on the map? Do you see both vector and raster data? Make a note of the vector data and raster data on the map.
    b. At the bottom of the window is a toolbar with tools to display Layer List, Legend, Measurement, Print, Bookmark, About, and Draw. Click on the tool to display Layer List. A window will open that shows all the layers with a checked box displayed on the map. At the bottom of the Layer List, you have Topographic Basemap and 2018 Aerial. Make sure that 2018 Aerial is checked and the Topographic Basemap is unchecked. Check and uncheck a few of the boxes and see the changes in your map. You may have to zoom out to see the changes when you check/uncheck the boxes.
    c. Check the box for Topographic Basemap, Major Contours and Minor Contours. What can you say about the elevation of the area where H-E-B is located?
    d. Check the box for Transportation to display all the roads on your map. How does the representation of roads by this layer compare with the roads

displayed on the 2018 Aerial photograph? Do they match up with each other?

e. Uncheck the box for Topographic Basemap and keep the box for 2018 Aerial checked. Explore the Measure tool. This tool gives you three options—Measure Area, Measure Distance, and Locate—that will give you latitudes and longitudes. Click on the Measure Area feature and measure the area of the parking lot of H-E-B. From the Units drop-down menu, select either sq. meters or sq. feet. Using the Measure Distance feature, measure the distance from the entrance of the grocery store to the end of the parking lot in feet or in meters. Using the Locate feature, record the latitudes and longitudes at the four corners of the parcel.

f. You can click on any of the parcels shown on the map and a pop-up window will open showing all the attribute information for that specific parcel. Explore the area around H-E-B. What can you say about the area? Is it a residential area or a commercial area? What kind of businesses, if any, are located in the area?

11. Check your city or county government to see if they have a similar online GIS website. What kind of layers are available for you to visualize? What specific tools are available to perform tasks similar to those in question number 10?

Key Terms

adjacency (p. 58)
connectivity (p. 59)
containment (p. 59)
continuous surface view (p. 53)
data model (p. 53)
DEM (p. 68)
digital number, or pixel value (p. 64)
discrete object view (p. 53)
DLG (p. 68)
georelational data model (p. 61)
GeoTIFF (p. 68)
GPX (p. 68)
IMG (p. 68)
KML (p. 68)
line (p. 54)
object-based vector model (p. 62)
point (p. 54)
polygon (p. 55)
raster data model (p. 63)
spaghetti data model (p. 58)
topological data model (p. 58)
topology (p. 58)
VCT (p. 68)
vector data model (p. 53)
vector objects (p. 53)

References

Jensen, John, and Ryan Jensen. 2013. *Introductory Geographic Information Systems.* Boston: Pearson.

Mukherjee, Falguni, and Deepika Singh. 2020. "Assessing Land Use–Land Cover Change and Its Impact on Land Surface Temperature Using LANDSAT Data: A Comparison of Two Urban Areas in India." *Earth Systems and Environment* 4: 385–407. https://doi.org/10.1007/s41748-020-00155-9.

Schmitz, Andy. 2012. *Essentials of Geographic Information Systems.* Arlington, VA: Saylor Academy.

Shellito, Bradley. 2018. *Introduction to Geospatial Technologies.* New York: W. H. Freeman–Macmillan Learning.

Attribute Data 4

Learning Objectives

- Explain what aspatial data is and its role in GIS
- Explain what an aspatial data model is in GIS
- Explain what a database and database management system are
- Articulate what a relational database management system is and how it is used in a GIS
- Describe the key characteristics of a relational database management system that is utilized in a GIS
- Articulate what cardinality is in a relational database management system and how it impacts GIS operations
- Discuss types of data and explain the differences between each type
- Identify different functions in a GIS attribute table and be able to execute basic operations using an attribute table

Representing Aspatial Data and Information

When viewing a GIS map in a digital medium, whether on your computer screen or your mobile device, your map is displaying information that is stored somewhere in the background and connected to the geospatial layers displayed on the map. Let's go back to the example of the Thai restaurant discussed in chapter 1. When looking for a Thai restaurant, you are obviously interested in a restaurant with good ratings. Also, you would like to select a restaurant that offers delivery services to your area. And most importantly, you would like a restaurant within 10 miles of your apartment. When you search for Thai restaurants, your map shows you all the Thai restaurants in your area. Now you can select each restaurant to review their ratings, visit the restaurant website, and check their hours, the menu, and whether the restaurant provides delivery options.

All of this information is nonspatial in nature and represents the attributes of individual restaurants that are displayed as points on your map, with specific symbology. This attribute information is very important in a GIS. It enables one to analyze geospatial data and make decisions, like you would decide which restaurant to order your food from based on the information provided. Many important decisions are made based on careful analysis of attribute information. A GIS has a unique ability to answer "what," "where," "why," and "how much" questions. This ability sets GIS apart from other information systems.

The answer to the "where" question relates to the location of the feature or phenomenon, discussed in chapter 2. Whereas answers to "what,", "why," and "how much" questions depend on detailed, and at times complex, analysis of information about a feature or phenomenon's attributes. This attribute information is not visible when you view your map at first. This is because the attribute information is stored as tabular data in what is called an attribute table. Every geospatial data that is displayed in a GIS has an associated attribute table. This table stores additional information about that specific geospatial data.

Can you recollect our discussion on data models from chapter 3? What is a data model? It is an abstract method for representing real-world features, phenomena, and their characteristics. It determines how data is stored, analyzed, extracted, retrieved, and displayed. We already know that every geospatial data has two components—a *spatial component* and an *aspatial component*—with a link between the two components, connecting attribute information to location information. This not only provides a GIS its unique analytical capabilities, but also enables it to spatially analyze geospatial data and map and display the results for visualization purposes. This is where other information systems and database systems lag behind a GIS. Accordingly, a GIS consists of two data models: one for the spatial component and one for the aspatial component (figure 4.1). We discussed data models for the spatial component of a GIS in chapter 3. In this chapter, we discuss data models for the aspatial component of a geospatial data.

There is no doubt that in today's world, everything is driven by data and information. No matter what discipline or background one comes from, data and information are the key drivers of all business processes and applications. The same holds true for a GIS. One can use a GIS to answer a myriad of questions. For instance, which is the closest park to my neighborhood? How many neighborhoods in the area have a tennis court? Which neighborhood is zoned to a specific school district? What is the average precipitation in an area during a particular season? What is the average lot size or the average home price in a specific neighborhood? What is the five-year trend in home prices in an area? Is there a correlation between incidence of a particular disease and income or race? Are certain areas or neighborhoods witnessing more crime or disease?

Answers to any of these questions can be derived from information that is obtained by processing and analyzing data. Therefore, in order for a GIS to perform any kind of analysis, it is important that it has access to data that is structured and organized in a certain manner, the *aspatial data model.* With widespread availability of the means to create digital data, large amounts of data are being generated and created in a number of different ways and in several different formats. GIS is increasingly used to answer complex questions that involve analyzing information obtained from processing large amounts and types of data. Data that is created and owned by multiple departments or agencies. Data that can be maintained and updated on a regular basis with ease, in order to derive the most updated information. Data that must be accessed by multiple users with different access privileges. Data with high performance capabilities that have the ability to maintain linkages with multiple applications. Hence, a system that is versatile enough to encapsulate these requirements is needed. A database management system (DBMS) provides such versatility and is the data model for the aspatial component of geospatial data.

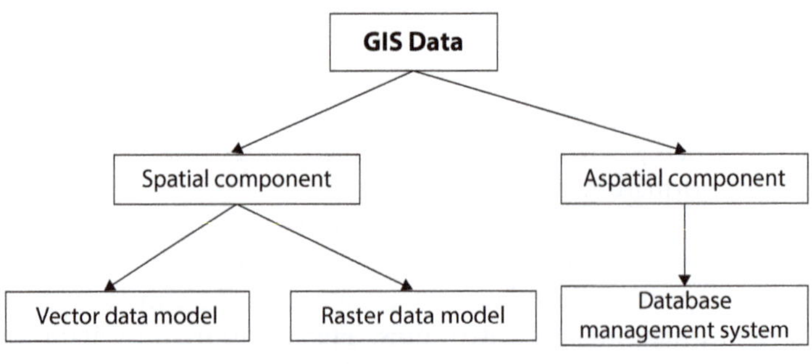

FIGURE 4.1 Geospatial data models.

Database Management Systems

Databases and database management systems have originated within the discipline of computer science. However, their application and use is not limited to computer science. DBMSs have widespread utility and are used by businesses, government agencies, universities, banks, or any organization that needs tabular data in a digital format to be used by a computer.

A **database** is a standardized collection of data stored in a computer system. It consists of a collection or series of files that store data. The data is stored in the form of rows and columns, and the file is called a **table**. A DBMS is a specialized information system for managing multiple databases. It is a versatile system designed to create, store, manipulate, retrieve, and manage data in a database. A GIS integrates several functionalities of a DBMS. Additionally, it contains specific geographic information for each data, such as its shape and location. Thus, in case of GIS, the DBMS manages databases that store data corresponding to a specific location or place in an attribute table. An attribute table is a table that stores attributes of map features and is associated with locational data. For instance, the attribute table shown in figure 4.2 consists of information for counties in Alabama. Columns in the attribute table contain the name of the county, name of the state, state abbreviation, state FIPS, and county FIPS.

Different types of DBMSs exist based on their specific data handling capabilities. DBMSs have evolved considerably since their inception in the 1960s, when the system would only handle individual databases that stored data formatted in a specific fashion, such as hierarchical or network DBMS. Several structures of DBMS have emerged since then such as relational, object oriented, distributed, NoSQL, open source, cloud, multimodal, and JSON. The most prevalent system is a relational DBMS that became a dominant system in the 1980s. It is the most common system used in a GIS.

A **relational database management system (RDBMS)** is a system designed to manage a relational database. A **relational database** consists of multiple tables where each table shares a common link with another table by means of a common attribute or an identifier, called the *key*. Let's take the example of your university. Your university has many division offices. Each university division has several departments under it. For instance, the division of academic affairs will have all colleges under it, and each college consists of many departments. The university also has other department offices housed under specific university divisions, such as the admissions office, bursar's office, registrar's office, financial aid, health center, international students' office, student affairs, and information technology. Most of these departments must store and maintain student records. However, each department needs to work with specific student record data.

	Name	State Name	State Abbreviation *	State FIPS	County FIPS
1	Autauga County	Alabama	AL	01	001
2	Baldwin County	Alabama	AL	01	003
3	Barbour County	Alabama	AL	01	005
4	Bibb County	Alabama	AL	01	007
5	Blount County	Alabama	AL	01	009

FIGURE 4.2 Attribute table showing information for counties in Alabama.

ArcMap interface courtesy of Esri, Inc. Data Sources: Esri; U.S. Department of Commerce, Census Bureau; U.S. Department of Commerce (DOC), National Oceanic and Atmospheric Administration (NOAA), National Ocean Service (NOS), National Geodetic Survey (NGS)

For example, for the financial aid office, data pertaining to students who have applied for financial aid is relevant, whereas for the office of international students, data pertaining to international students who have applied for admission to the university and are enrolled in classes is relevant. However, both offices must also keep track of student GPA, data that is stored and maintained by respective departments, colleges, the registrar's office, and other university departments. Think about how many students attend your university. All university departments must store and maintain records of every student enrolled in the university in their respective databases. In this scenario, the size of the database is very large, consisting of several files with huge amounts of data. Moreover, not everyone must have access to all student data in the database due to privacy reasons. An RDBMS, in this case, will enable each department to store and maintain specific data pertaining to student records following appropriate protocols and retrieve data from different databases that belong to other divisions or departments but are related by means of a unique identifier or key, which in this case would be the student ID.

Take another case of the tables shown in figures 4.3 and 4.4. These tables contain specific information pertaining to US states. They are two separate tables

STATE_NAME	STATE_FIPS	STATE_ABBR	POP2010	POP10_SQMI
Alabama	01	AL	4779736	92.5
Alaska	02	AK	710231	1.2
Arizona	04	AZ	6392017	56.1
Arkansas	05	AR	2915918	54.8
California	06	CA	37253956	235.6
Colorado	08	CO	5029196	48.3
Connecticut	09	CT	3574097	720.2
Delaware	10	DE	897934	454.5
District of Columbia	11	DC	601723	9675.6
Florida	12	FL	18801310	331.6
Georgia	13	GA	9687653	164.7
Hawaii	15	HI	1360301	211.6
Idaho	16	ID	1567582	18.8
Illinois	17	IL	12830632	227.7
Indiana	18	IN	6483802	179.2
Iowa	19	IA	3046355	54.1
Kansas	20	KS	2853118	34.7
Kentucky	21	KY	4339367	107.4
Louisiana	22	LA	4533372	96.9
Maine	23	ME	1328361	40.9
Maryland	24	MD	5773552	583.2
Massachusetts	25	MA	6547629	808
Michigan	26	MI	9883640	170
Minnesota	27	MN	5303925	62.9
Mississippi	28	MS	2967297	62.2
Missouri	29	MO	5988927	85.9
Montana	30	MT	989415	6.7
Nebraska	31	NE	1826341	23.6
Nevada	32	NV	2700551	24.4
New Hampshire	33	NH	1316470	142.1
New Jersey	34	NJ	8791894	1164
New Mexico	35	NM	2059179	16.9
New York	36	NY	19378102	398.9

FIGURE 4.3 Table of US states with 2010 population data.

ArcMap interface courtesy of Esri, Inc. Data Sources: Esri, TomTom, U.S. Department of Commerce, U.S. Census Bureau

USA-States

Field: Add Calculate Selection: Select By Attributes Zoom Switch Clear Delete

	Shape	FID_	STATE_NAME	STATE_FIPS	WHITE	BLACK	AMERI_ES	ASIAN	HAWN_PI	HISPANIC
1	Polygon	0	Alaska	02	473576	23263	104871	38135	7409	39249
2	Polygon	0	California	06	21453934	2299072	362801	4861007	144386	14013719
3	Polygon	0	Hawaii	15	336599	21424	4164	525078	135422	120842
4	Polygon	0	Idaho	16	1396487	9810	21441	19069	2317	175901
5	Polygon	0	Nevada	32	1786688	218626	32062	195436	16871	716501
6	Polygon	0	Oregon	41	3204614	69206	53203	141263	13404	450062
7	Polygon	0	Washington	53	5196362	240042	103869	481067	40475	755790
8	Polygon	0	Arizona	04	4667121	259008	296529	176695	12648	1895149
9	Polygon	0	Arkansas	05	2245229	449895	22248	36102	5863	186050

FIGURE 4.4 Table of US states with demographic data.

ArcMap interface courtesy of Esri, Inc. Data Sources: Esri, TomTom, U.S. Department of Commerce, U.S. Census Bureau

and therefore two separate files. In a relational DBMS, these two tables will share a common link by means of a key. In this case, the key is state ID, stored in the tables as STATE_FIPS.

An RDBMS has two types of keys: a primary key and a foreign key. The primary key is a unique identifier used to link one database with another. This will be a specific column in a table that uniquely identifies records in each row to link it with another table. The primary key corresponds to a specific column in another table, called the foreign key. The column STATE_FIPS in the table in figure 4.3 is the primary key, and the column with the same name in figure 4.4 is the foreign key. However, column names need not be the same. It is the data contained in the column that must be unique to link the two tables. Take the case of a university database that stores and maintains records of different university divisions and departments. Let's consider the two tables shown in figure 4.5. Each table is stored and maintained by two different departments.

Let's say the table on the top is one of the files in a database by department A in division 1, and the table at the bottom is a file in another database that belongs

PRIMARY KEY

Student_name	Student_id	Major	Minor	Department	College	GPA
Smith, John	15935709	Geography	Geology	Env-Science	Science	3.6
Johnson, Dave	25848602	History	Sociology	History	Humanities	2.98
Cooper, Allison	11185209	Biology	GIS	Env-Science	Science	3.25
Williams, Charles	35795102	Statistics	Math	Math-Statistics	Science	2.35
Phillip, Harold	35620808	Chemistry	Physics	Chemistry	Science	3.0
Hernandez, Luisa	82274603	Geology	GIS	Env-Science	Science	2.0
Coffey, Mike	25430809	English	Philosophy	English	Humanities	2.50
Seaman, Pearl	82329710	Mass Communicati ons	Art	Art	Arts	3.75

FOREIGN KEY

Name	Id_number	Application number	Approved	Department	College
Neilson, Sam	10732906	NS_906253	Y	Chemistry	Science
Hernandez, Luisa	82274603	HL_603467	Y	Env-Science	Science
Krishnan, Neel	84956605	KN_605249	Y	Accounting	Business
Seaman, Pearl	82329710	SP_710861	N	Art	Arts
Morgan, Mike	26784302	MM_30244	N	Sociology	Humanities
Cooper, Allison	11185209	CA_209656	Y	Env-Science	Science
Williams, Charles	35795102	WC_102956	N	Math-Statistics	Science

FIGURE 4.5 Primary key and foreign key.

to department B in division 2. Both of the tables store and maintain data relevant to individual departments. One table stores data related to student major, minor, department, college, and GPA, while the other table stores data pertaining to student grant applications, with the application number, whether the application was approved or not, and the applicant's department and college. Let's say the university has undertaken a study that examines student performance and would like to assess how students who were approved for grants are performing in their classes. One way to do that would be to look at student GPAs. So the best approach would be to link the two tables. In order to do that, you would need a common identifier or key. In this case, student ID would be a unique identifier.

In the table at the top of figure 4.5, student ID is stored in a column named Student_id, and in the table at the bottom, student ID is stored in a column named Id_number. Here, Student_id is the primary key and Id_number is the foreign key. Although the columns in both of the tables have different names, it is the content of the columns that uniquely identifies each student and can link student records from one table to another. If the primary key and the foreign key have the same field name in each database, they are called the common key.

In an RDBMS, two tables that have at least one common attribute or key can be combined into a single table using the key. For instance, the table in figure 4.3 contains 2010 data for the population of US states, whereas the table in figure 4.4 contains 2010 demographic data of US states. Let's say that we would like to have a single table that contains both population and demographic data of US states. In this case, the two separate tables can be combined into a single table using the key STATE_FIPS. Now the new table shown in figure 4.6 contains 2010 population and demographic data in a single table.

The new table can now be used to create new data. For instance, let's say we would like to calculate the difference between male and female populations in US states to examine which states have a higher male population and which states have a higher female population and by how much. You can do that by adding a new column to your table and calculating the difference between the two population numbers, as shown in figure 4.7.

You can also use the table to retrieve data for a specific US state. Most RDBMSs use a special programming language called structured query language (SQL) to access the database and communicate with data. (We will cover query later in this chapter.) For instance, if you are interested in retrieving data for a specific US state from the table shown in figure 4.7, you can do so by querying the table using SQL. Similarly, let's say the university is particularly interested in students with a GPA of 2.5 or lower. Using the tables in figure 4.5, a query for student GPA will give you results for all students with a GPA of 2.5 or lower, with additional information such

Field: Add Calculate Selection: Select By Attributes Zoom To Switch Clear

	Shape	FID_	STATE_NAME	STATE_FIPS	POPULATION	POP_SQMI	WHITE	BLACK
1	Polygon	0	Alaska	02	744733	1.3	473576	23263
2	Polygon	0	California	06	39611295	250.5	21453934	2299072
3	Polygon	0	Hawaii	15	1461211	227.3	336599	21424
4	Polygon	0	Idaho	16	1714694	20.5	1396487	9810
5	Polygon	0	Nevada	32	2994047	27.1	1786688	218626
6	Polygon	0	Oregon	41	4122334	42.5	3204614	69206

FIGURE 4.6 Table with population and demographic data of US states.

ArcMap interface courtesy of Esri, Inc. Data Sources: Esri, TomTom, U.S. Department of Commerce, U.S. Census Bureau

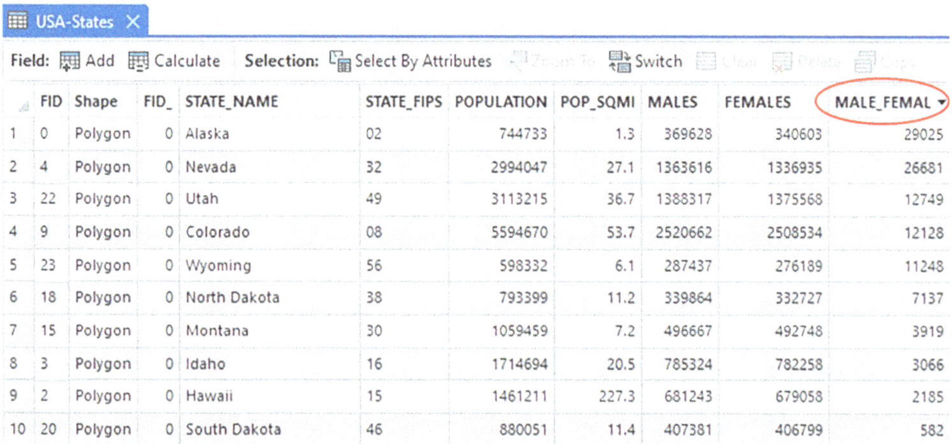

	FID	Shape	FID_	STATE_NAME	STATE_FIPS	POPULATION	POP_SQMI	MALES	FEMALES	MALE_FEMAL
1	0	Polygon	0	Alaska	02	744733	1.3	369628	340603	29025
2	4	Polygon	0	Nevada	32	2994047	27.1	1363616	1336935	26681
3	22	Polygon	0	Utah	49	3113215	36.7	1388317	1375568	12749
4	9	Polygon	0	Colorado	08	5594670	53.7	2520662	2508534	12128
5	23	Polygon	0	Wyoming	56	598332	6.1	287437	276189	11248
6	18	Polygon	0	North Dakota	38	793399	11.2	339864	332727	7137
7	15	Polygon	0	Montana	30	1059459	7.2	496667	492748	3919
8	3	Polygon	0	Idaho	16	1714694	20.5	785324	782258	3066
9	2	Polygon	0	Hawaii	15	1461211	227.3	681243	679058	2185
10	20	Polygon	0	South Dakota	46	880051	11.4	407381	406799	582

FIGURE 4.7 Table with new data.

ArcMap interface courtesy of Esri, Inc. Data Sources: Esri, TomTom, U.S. Department of Commerce, U.S. Census Bureau

as their major, minor, college, department, and whether their grant application was approved or not.

Thus, an RDMBS enables a user the means to retrieve data based on specific criteria, student GPA in case of the university database in figure 4.5 or state name in case of the database file in figure 4.7. It also allows individual files to be created, maintained, and stored separately. For instance, a university department only needs to store and maintain data that is relevant to their department. And when needed, separate tables can be linked using student ID and the necessary information can be retrieved. An RDBMS also allows multiple users to edit the database simultaneously. If the data in one file in a database changes, only that file needs to be updated. There is no need to update every file in every database. Thus, an RDBMS provides an efficient and flexible approach to process data and manage databases. It also enables data protection and security so that data can be safely accessed and processed, whereby specific users can be authorized to access certain data or make changes. This is particularly useful in an environment involving large databases with multiple users.

RDBMS and GIS

A GIS includes the functionality of an RDBMS and integrates it to work with geographic data. Therefore in a GIS, an RDBMS manages databases that store geographic data. The geographic data is stored and maintained in attribute tables. A GIS can also work with non-geographic data files in the database. These are called standalone tables because they are not associated with geographic data. However, standalone tables must be linked with an attribute table for spatial analysis and visualization purposes.

In an attribute table, attributes are stored as column headers and records are stored as rows (figure 4.8). The contents of an attribute for each individual record is its *value*. The intersection of a row and a column is unique and is called a *cell*. A cell holds individual attribute values.

Several conditions of an RDBMS are particularly relevant in a GIS. Each column in a table represents the data for a single attribute for the dataset. Every row in a table is unique, and a table must not have duplicate rows. Every cell in a table contains one and only one value, and all values in a column pertain to the same specific

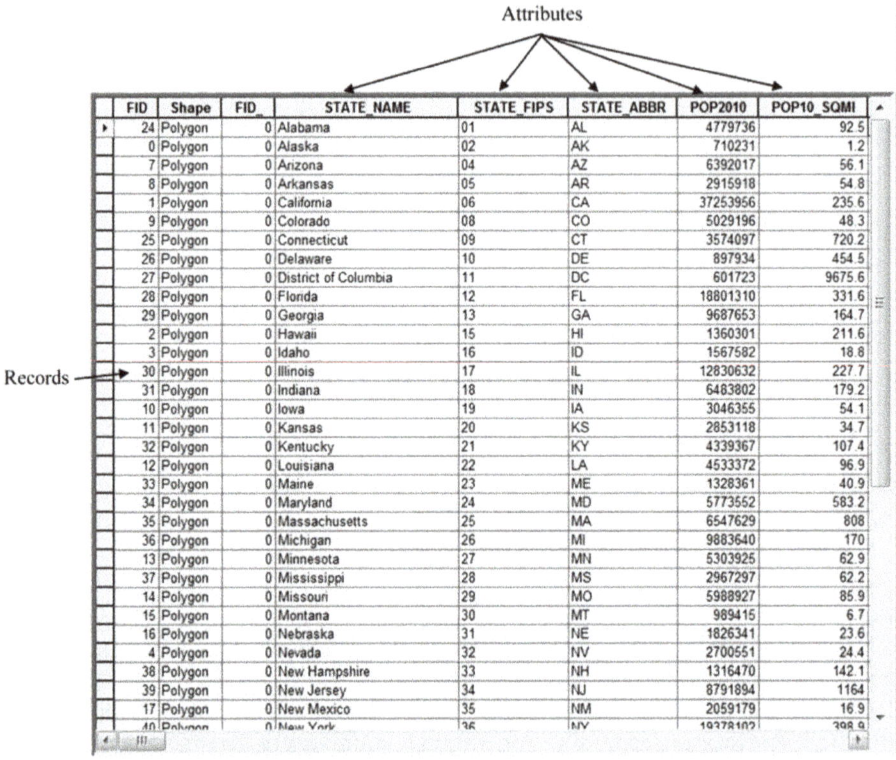

	FID	Shape	FID_	STATE_NAME	STATE_FIPS	STATE_ABBR	POP2010	POP10_SQMI
▸	24	Polygon	0	Alabama	01	AL	4779736	92.5
	0	Polygon	0	Alaska	02	AK	710231	1.2
	7	Polygon	0	Arizona	04	AZ	6392017	56.1
	8	Polygon	0	Arkansas	05	AR	2915918	54.8
	1	Polygon	0	California	06	CA	37253956	235.6
	9	Polygon	0	Colorado	08	CO	5029196	48.3
	25	Polygon	0	Connecticut	09	CT	3574097	720.2
	26	Polygon	0	Delaware	10	DE	897934	454.5
	27	Polygon	0	District of Columbia	11	DC	601723	9675.6
	28	Polygon	0	Florida	12	FL	18801310	331.6
	29	Polygon	0	Georgia	13	GA	9687653	164.7
	2	Polygon	0	Hawaii	15	HI	1360301	211.6
	3	Polygon	0	Idaho	16	ID	1567582	18.8
	30	Polygon	0	Illinois	17	IL	12830632	227.7
	31	Polygon	0	Indiana	18	IN	6483802	179.2
	10	Polygon	0	Iowa	19	IA	3046355	54.1
	11	Polygon	0	Kansas	20	KS	2853118	34.7
	32	Polygon	0	Kentucky	21	KY	4339367	107.4
	12	Polygon	0	Louisiana	22	LA	4533372	96.9
	33	Polygon	0	Maine	23	ME	1328361	40.9
	34	Polygon	0	Maryland	24	MD	5773552	583.2
	35	Polygon	0	Massachusetts	25	MA	6547629	808
	36	Polygon	0	Michigan	26	MI	9883640	170
	13	Polygon	0	Minnesota	27	MN	5303925	62.9
	37	Polygon	0	Mississippi	28	MS	2967297	62.2
	14	Polygon	0	Missouri	29	MO	5988927	85.9
	15	Polygon	0	Montana	30	MT	989415	6.7
	16	Polygon	0	Nebraska	31	NE	1826341	23.6
	4	Polygon	0	Nevada	32	NV	2700551	24.4
	38	Polygon	0	New Hampshire	33	NH	1316470	142.1
	39	Polygon	0	New Jersey	34	NJ	8791894	1164
	17	Polygon	0	New Mexico	35	NM	2059179	16.9
	40	Polygon	0	New York	36	NY	19378102	398.9

FIGURE 4.8 Attribute table.

ArcMap interface courtesy of Esri, Inc. Data Sources: Esri, TomTom, U.S. Department of Commerce, U.S. Census Bureau

attribute. For instance, in the attribute table shown in figure 4.8, all values in column STATE_NAME will pertain to the attribute state names, and all values in column STATE_FIPS will pertain to the attribute state FIPS, which is a unique identification code for each state. The column STATE_NAME will not have a single data for STATE_FIPS. That data is stored in the column STATE_FIPS. Each cell holds individual attribute values that are unique to a specific row and column. The intersection of column POP2010 and the seventh row in the table shown in figure 4.8 with FID 25 is 3574097, which is the 2010 population data for Connecticut.

In a GIS, separate tables are linked in two ways, using either a join operation or a relate operation. In order to understand the two operations, we must first understand how the relationship between two separate tables is established. This relationship between two separate tables is called **cardinality**, which defines the *R*, or relational, in the RDBMS. When thinking about cardinality between two tables, think in terms of a source table, a destination table, and the direction in which the association between the two tables is established. Cardinality is always defined between a destination table and a source table. In other words, the destination table is the point of reference. Four types of cardinality can be defined between a destination table and a source table: one to one, many to one, one to many, and many to many.

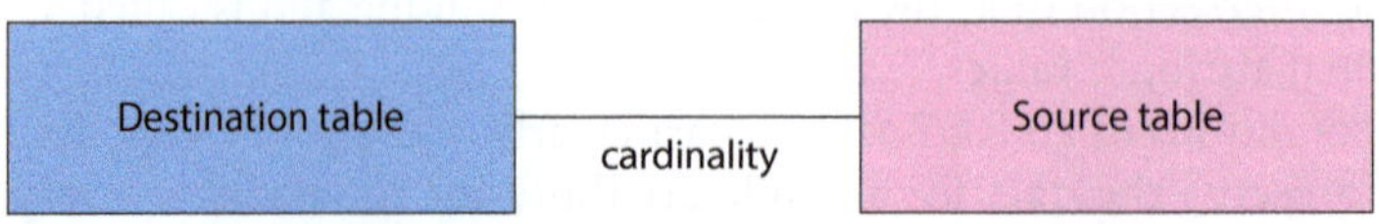

FIGURE 4.9 Cardinality in RDBMS.

In a one to one relationship, one record in the destination table corresponds to only one record in the source table. Take the example of a database that has one table with information for all US states and another database with a separate table with information about state capitals. Here the table with information for US states is the destination table, and the table with information for state capitals is the source table. Each state in the states information table will have only one matching record for a state capital in the capital's information table. Thus, each record in the state table corresponds to only one record in the state capital table (figure 4.10).

In a many to one relationship, multiple records in the destination table will match only one record in the source table. Take the example of US cities and states. Let's consider a database where you have one table with information for US states and another database with a table with information for US cities. Here the US cities table is the destination table, and the US states table is the source table. Multiple cities belong to one state. Thus, multiple records in the US cities table will be associated with one record in the states table (figure 4.11).

In a one to many relationship, each record in the destination table will match multiple records in the source table. Let's again look at the example of a database with a table that has information for US cities and a separate table with information

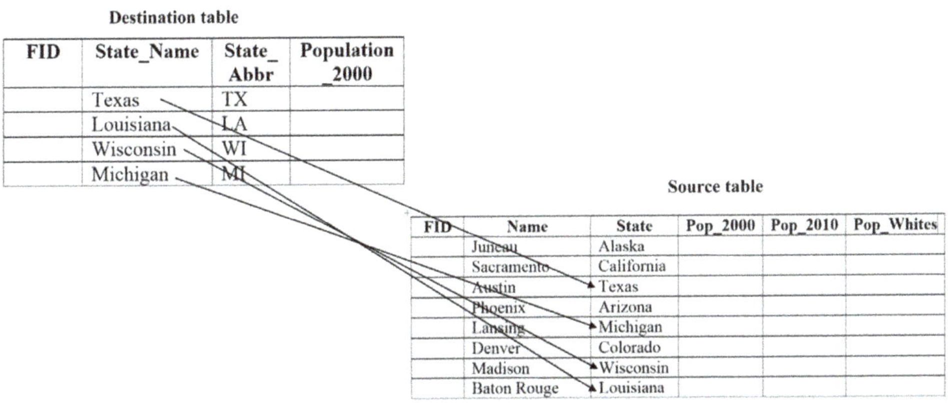

FIGURE 4.10 One to one cardinality.

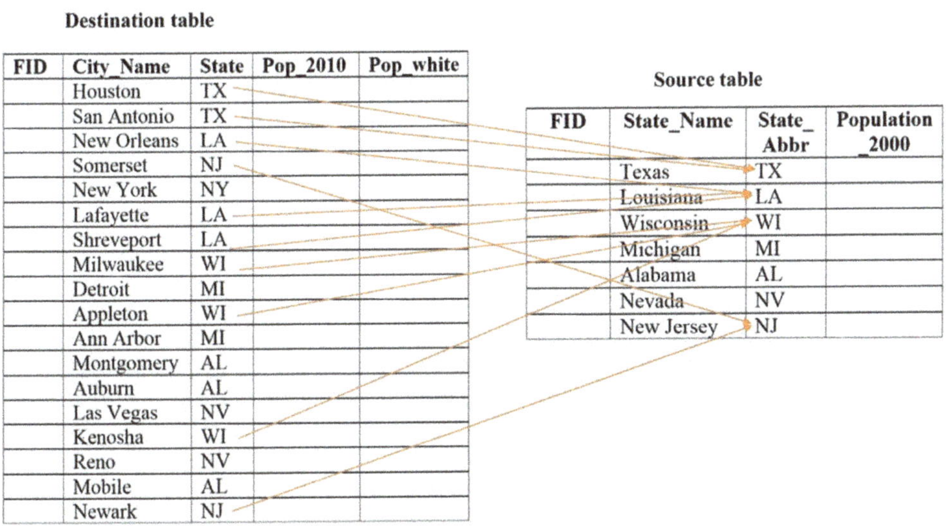

FIGURE 4.11 Many to one cardinality.

for US states. Consider that the table for US states is the destination table, and the table for US cities is the source table. Each US state has multiple cities in it. Thus, each state in the destination table will correspond to multiple cities in the source table (figure 4.12).

In a **many to many relationship**, multiple records in the destination table will match multiple records in the source table. Take the case of a university database that has one table with student records and another table with classes offered at the university. Each class has multiple students enrolled in it, and most students are enrolled in multiple classes. Multiple students are thus enrolled in multiple classes. Therefore, multiple records in the destination table will match multiple records in the source table (figure 4.13).

In a GIS, in order to link two separate tables using a join operation, the rule of joining must be met. Per the rule of joining, there must be one and only one record in the source table for each record in the destination table. This rule is met in a one to one cardinality and in a many to one cardinality. Therefore, a join operation must only be used when the relationship between two tables is one to one or many to one.

Join is a temporary relationship between two tables in a GIS using RDBMS. The join operation uses a primary key and a foreign key. It appends fields of one table to fields in another table using these keys, which is a common attribute/field in both of the tables. In a join function, the direction in which the two tables are linked is very important, since it determines the cardinality between the two tables. The table that contains the information that will be appended is the **source table**. The table that receives the information is the **destination table**. A join function takes the records

Destination table

FID	State_Name	State_Abbr	Population_2000
	Texas	TX	
	Louisiana	LA	
	Wisconsin	WI	
	Michigan	MI	
	Alabama	AL	
	Nevada	NV	
	New Jersey	NJ	

Source table

FID	City_Name	State	Pop_2010	Pop_white
	Houston	TX		
	San Antonio	TX		
	New Orleans	LA		
	Somerset	NJ		
	New York	NY		
	Lafayette	LA		
	Shreveport	LA		
	Milwaukee	WI		
	Detroit	MI		
	Appleton	WI		
	Ann Arbor	MI		
	Montgomery	AL		
	Auburn	AL		
	Las Vegas	NV		
	Kenosha	WI		
	Reno	NV		
	Mobile	AL		
	Newark	NJ		

FIGURE 4.12 One to many cardinality.

Destination table

FID	LAST_NAME	FIRST_NAME	Student_id
	Smith	Evan	
	Seaman	Matthew	
	Ray	Douglas	
	Coffey	Charmin	
	Morgan	Michael	
	Smith	Jeffery	
	Roy	Shaun	

Source table

FID	Course	Course_title	Department	College
	HIST 101			
	MATH 203			
	GEOL 203			
	GEOG 246			
	BIOL 334			
	CHEM 204			
	CHEM 334			

FIGURE 4.13 Many to many cardinality.

from the source table and appends them to the corresponding records in the destination table, temporarily combining the source table and the destination table into a single database. In figure 4.14, the destination table stores information about US cities and the source table stores information for US states. Both tables have a common field/attribute, state name abbreviation, stored under the column heading "State" in the destination table and "State_Abbr" in the source table. State name abbreviation acts as the key to join the two tables.

In a join function, the original data stored in the source table and in the destination table remain unchanged. A join between two tables can be removed when it is no longer needed. A join function is particularly useful when you have a standalone table that is rich with detailed data, but you cannot map it because it is not associated with geographic data. In such a case, joining a standalone table with an attribute table will bring the rich data from the standalone table to the attribute table by appending corresponding records and allow you to spatially analyze and map the data in the standalone table. In this case, the attribute table will be the destination table and the standalone table will be the source table.

When the cardinality between two tables is one to many or many to many, the rule of joining is not met. In those cases, a relate function is used in GIS to link

Destination table

FID	City_Name	State	Pop_2010	Pop_white
	Houston	TX		
	San Antonio	TX		
	New Orleans	LA		
	Somerset	NJ		
	New York	NY		
	Lafayette	LA		
	Shreveport	LA		
	Milwaukee	WI		
	Detroit	MI		
	Appleton	WI		
	Ann Arbor	MI		
	Montgomery	AL		
	Auburn	AL		
	Las Vegas	NV		
	Kenosha	WI		
	Reno	NV		
	Mobile	AL		
	Newark	NJ		

Source table

FID	State_Name	State_ Abbr	Population _2000
	Texas	TX	
	Louisiana	LA	
	Wisconsin	WI	
	Michigan	MI	
	Alabama	AL	
	Nevada	NV	
	New Jersey	NJ	

FID	City_Name	State	Pop_2010	Pop_white	FID	State_Name	State_Abbr	Population_2000
	Houston	TX				Texas	TX	
	San Antonio	TX				Texas	TX	
	New Orleans	LA				Louisiana	LA	
	Somerset	NJ				New Jersey	NJ	
	New York	NY				New York	NY	
	Lafayette	LA				Louisiana	LA	
	Shreveport	LA				Louisiana	LA	
	Milwaukee	WI				Wisconsin	WI	
	Detroit	MI				Michigan	MI	
	Appleton	WI				Wisconsin	WI	
	Ann Arbor	MI						
	Montgomery	AL						
	Auburn	AL						
	Las Vegas	NV						
	Kenosha	WI						
	Reno	NV						
	Mobile	AL						
	Newark	NJ						

FIGURE 4.14 Joining two tables using a common field or key.

two separate tables. A relate function also uses a key to connect two separate tables, but the tables remain physically separate. However, when a record is selected in one table, corresponding records in the related table are selected. In the many to many cardinality for the table in figure 4.13, you would use multiple relates to examine the relationship between students and classes. A one to many relate from the students table to the classes table will give you a list of classes for each student. When you select a student record in the destination table, all classes the student is enrolled in will be selected in the source table. So if you select Evan in the destination table in figure 4.13, HIST 101, GEOG 246, and CHEM 334 will be selected in the source table. Those are the classes Evan is enrolled in. Similarly, another one to many relate from the classes table to the students table will give you a list of students enrolled in each class.

Types of Data

A data can be either qualitative or quantitative in nature. Every data falls into one of these two categories. Quantitative data is any data that is described using numbers, while qualitative data is any data that is described using text. Both quantitative and qualitative data come in different types or formats. Type of data is an important part of discussion on databases. The type of attribute data will determine how it is stored in a database and what types of functions and operations can be performed using an attribute table. Attribute values in a GIS can be qualitative or quantitative in nature. In any database, all data, whether qualitative or quantitative, are of four types: nominal, ordinal, interval, and ratio.

Nominal data is any data that can identify or label an object. Values for nominal data uniquely identify an object. For instance, names of cities or states are nominal data, as they uniquely identify a city or a state. Similarly, student ID, driver's license number, and parcel ID are also examples of nominal data, as the data values uniquely identify a student, a driver, or a parcel. Nominal data values can be described using text or numbers. If nominal data values are described numerically, any kind of mathematical computation with the data values does not carry any meaning.

Ordinal data is any data that is ranked on a numeric scale or hierarchy. It represents which data is first in a hierarchy, which data is second, and so on; for instance, the top five colleges in your region or the best neighborhoods in your area. Student grades of A, B, C, D, or F are also ordinal data. Ordinal data only ranks data in a hierarchy. It does not represent by how much a data ranks higher than the next data in the hierarchy.

Interval data is numeric data measured on a regular scale with no fixed zero point. Interval data is not related to a meaningful zero point. In other words, a value of zero does not mean absence of data. For instance, temperature measured in degrees Celsius or Fahrenheit is interval data. A temperature of zero degrees does not mean complete absence of temperature. Temperature can be zero degrees and can fall below zero degrees. Similarly, elevation is interval data. Because interval data does not have a fixed zero point, the data can be added or subtracted. However, dividing or multiplying interval data will be meaningless.

Ratio data is any data that has a meaningful fixed zero point. In other words, a value of zero indicates absence of data. For instance, zero precipitation would indicate absence of rain. Similarly, cost of attending college and monthly salary of an employed individual are examples of ratio data. Ratio data can be added, subtracted, divided, and multiplied.

TABLE 4.1 Field Data Types for Attribute Tables

Field Type	Application	Examples
Short integer	Whole numbers. Numeric values without fractions. Values can range from 32,768 to 32,767.	612 61275
Long integer	Whole numbers. Numeric values without fractions. Values can range from −2,147,483,648 to 2,147,483,647.	235001 326851
Float	Fractional numbers with eight significant digits after decimal point. Values range from approximately −3.4E38 to 1.2E38.	15.678
Double	Floating point values with 16 significant digits after decimal point. Values range from approximately −2.2E308 to 1.8E308.	4.6904351234
Text	Series of alphanumeric symbols	Houston
Date	Dates, times, or dates and times Default format: mm/dd/yyyy, hh:mm:ss	12/06/1980 02:35:45
BLOB	Binary large object. Data is stored as a long sequence of binary numbers.	Annotations, dimensions, items such as images, multimedia, and bits of code
Object identifier	ObjectID field maintained by ArcGIS. Unique identifier for each row in a table. Key ArcGIS functions depend on this field and the field must be unique.	OID, ObjectID
Global identifier	Stores registry style strings consisting of 36 characters. Uniquely identifies a feature or table row within a geodatabase and across geodatabases.	GUID

Source: ArcGIS Help, https://desktop.arcgis.com/en/arcmap/latest/manage-data/geodatabases/arcgis-field-data-types.htm

Different GIS systems allow different data types to input for attribute values. For instance, in ArcGIS, numbers can be stored in four numeric data types: short integer, long integer, float, or double. Other data types include text, date, BLOB, object identifiers, and global identifiers. Each of the data types has unique properties and works for different types of data, as summarized in table 4.1.

Functions in an Attribute Table

An attribute table in a GIS provides several functionalities that are important for data analysis. One of the most basic functions is the ability to add data. This function is very important in order to input results of any kind of computation utilizing attributes present in the table. We often find ourselves using this function in a GIS, particularly when the table has detailed information about a feature. When you add a field, the first thing you must do is specify the type of data that will be stored. Data type can be any of the options in table 4.1. One must think carefully about the new field values that will be added when specifying data type. Once a field is added to the table, data type cannot be changed. Moreover, depending on the type of data, you may or may not be able to perform certain operations using the field values. For instance, if you select "text" as your data type, you will not be able to enter numeric values in that column or perform any mathematical computation using the field values. Similarly, if the new data to input is dates, you must select "date" as your data type or else you will not be able to enter your data in the correct date format.

The next important decision is the length of the new field, as the length of a field allocates file space. In large databases with many records, this may hamper data processing and is an important consideration.

Just like a field in a table can be added, a field can also be deleted. You may want to delete a field if there is data redundancy in a database. This is an important part of the data preparation stage before actual analysis can be done. When obtaining data from multiple sources, there is always a chance that you will have a database with fields that are not necessary or have multiple fields with the same data. These unnecessary fields must be deleted, as they add to file storage space and data processing.

After adding a new field, the next step is to add field values. This can be done in two ways. One is where you simply click a specific field record and type in the value. Another option is to use the field calculator, available in an attribute table. The field calculator is particularly useful in calculating specific values and adding those values to multiple fields. For instance, let's say you have a column with population data and another column with area. You want to calculate population density. You can add a new field, "Population Density," to your table and use the field calculator to calculate population density using the two columns to input the formula:

$$\textit{Population density} = \frac{\textit{Population}}{\textit{Area}}$$

An important feature in a database is the ability to query a database table. This is a valuable tool in GIS. A query tests specified conditions by a user and provides an output. In the case of a GIS, the output is shown in the table and on the map, if querying an attribute table. There are many utilities of queries in GIS, from data exploration to advanced data analysis. In GIS, two types of queries are possible. If you recollect, a geospatial data has two components: a spatial component and an aspatial component. Accordingly, you can query using both components (figure 4.15). A query for the aspatial component is conducted on the database table (attribute table or standalone table), and a query for the spatial component is conducted on the spatial layers. In order to query attributes in a table, a programming language called Structured Query Language (SQL) is used. SQL is also used by many commercial database platforms such as Oracle, Microsoft SQL Server, Microsoft Access, and DB2. In order to query spatial layers, spatial location is used.

An **attribute query** will test a specified condition using records present in a table and give an output whose values meet the specified condition. Attribute data query follows Boolean algebra where conditions are specified using SQL query statements

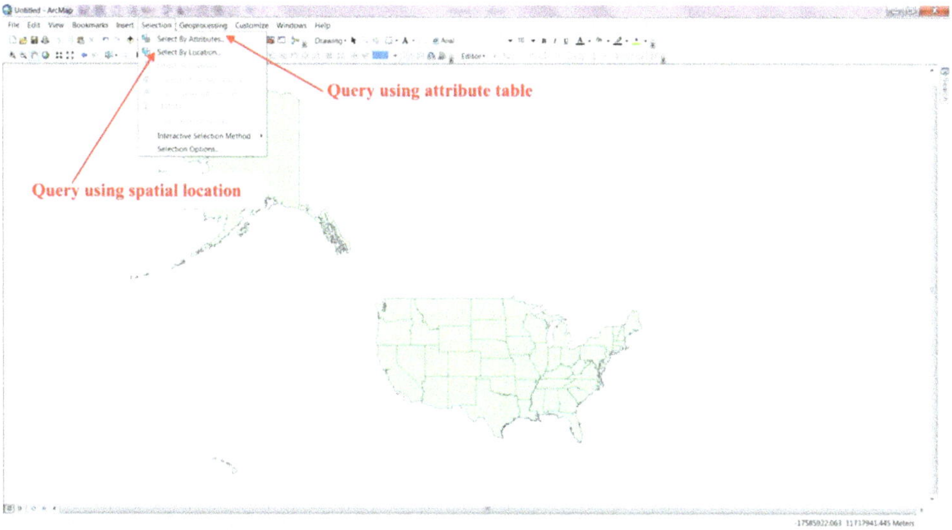

FIGURE 4.15 Two types of queries in GIS shown in ArcMap.

ArcMap interface courtesy of Esri, Inc. Data Sources: Esri, TomTom, U.S. Department of Commerce, U.S. Census Bureau

in a text-based interface called a *command line interface* (figure 4.16). Query statements are written as an expression that tests specified conditions between two or more attributes using two operands and a logical operator. An attribute query uses one of the relational operators of equal (=), not equal (< >), greater than (>), greater than or equal to (>=), less than (<), and less than or equal to (<=) to build a query. For example:

*SELECT *FROM USA_States WHERE "STATE_NAME" = 'TEXAS'*

Here the query is being conducted on an attribute table that is part of a spatial layer named USA_States. STATE_NAME is one of the attributes in the table. STATE_NAME and TEXAS are the operands that are connected by a logical operator =.

*SELECT *FROM students WHERE "LAST_NAME" = 'SMITH'*

Here the query is being conducted on a table named "students." LAST_NAME is one of the attributes in the table. LAST_NAME and SMITH are the operands that are connected by a logical operator =.

*SELECT *FROM USA_States WHERE "POP10_SQMI" >= '25'*

Similarly, here the query is being conducted on a table named USA_States. STATE_NAME is one of the attributes in the table. STATE_NAME and TEXAS are the operands that are connected by a logical operator >=.

A simple query uses one operator and one field and is based on one criterion. However, simple queries can be connected to build a compound query and test multiple criteria. Two or more query expressions testing more than one condition can be connected by using Boolean connectors, which are AND, OR, XOR, and NOT. Boolean connectors test for the specified criteria and give an output of records that meet those criteria. It is important to understand how Boolean connectors work, as they impact the output of your query expression. Particularly the AND and OR connectors can be confusing. The AND connector will select records that meet specified conditions of both expressions connected by the AND

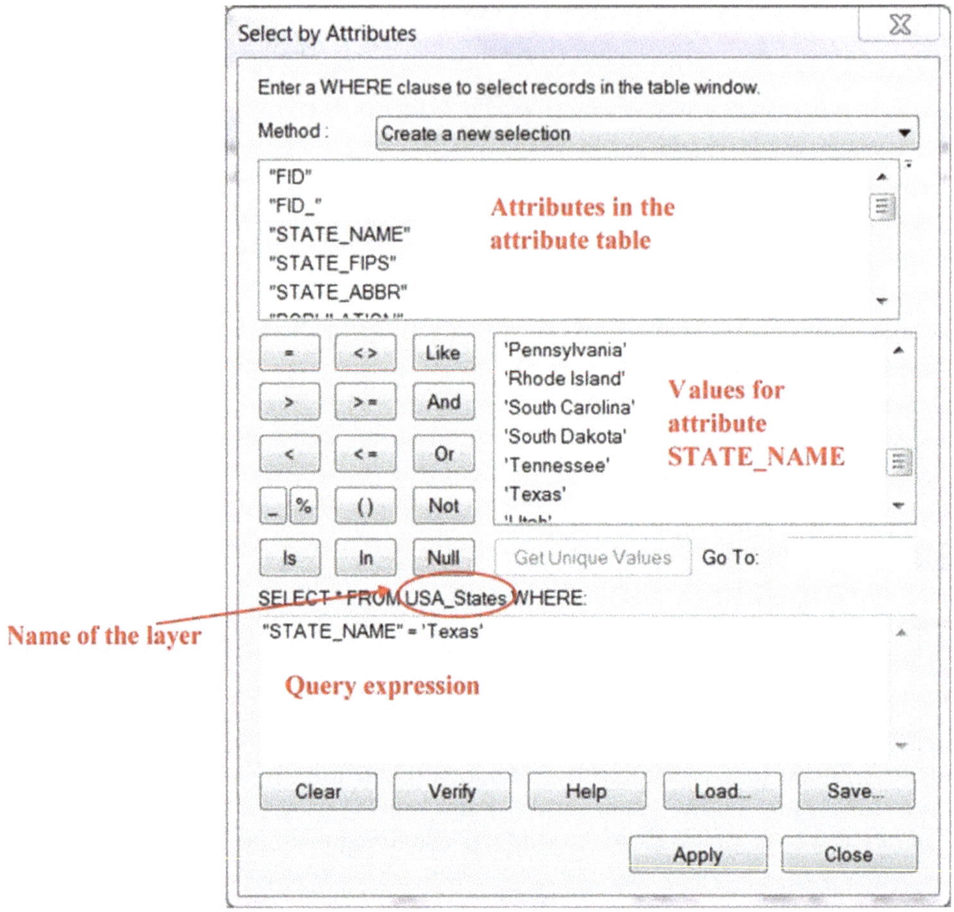

FIGURE 4.16 Attribute query interface. In this example, an attribute query can be used to select features based on a specified condition from a file containing US states information.

ArcMap interface courtesy of Esri, Inc.

operator, whereas the OR operator will select records that meet the specified criteria in either of the two expressions connected by the OR operator. The diagrams in figure 4.17 explain how the Boolean connectors work.

Consider the following expressions connected by a Boolean connector:

*SELECT *FROM students WHERE “LAST_NAME” = ‘SMITH’ OR ‘GPA’ = 3.0*

Here OR is a Boolean connector connecting two query expressions. This expression will select records from a student table identifying all students that have last name of Smith and all records identifying all students that have a GPA of 3.0. The output will consist of records where student last name is Smith but GPA may or may not be equal to 3.0 and student records where student GPA is equal to 3.0 but last name may or may not be Smith.

*SELECT *FROM USA_States WHERE “POP10_SQMI” > 25 AND “POP10_SQMI” < 150*

Here the expression will select records from the US states table all states that have a population density greater than 25 and less than 150, based on 2010 population.

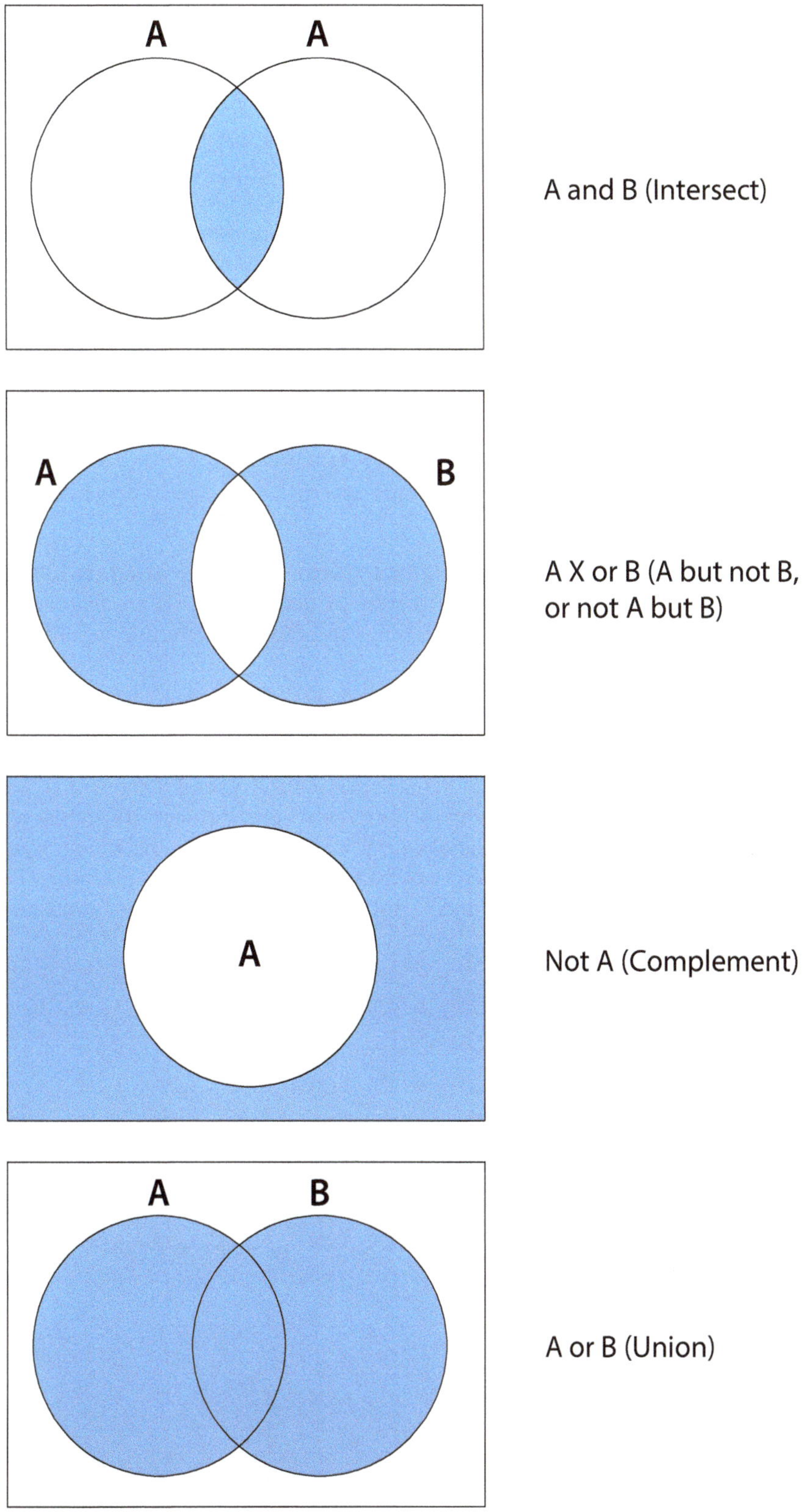

FIGURE 4.17 Boolean connectors AND, XOR, NOT, and OR

In a GIS, the output of an attribute query is selected in the attribute table as well as on the map, as shown in figure 4.18. The output of the query expression *SELECT *FROM USA_States WHERE "POP10_SQMI" > 25 AND "POP10_SQMI" < 150* are the 22 states that are selected in the attribute table and also on the map.

The output from an attribute query can be exported to create a new spatial layer or a new standalone table. Selected features from the query output can be exported to create a new spatial layer with its own attribute table, or only the selected records from the attribute table can be exported to create a new standalone table. This is again a useful feature when you have data from multiple external sources and would like to remove unnecessary or redundant data. This feature can also be used to create new data from a subset of a large data.

Attribute queries are not very useful for geographic search, where the specific condition for query is based on geography, location, or some sort of a spatial relationship. For example, a query that specifies to select

"*all houses within a 1 mile radius of a major freeway*"

"*house situated within a certain ISD*"

"*select cities that intersect rivers*"

cannot be tested from an attribute table because they are based on relationship between two features. Houses and freeway or house and ISD are features that would be on a map, and to evaluate a relationship between them one would need to conduct a query using spatial layers.

In a GIS, queries can be performed on spatial layers and selected based on their spatial location or spatial relationship with other layers. Spatial queries can be done in two ways. One way would be interactively by using the "Select feature" tool and drawing a box around features. This is particularly useful if the size of your dataset is small or if you are familiar with the area where features of interest are located. Another way to conduct a spatial query would be to test specific conditions based on spatial relationship between two spatial layers or features at a time. This is done by using the "Select by location" tool (figure 4.19). This tool does not use SQL

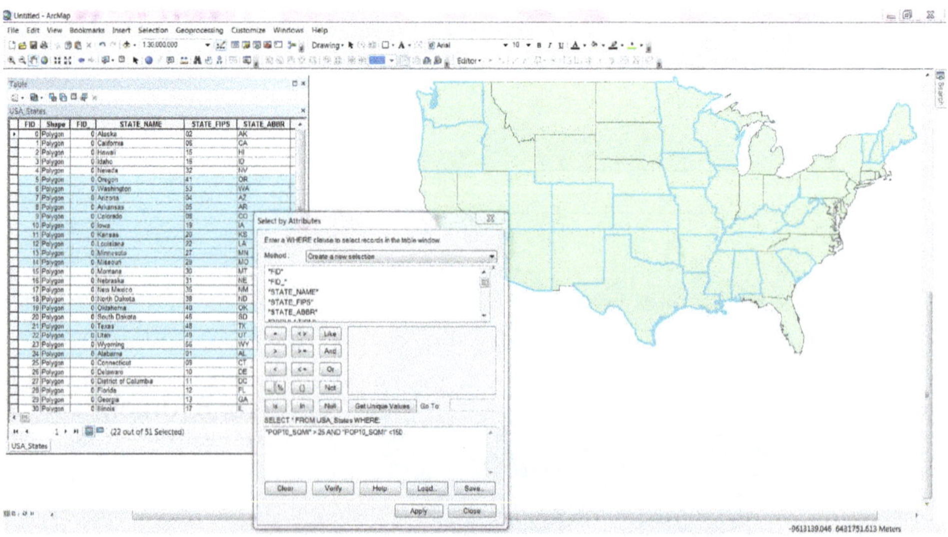

FIGURE 4.18 Output of 22 states from attribute query selected in attribute table and map.

ArcMap interface courtesy of Esri Inc. Data Sources: Esri, TomTom, U.S. Department of Commerce, U.S. Census Bureau, U.S. Department of Commerce, National Oceanic and Atmospheric Administration (NOAA), National Ocean Service (NOS), National Geodetic Survey (NGS)

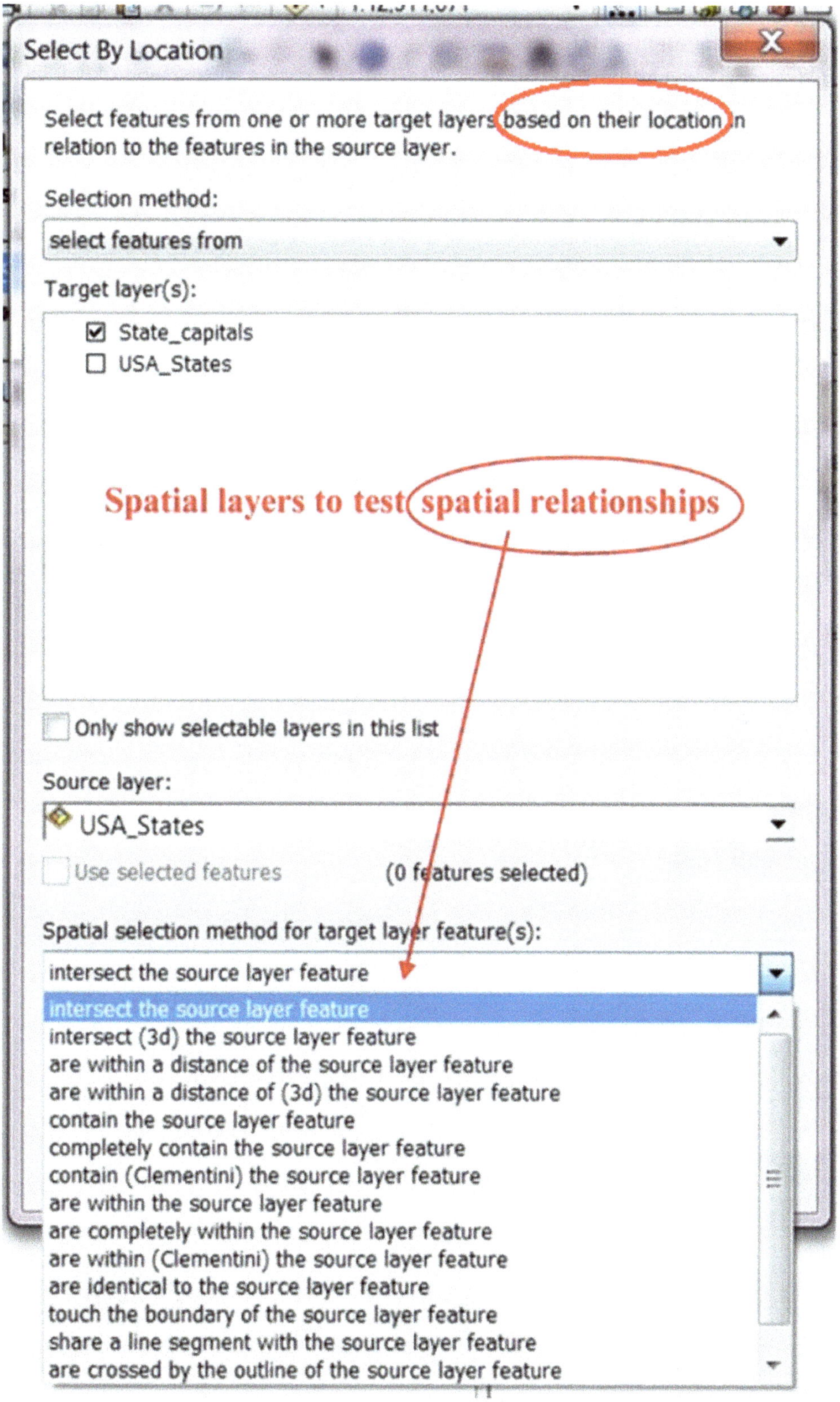

FIGURE 4.19 Spatial query interface in ArcGIS.

ArcMap interface courtesy of Esri, Inc.

for constructing query statements. Instead, a query statement is constructed by putting together options within the "Select by location" tool using traditional sentences. Spatial query in a GIS uses different operators to test spatial relationships of containment, proximity, and intersection between two spatial layers. Spatial queries result in either locating sets of spatial layers or building new spatial layers. The output from a spatial query is selected on the map, and the corresponding records in the attribute table are also selected. Selected features from the query output can be exported to create a new spatial layer with its own attribute table, or selected records from the attribute table can be exported to create a new standalone table.

Questions and Activities

1. From your everyday experience, pick any five examples of database use. Do you see any use of a DBMS in your examples? Explain.
2. What is a DBMS? Explain the properties of an RDBMS and its benefits for an organization.
3. Tables 4.2, 4.3, 4.4, and 4.5 are tables in a relational database management system that are typically used in the geographic information systems of local governments. Use their relational characteristics to find the *sales price of the property zoned R-2 in the Census block that has an average household income greater than $40,000.* Also, find the polygon ID needed to display the parcel on the map.

   Sale Price = $__________ *Polygon ID* = __________
4. What is the cardinality between each pair of tables:
   a. departments to instructors
   b. staff to university
   c. county officials to county
   d. state to state capitols
   e. school district to schools

TABLE 4.2 US Census of Population

Tract/Block ID	Avg. Household Income
52701212	$40,310
52701213	$32,920
52701214	$33,814
52701215	$24,214

TABLE 4.3 Property Data

Parcel ID	Map/Block/Lot ID	Zoning	Land Use	Tract/Block ID
123456789	989-00000-0596	R-1	110	52701212
112233445	989-00000-0414	R-2	120	52701212
876543210	985-00000-0003	R-2	120	52701213
563865243	985-00000-1530	C	210	52701213

TABLE 4.4 Register of Deeds Data

Parcel ID	Assessed Value	Sale Price	Owner
123456789	$82,610	$126,300	H. Jones
112233445	$93,415	$175,035	W. Smith
876543210	$75,050	$97,545	F. Nelson
563865243	$99,560	$189,760	S. Fishman

TABLE 4.5 Parcel Map

Map/Block/Lot ID	Polygon ID	Area
989-00000-0596	193	0.95
989-00000-0414	204	1.20
985-00000-0003	154	0.56
985-00000-1530	176	1.15
979-00000-0254	132	0.03

5. Explain the difference between an attribute table and a standalone table. If you have an attribute table and a standalone table and want to link the two tables using a join function in GIS, what will be your source table and what will be your destination table? Why?
6. What is a source table and a destination table? Explain the importance of designating tables as source and destination in an RDBMS.
7. Attribute data can be classified into four classes. For each of the following types of data, state whether it is nominal, ordinal, interval, or ratio. Explain why.
   a. precipitation measured in inches
   b. your possible grade in this grade: A, B, C, or D
   c. your driver's license number
   d. names of counties in the state of Texas
   e. temperature measured in degree Celsius
   f. house number
   g. street address
   h. land use category
   i. region's top 10 places to visit
   j. hospitals in your local area
8. Explain the types of queries you can perform in a GIS. How are they different or similar to one another?
9. Compare a join function to a relate function in a GIS.

Key Terms

attribute query (p. 86)
attribute table (p. 73)
cardinality (p. 80)
column header (p. 79)
common key (p. 78)
database (p. 75)
database management system (DBMS) (p. 74)
destination table (p. 82)
foreign key (p. 77)
interval data (p. 84)
join (p. 82)
many to many relationship (p. 82)
many to one relationship (p. 81)
nominal data (p. 84)
one to many relationship (p. 81)
one to one relationship (p. 81)
ordinal data (p. 84)
primary key (p. 77)
qualitative data (p. 84)
quantitative data (p. 84)
ratio data (p. 84)
relate function (p. 83)
relational database (p. 75)
Relational Database Management System (RDBMS) (p. 75)
source table (p. 82)
standalone table (p. 79)
structured query language (SQL) (p. 79)
table (p. 75)

References

ArcGIS Desktop Help. n.d. "Databases and ArcGIS." Accessed October 11, 2021. https://desktop.arcgis.com/en/arcmap/latest/manage-data/databases/databases-and-arcgis.htm.

Huxhold, William. 1991. *An Introduction to Urban Geographic Information Systems.* New York: Oxford University Press.

Jensen, John, and Ryan Jensen. 2013. *Introductory Geographic Information Systems.* Boston: Pearson.

Price, Maribeth. 2016. *Mastering ArcGIS.* New York: McGraw Hill Education.

Schmitz, Andy. 2012. *Essentials of Geographic Information Systems.* Arlington, VA: Saylor Academy.

Geospatial Data 5

Learning Objectives

- Explain the key characteristics of geospatial data
- Identify different sources of geospatial data and understand how to use them
- Identify the types of data and data formats available from different agencies
- Understand different field methods and technologies used to create geospatial data
- Explain the different segments of GPS technology and how they work
- Describe the sources of error in GPS data and the technology to improve the accuracy of GPS receivers
- Explain the process of remote sensing
- Discuss remote sensing data collection methods, including aerial photography and satellite imaging
- Classify types of aerial photos
- Explain and classify earth orbiting satellites based on their properties
- Explain geocoding techniques to create digital data, including digitizing and address matching

What Is Geospatial Data?

Geospatial data is an important component of any working GIS. In chapter 1, we discussed that the key components of any working GIS include hardware, software, data, people, and appropriate methods. You may have the most advanced cutting-edge hardware, the most sophisticated and latest version of GIS software, and the best team of GIS technicians, but without appropriate data none of these can be put into action. An effective and successful GIS implementation requires the presence of *appropriate* geospatial data—GIS projects require *specific* types of data, and several characteristics of GIS data set them apart from other types of data.

First, GIS data consists of a map component, which can be visualized, and a tabular component that enables a user to analyze the data and perform queries and other functions. GIS datasets, whether vector or raster, can be very large. Take the example of parcel data, transportation data, or infrastructure data maintained by a city, county, or municipality. Such datasets can have thousands of records. Thus, the GIS data will have thousands of features and thousands of attributes. Similarly, as discussed in chapter 3, raster datasets tend to be very large, and multiple large datasets may require special consideration for system requirements.

Second, GIS implementations do not occur in silos. Rather, they always occur in a multi-departmental and multiuser environment where multiple users work with the data. One department may be responsible for collecting, creating, or maintaining the data but ultimately multiple users have access to the data to use it for their day-to-day activities. GIS datasets are a representation of the real world, which is dynamic and ever-changing. Accordingly, GIS data must be updated frequently. In a multiuser environment it is impractical to provide each user a copy of the dataset every time it is updated. A simple solution followed by modern work environments is to store the GIS dataset on a central server, where different users will have different types of access privileges.

Third, any GIS project will involve geospatial datasets derived from multiple sources in different formats. It is rare that datasets used in a project have all been derived from a single source. As we will cover in this chapter, GIS datasets can be obtained from several sources, including public agencies, state and federal government agencies, private corporations, online resources, government satellite programs, aerial photographs, GPS units, scanners, unmanned aerial vehicles, or smartphones. Geospatial datasets are also derived in different formats, including tables, paper maps, images, or web pages. These data are further derived in different file formats using different projections, datums, and coordinate systems. Moreover, with advances in technology and data-sharing protocols, geospatial data is increasingly available for use over the internet (for example, Esri's ArcGIS Online and ArcGIS Living Atlas of the World). Nonetheless, geospatial datasets derived from multiple sources in a variety of formats must work together in sync with other datasets.

The next time you look at a digital map on your computer screen or smartphone, take note of the datasets represented on the map and their data sources. For example, the map shown in figure 5.1 displays current wildfires in the United States derived from Esri's ArcGIS Living Atlas of the World. Look at the bottom right corner of the map and you will notice that multiple sources are attributed for the map.

Many such maps and datasets are freely available to view or download from Esri's ArcGIS Living Atlas of the World, a great online resource for GIS data and maps. These datasets and maps have been created by curating datasets from multiple sources available in different formats.

Today, we are able to display and view a variety of maps and datasets because some agency or individual created those datasets and made them accessible. Remember, when we talk about GIS data we are referring to data that exists in a digital format to be viewed. Ultimately, when we talk about geospatial data, there are three scenarios: the data exists in digital format; the data exists but not in digital format and must be digitized so that it can be utilized in a GIS; or the data does not exist in either digital or analog form and must be created. GIS data may also be available in digital form but be out of date or in an inappropriate format or unsuitable for a GIS application. In such cases too, data must be created.

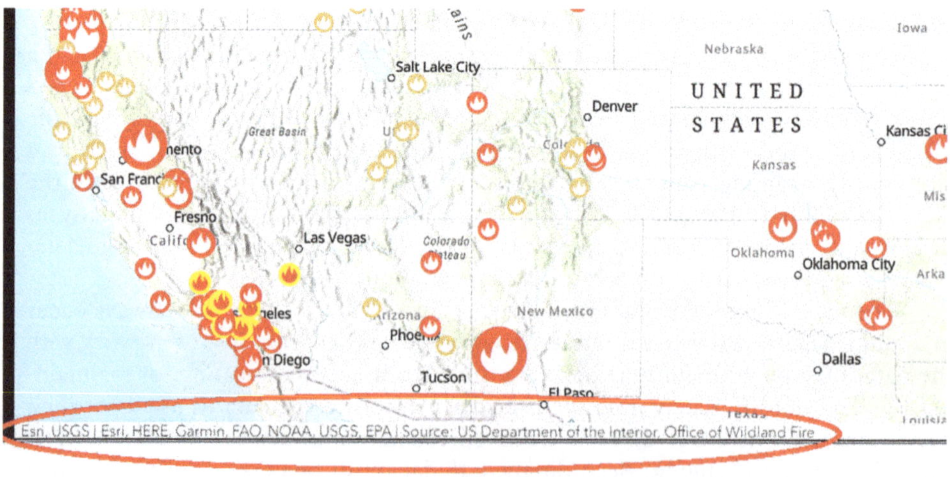

FIGURE 5.1 Esri's ArcGIS Living Atlas of the World displaying current wildfires in the United States.
ArcGIS Living Atlas of the World interface courtesy of Esri, Inc.

GIS data can be created or procured in different ways. One way is to create geospatial data in digital form by transforming analog data into digital format, a process called geocoding. Examples of geocoding include scanning a paper map, digitizing features from a paper map, and address matching. Another way geospatial data is created in a digital form is by using global positioning systems, aerial photography, or remotely sensed data derived from satellite images.

Obtaining appropriate geospatial data is the most crucial part of any GIS implementation project and consumes large amounts of time, money, and labor. However, once appropriate geospatial data is obtained, it can be reused in a GIS for multiple projects as many times as needed until the data must be updated. It is rare for a contemporary GIS project to commence with no geospatial data at all. Today we have access to vast amounts of useful geospatial data because of the many agencies with large geospatial data distribution programs. These agencies create and disseminate geospatial data to the public that can be readily used in a GIS and form a crucial part of effective GIS use and implementation. We will cover some of the agencies that maintain such geospatial data distribution programs and the types of geospatial data disseminated by them. Thereafter we will cover different techniques for geospatial data creation.

Finding Geospatial Data

The first place to start looking for data is, unsurprisingly, the internet. However, just because vast amounts of data are easily located on the web does not necessarily mean the data will be useful for a project on hand. Searching for and finding appropriate data is a complex process. Online sources of geospatial data are dynamic, as are the processes to make geospatial data available on the web. New data types and file formats are available more rapidly today than in the past, making available data quickly obsolete. Every website follows its own protocol to visualize or download data, websites change their addresses, and many websites do not work as expected. Websites may also be missing key information about the data, making it difficult to ascertain if it is current or if the specifications of the dataset meet widely accepted standards. Therefore, everything that may be available on the internet to download may not be useful.

Moreover, searching for geospatial datasets requires a user to be technically savvy, as they may be available to download in different formats, requiring the user to understand how to download, check, and use the data. Users must know whether they are accessing both parts of the geospatial data—the map component and the tabular component. Thus, searching for geospatial data on the web may seem easy, but it is similar to finding your way through a labyrinth.

The good news is that many government agencies in the United States create and distribute geospatial data and maps to the public. Each agency creates a specific type of geospatial data and it is best to start your search with one of those agencies, particularly the ones discussed here that have been used the most by GIS professionals.

United States Geological Survey (USGS)

The United States Geological Survey (USGS), part of the Department of the Interior, is a key federal agency that supplies geospatial data covering land features, topographical information, and man-made features of the United States.

The agency supplies both vector and raster datasets, made available to the public for free to download. Vector datasets available from the USGS include the national hydrography dataset, watershed boundary dataset, governmental boundary units, transportation, structures, elevation contours, and geographic names. Raster datasets include digital elevation models (DEMs), orthoimagery, and land cover. Lidar Point Cloud data is also available over many areas of the country (https://www.usgs.gov/products/data-and-tools/gis-data). All of these datasets are available in formats that can be used in any GIS software.

USGS geospatial datasets are available online through the National Map (figure 5.2), the foundational platform of the agency's National Geospatial Program. **The National Map** is a collaborative effort between the USGS and other federal, state, and local government agencies to disseminate topographic information and geospatial data of the United States to the public. Topographic data and geospatial data available through the National Map are also packaged into downloadable products that are available from the USGS in multiple formats.

The National Map (https://apps.nationalmap.gov/viewer) thus functions as a one-stop source for free downloadable digital geospatial data for the entire country. Data is downloaded by using the "Data Download" menu on top of the page, providing a list of all geospatial data and maps that the USGS distributes through the platform. Users can search for a specific location by selecting a dataset from the menu and using the "Search Products" menu. This will display the datasets for all the regions that are available through the National Map. A user may then select geospatial data of any region and add it to the cart. Depending on the selection, data are available in different formats, scales, and resolution. For instance, geospatial data can be downloaded in shapefile and geodatabase file formats at two scales, national and state.

An important digital geospatial data available through the National Map is digital elevation data of the United States, available as part of the USGS's 3DEP Program, a component of the USGS National Geospatial Program. 3DEP products available through the National Map include digital elevation data and Lidar Point Cloud data. Data for 3DEP is derived from several different sources. Elevation data available through the National Map include Digital Elevation Models (DEM), Digital Surface Models (DSM), Orthorectified Radar Images (ORI), and Lidar Point Cloud (LPC)

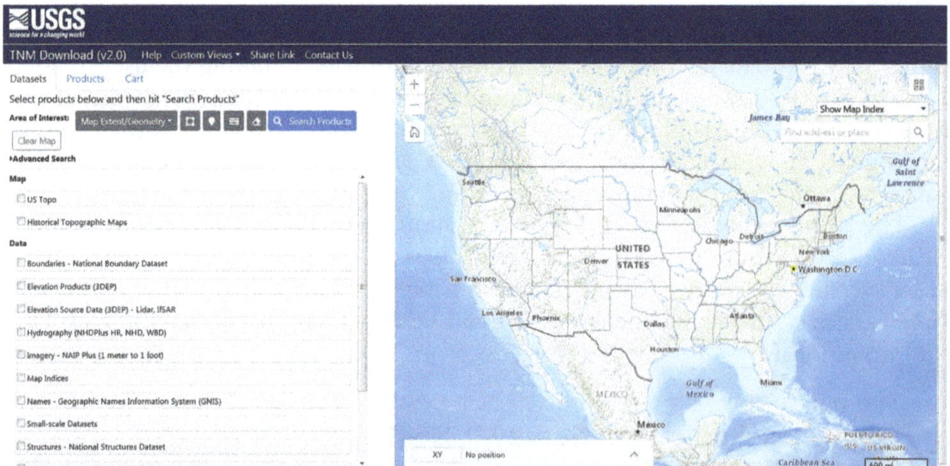

FIGURE 5.2 The National Map.

The National Map interface courtesy of USGS

data in LAS, LAZ, and TIFF file formats. DEMs are available at 1 arc second (about 30 meters), 1/3 arc second (about 10 meters), 1/9 arc second (about 3 meters), and 1-meter resolutions in GeoTIFF and IMG file format. DEMs are raster representation of elevation data, created and distributed by the USGS. It is a widely used elevation data and can be read directly by most GIS packages. For instance, figure 5.3 shows a DEM at 1/3 arc second spatial resolution in GeoTIFF format covering Montgomery, Walker, and San Jacinto Counties in Texas on the National Map.

Other important resources available through the National Map are topographic maps of the United States. US Topo maps can be downloaded at 7.5-minute resolution quad sections in GeoPDF file format. A topographic map is used to represent topography of the land and corresponding terrain features. The US Topo maps distributed by the USGS through the National Map are layered maps that include orthoimagery, roads, railroads, airports, grids, geographic names, elevation contours, hydrography, and contour boundaries. Figure 5.4 shows a US Topo map of Spring, Texas, downloaded in a GeoPDF file format from the National Map.

Historical topographic maps of the United States are also available for users to download through the National Map. These are US topographic maps dating back to 1882 and are part of the USGS's Historical Topographic Map Collection (HTMC). The maps are available in 11 different resolutions in GeoTiff and GeoPDF file formats. The USGS's TopoView application is a useful resource for accessing topographic maps, including those from the US Topo series as well as historical topographic maps.

A US Topo map contains a lot of other information in the white border surrounding the map: the datum, projection, coordinate system, data sources for all the datasets layered on the map, scale, and date when the map was created. As is evident from the map, data contained in the US Topo maps are derived from different government agencies. For instance, the imageries included in the US Topos are orthophotos produced by the National Agriculture Imagery Program (NAIP). An orthophoto is a geometrically corrected aerial photo produced using a process known as orthorectification. This process removes distortions in an aerial photo caused by camera tilt and terrain relief and gives the aerial photo uniform scale, producing an orthophoto. Orthophotos are used to produce digital orthophoto

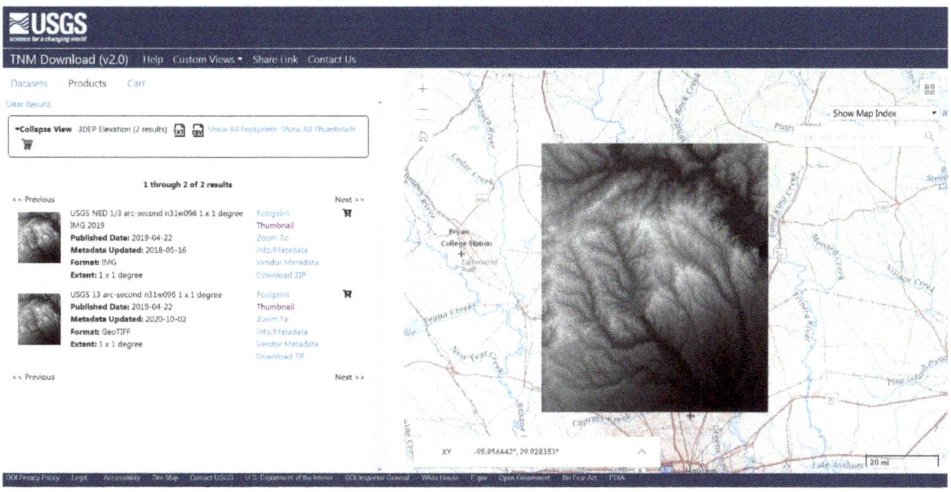

FIGURE 5.3 DEM at 1/3 arc second spatial resolution covering Montgomery, Walker, and San Jacinto Counties in Texas on the National Map.

The National Map interface courtesy of USGS

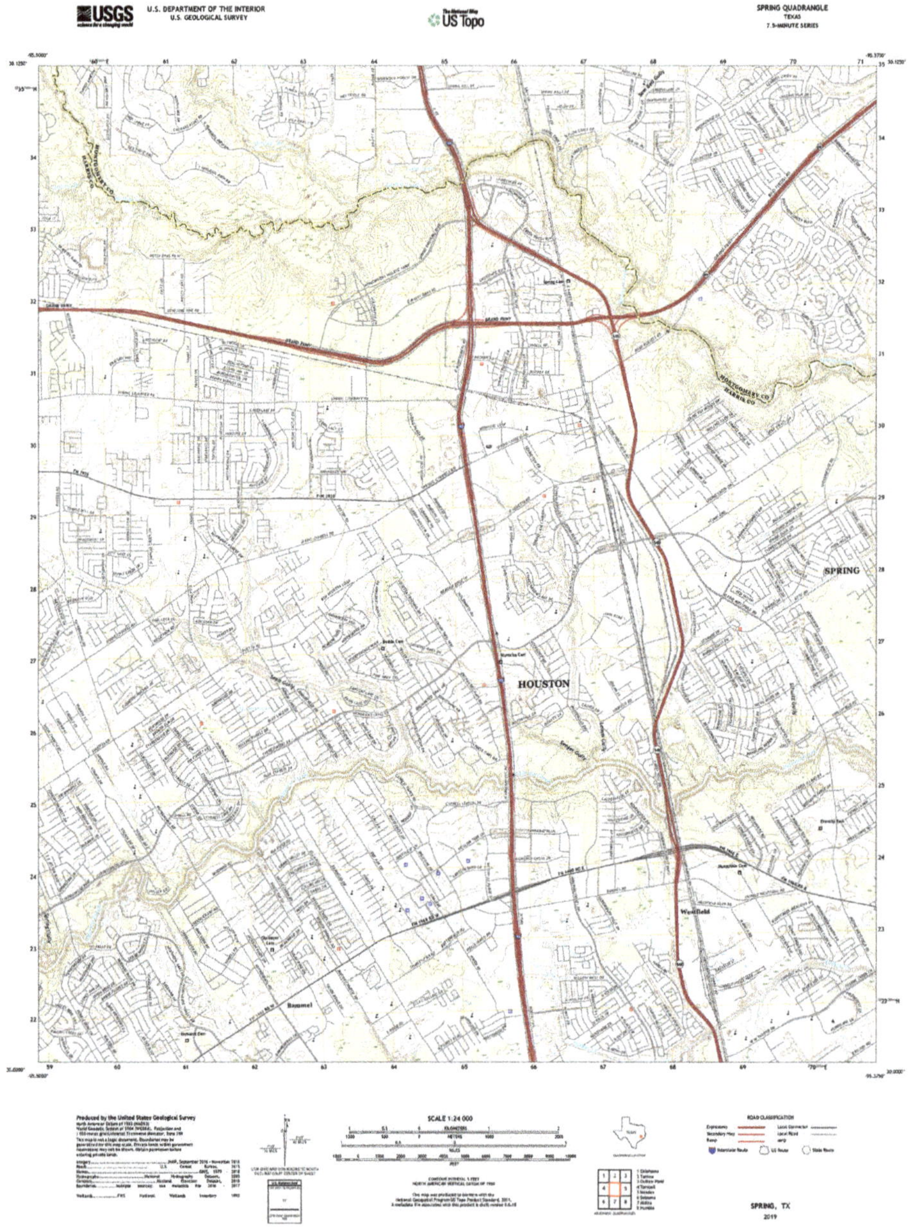

FIGURE 5.4 7.5-minute quad section of a US Topo map of Spring, Texas, downloaded in a GeoPDF file format from the National Map.

Credit: USGS

quadrangles (DOQs), as shown in figure 5.5, a 1996 DOQ of the northern part of Port Arthur on the Texas-Louisiana border, downloaded from USGS Earth Explorer.

DOQs are orthophotos that cover an area of 3.75 minutes of latitude by 3.75 minutes of longitude on the ground, an area equivalent to one-fourth of a 7.5-minute USGS topographic map. DOQs are georeferenced so that every pixel on the photo can be accurately located. A DOQ is therefore a digital image of an aerial photo that incorporates the image characteristics of a photo with the georeferenced qualities of a map. As a result, DOQs match with other geospatial datasets in a GIS and are an excellent source of data.

FIGURE 5.5 1996 DOQ of the northern part of Port Arthur on the Texas-Louisiana border downloaded from USGS Earth Explorer.

Credit: USGS, Earth Explorer

The USGS produced two types of DOQs from 1987 to 2006, 3.75-minute DOQs and 7.5-minute DOQs. DOQs were incorporated into the National Map as a base layer until recently. USGS DOQs produced through 2006 can be accessed through either USGS Earth Explorer or the USGS's Global Visualization Viewer (GloVis), a portal to access publicly available remote sensing data created by federal agencies. Figure 5.6 shows a 1998 DOQ of Port Arthur on the Texas-Louisiana border displayed on the USGS GloVis portal.

Recently DOQs were replaced in the National Map by more recent orthophoto imagery derived from NAIP (figure 5.7), which can be downloaded in JPEG file format. The USGS and US Department of Agriculture through its NAIP program are the two important agencies in the United States that produce orthoimagery used in the creation of orthophoto maps.

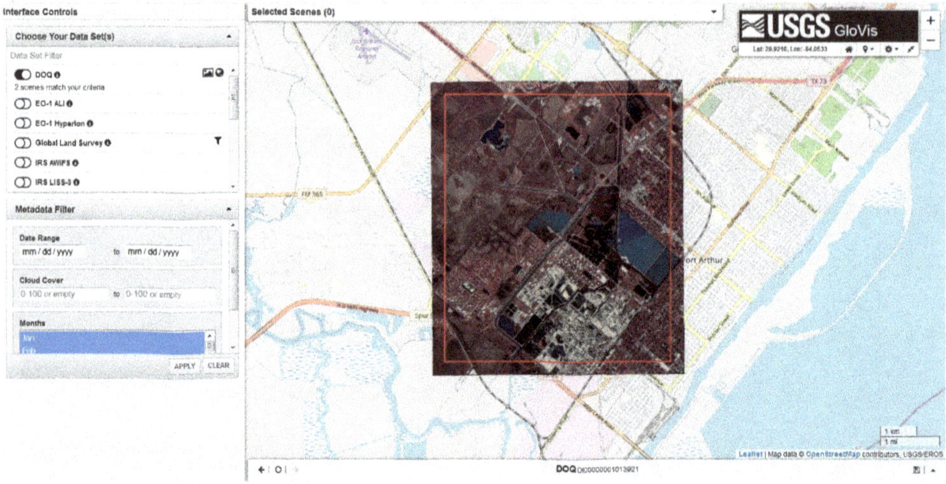

FIGURE 5.6 USGS GloVis portal showing a 1998 DOQ of Port Arthur on the Texas-Louisiana border.

GloVis interface courtesy of USGS

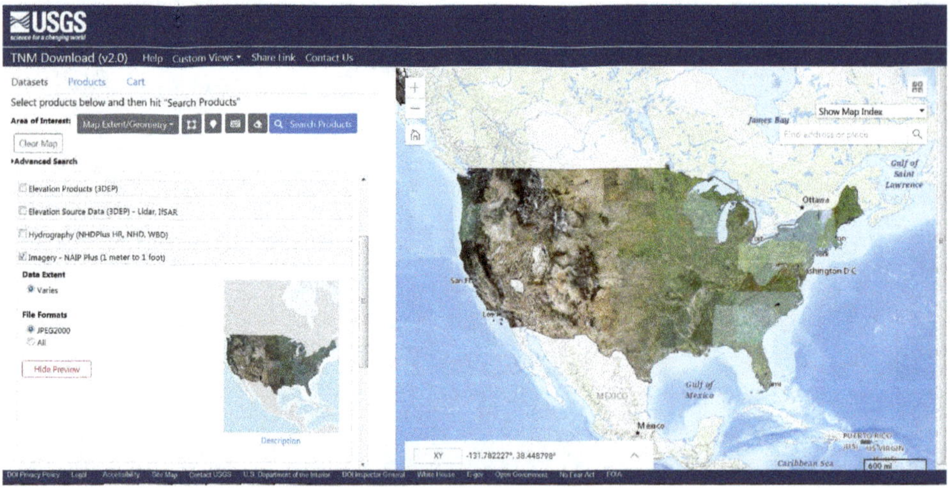

FIGURE 5.7 NAIP imagery on the National Map.

The National Map interface courtesy of USGS

An excellent source of data through the USGS is the **Earth Explorer**, a USGS data portal that provides access to a vast number and variety of land-related digital geospatial data that includes USGS DOQs produced through 2006; satellite image data such as Landsat, Sentinel, IKONOS, and OrbView3 imageries; radar data; digital line graphs; digital elevation models; aerial photos; land cover data; and other datasets. For all the datasets users can view their metadata, an important element of any good-quality geospatial data. **Metadata** is data about data that carries detailed information about the specific dataset such as when the dataset was created or procured and technical specifications of the data, datum, coordinates, and other information. Every geospatial data must be accompanied by metadata.

A useful vector data available in digital format through the Earth Explorer is Digital Line Graph (DLG). DLGs are digital representations of point, line, and area features from USGS topographic maps at the scale of 1:24,000, 1:100,000, and 1:2,000,000. DLGs are vector datasets representing hydrography such as river and creek features; boundary features for state, county, city, and national forests; transportation features that include roads, railroads, and highways; and features representing the US Public Land Survey System.

Once users have made their selection on the National Map or Earth Explorer, data is downloaded as a zipped file directly to their computer. Files must be unzipped before they can be used. In order to download data from the National Map or the Earth Explorer website, users must be knowledgeable of the different file formats available, have the technical know-how to uncompress the downloaded files, and be aware of the data extent and file format needed for their projects.

United States Census Bureau

Another government agency that provides key geospatial data is the US Census Bureau, the nation's primary provider of demographic and economic data. Geospatial data are available as TIGER/Line shapefiles, TIGER/Line geodatabase files, and KML files. Tabular data are also available to download from the US Census Bureau website that contain a variety of detailed demographic and economic information at different geographic scales. These are standalone tables that can be linked to TIGER/Line shapefiles or geodatabase files downloaded from the Census Bureau website.

An important dataset created by the US Census Bureau and available to the public for free are TIGER/Line files. Today, when your smartphone or other device is conveniently able to give you directions from point A to point B, it is largely due to the origin of this file and its corresponding digital geographic database. **TIGER** stands for **Topologically Integrated Geographic Encoding Referencing**. It is a system and digital geographic database that was created by the Census Bureau to modernize their mapping process for the decennial census and create a single source of good-quality geographic data. The system used street segments as the building blocks for their database to generate street-level address maps, generating the first street centerline data throughout the entire United States.

Creating the TIGER system was a massive collaborative effort involving the USGS. The system was created for the 1990 census and was first released in 1989. The TIGER system provided the public the very first nationwide digital map of roads and boundaries covering the entire United States at the street level and played a key role in GIS analysis. TIGER/Line files were released by the Census Bureau in ASCII format until 2007, when the format of TIGER/Line files was changed to shapefile. The shapefiles include point features, line features representing roads and hydrography, and polygon boundary features. They contain geographic identifiers that can be used to link the shapefiles with Census Bureau's tabular data.

The shapefiles themselves do not contain any demographic data from the census survey. Demographic data from the survey is available for download in tabular format from the US Census Bureau website (https://www.census.gov/geographies/mapping-files/time-series/geo/tiger-line-file.html). These contain detailed demographic and economic information collected as part of the census. The tables can be linked with TIGER/Line shapefiles and geodatabase files, giving a GIS user good-quality geospatial data rich with detailed attribute information. TIGER/Line datasets include political and census boundaries at state, county, census tract, block group, block, zip code, urban boundaries, tribal subdivision, congressional district, city, and school district level. In addition to their street centerline data, other TIGER/Line shapefiles are also available that include primary and secondary roads, hydrology, railroads, landmarks, coastlines, and many more. TIGER/Line files can be used by any GIS software and have become an important source of vector data for geospatial data users. Figure 5.8 shows TIGER/Line census tracts and primary roads shapefiles of Texas.

The TIGER database is used as the primary roads base layer on the USGS's National Map and US Topo maps. Figure 5.9 shows a TIGER/Line shapefile of street centerline data of Montgomery County in Texas. The road highlighted in blue on the map is State Highway 105 that runs across the county. Street centerline shapefiles are composed of small segments of road, each associated with its own attribute information assigned to it. Although State Highway 105 is one road, it is composed of 182 line segments, as shown in figure 5.9.

The attribute information describes the characteristics of each line segment, such as road name, zip code, and address range. This information is essential for any application that references locations for pinpointing addresses. Files such as the street centerline TIGER/Line shapefile serve as a valuable reference basemap for applications that pinpoint location. So, when you input an address into your smartphone's mapping application, the address is matched against the values of the attribute information of a reference basemap. If a match is not found between the input address and the values of attribute information of the reference basemap, the mapping application will not be able to locate the address. It is also important that the basemap used for referencing purposes is highly accurate or the mapping application will not be able to pinpoint addresses or will navigate to an incorrect address or location. The next time a mapping application navigates you to an incorrect location, the chances are high that some of the values of the attribute information in the reference file are incorrect or missing.

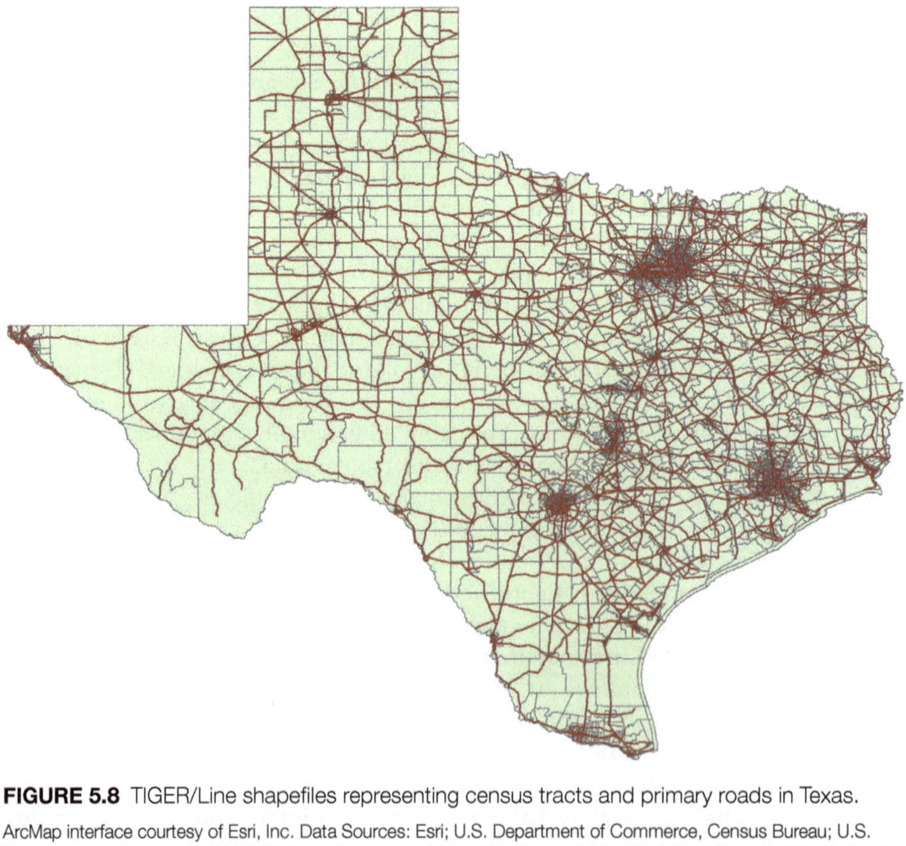

FIGURE 5.8 TIGER/Line shapefiles representing census tracts and primary roads in Texas.

ArcMap interface courtesy of Esri, Inc. Data Sources: Esri; U.S. Department of Commerce, Census Bureau; U.S. Department of Commerce (DOC), National Oceanic and Atmospheric Administration (NOAA), National Ocean Service (NOS), National Geodetic Survey (NGS), United States Census Bureau, 2022

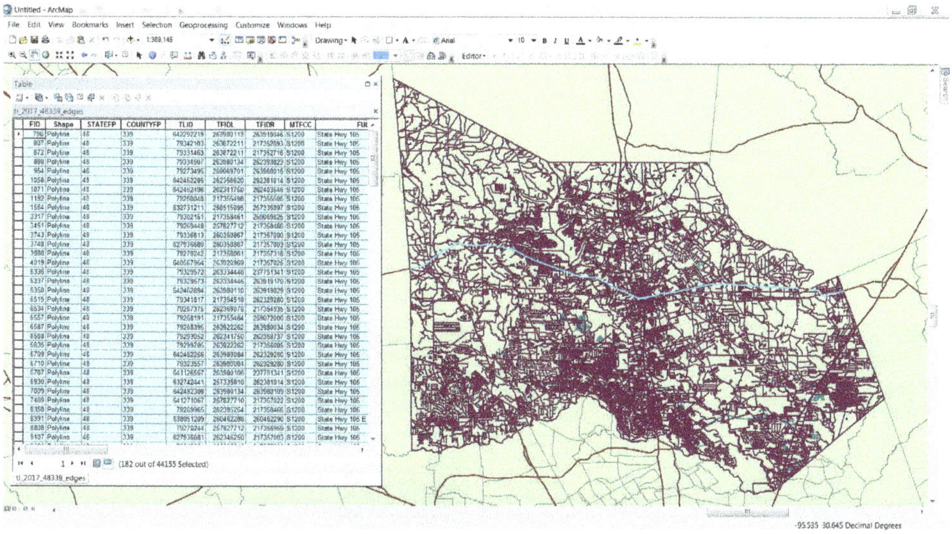

FIGURE 5.9 The State Highway 105 TIGER/Line shapefile is highlighted in blue on the map and the table shows the attribute information of the 182 line segments the State Highway 105 TIGER/Line shapefile is composed of.

ArcMap interface courtesy of Esri, Inc. Data Sources: Esri; U.S. Department of Commerce, Census Bureau; U.S. Department of Commerce (DOC), National Oceanic and Atmospheric Administration (NOAA), National Ocean Service (NOS), National Geodetic Survey (NGS), United States Census Bureau, 2022

Esri ArcGIS Living Atlas of the World

Another excellent source of geospatial data is Esri. Esri has a massive collection of very good-quality geospatial data, layers, and maps that it has curated from thousands of data sources the world over. This collection is available through two portals: Esri's location data browser and its Living Atlas of the World.

Esri location data have ready-to-use demographic data, behavioral and business data, places and movement data, high-quality imagery, and basemaps. Most of these datasets are available for the general public to download through the ArcGIS Living Atlas of the World.

The ArcGIS Living Atlas of the World is a collection of ready-to-use global geospatial data from Esri, a leading provider of GIS solutions including the ArcGIS suite of software. Datasets available through the Living Atlas of the World include imagery, basemaps, demographics and lifestyle data, landscape data, features representing boundaries and places, transportation data, earth observations, urban systems, oceans, and historical maps (https://livingatlas.arcgis.com). Users can download data and combine them with their own datasets to create maps, images, and apps or perform analysis. Maps and apps can be created in Map Viewer, Esri's interface solution to work with online maps, or in ArcGIS Desktop.

The Living Atlas represents a massive collection of geospatial data from Esri and thousands of private, government, and nongovernmental entities. Datasets fall into two main categories: authoritative and premium content. Authoritative datasets are created by Esri by bringing together quality datasets and layers from other organizations. These datasets and layers are high-quality, authoritative, and recommended by Esri, while also freely available to the public. Premium content data are curated by Esri from different data sources. In order to use premium content data, users must have a subscription to Esri's ArcGIS Online. The majority of the data available through the Living Atlas is free. However, users must have access to ArcGIS software to be able to view or use the datasets. Users have different options to browse and search for data (figure 5.10).

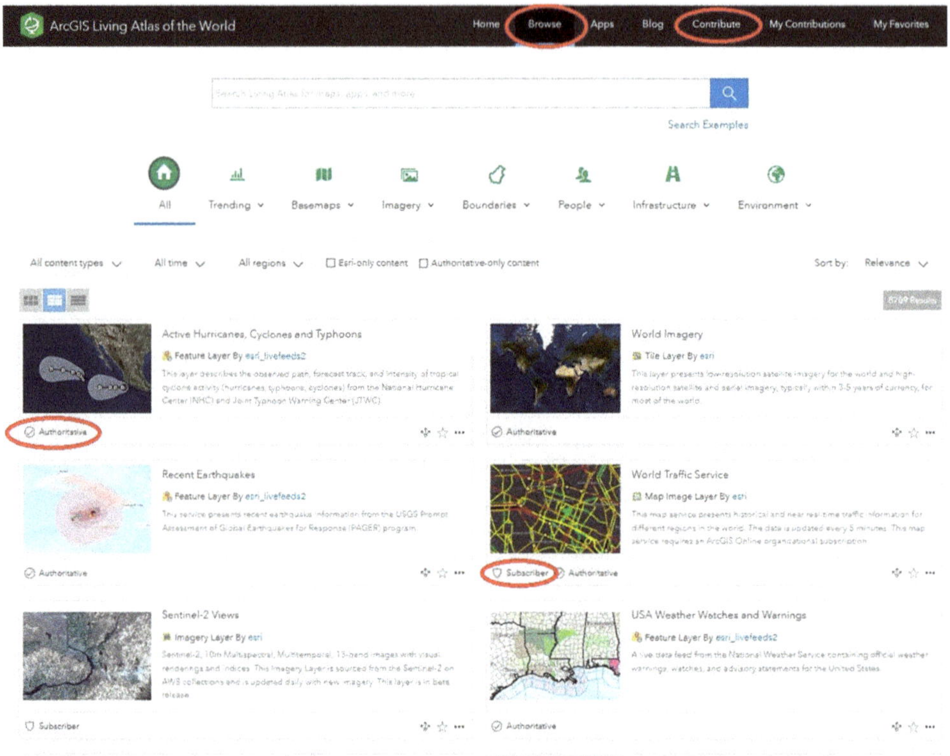

FIGURE 5.10 ArcGIS Living Atlas of the World.

ArcGIS Living Atlas of the World interface courtesy of Esri, Inc.

As you can see in figure 5.10, some of the datasets have a small circle with a checkmark inside it. Datasets with those signs are authoritative data that are recommended by Esri. Some other datasets have the subscriber sign, meaning users must have a subscription account with Esri's ArcGIS Online. The Living Atlas also has an option for community members to contribute their data, apps, and maps to the atlas.

Field Methods and Geospatial Data Creation

As you can see, a lot of geospatial data has already been created for you, and agencies like the USGS, US Census Bureau, and Esri, discussed in the previous sections, continue to create new datasets to update and add to their current collection. Geospatial datasets can be collected and created in a number of different ways. All geospatial data collection methodologies fall into two broad categories: in situ data collection and ex-situ data collection methods. In-situ data collection methods are conducted on the ground by a technician or by using a data collection device. Ex-situ data collection methods are conducted in a laboratory setting using GIS software and applications.

In-situ data collection can be conducted in two ways. First, a field-worker or researcher is present on-site to collect data in its true setting. For instance, a field-worker may go door-to-door to collect data, or a researcher may visit the study area to interview research subjects. Thus, data is collected in the geographic space being studied. Alternatively, in situ collection of geospatial data can be conducted using an instrument or device that collects data in its true setting in lieu of a human interaction; for instance, collecting rainfall data using a rain gauge or temperature data using a thermometer. Similarly, geospatial data can be collected in situ by using a global positioning system (GPS) technology or data can be collected by means of remote sensing techniques.

Global Positioning System (GPS)

Today, we are all well aware of GPS devices. They have become a part of our everyday activities to such an extent that we may use a GPS or GPS-based application without realizing it. For instance, if you are on a road trip you may need to exit to fill the gas tank or use the restroom on the way. During the course of your trip you may take a wrong exit and end up in a completely unknown location where you are unfamiliar with the street names and route numbers. What would you do in this situation to navigate yourself back to your planned itinerary? You must first be able to locate your current position. Thereafter, chances are the mapping application on your smartphone would recalculate the route and assist you (or the vehicle) in navigating to the new path. But what if you are in a situation where you have no cell phone reception? Can you imagine what you would do if you were lost in an unknown location with no cell phone reception and with the possibility of not knowing your current position? It can be a scary situation. However, the availability of geospatial technology and tools have made such situations extremely unlikely.

Even without cell reception, applications downloaded and installed on most smartphones can help users because of geospatial technology and the tools available today. Using this technology, one can determine their position on the earth's surface with relative accuracy and navigate their movements. A small handheld device can be used to record one's coordinates and movements. The output can also be plotted onto a mapping application. Most smartphones available today can also be used in lieu of a separate handheld device. The position and movements can be plotted onto a mapping application stored in most smartphones. This is possible because of a technology that we are all very familiar with, a **Global Positioning System (GPS)**.

GPS devices have become ubiquitous. Gone are the days when they were only available in special electronic stores. Today one can purchase GPS devices in sporting goods stores, home improvement stores, and retail merchandizing stores. GPSs have become a part of standard gear used by bikers, hikers, trekkers, campers, golfers, and travelers. GPS receivers come loaded in special watches, tablets, cars, and smartphones. Using a smartphone's GPS technology for navigation purposes has become common practice today. No matter where you are on the earth's surface, you can use a GPS device for free.

GPS devices that we are all very familiar with use Global Positioning System technology. GPS is a type of a **global navigation satellite system (GNSS)**. The GNSS technology was developed by the US Department of Defense and first launched into orbit in 1978. The 1990 Gulf War was the first war to use GPS technology. GPS is a US government–owned utility operated and maintained by the United States Space Force. A GPS system consists of three components. Any GPS device relies on these three components working in sync with one another to pinpoint locations on Earth. They are a space segment, a control segment, and a user segment.

The space segment of a GPS system consists of a constellation of GPS satellites orbiting the earth in fixed paths. These satellites transmit radio signals to users on earth. The satellites in the GPS constellation orbit the earth in six orbital planes. Each of the six orbital planes surrounding the earth are equally spaced, 60° apart. Each of the orbital planes contain four satellites at any given time, creating a 24-slot core satellite constellation (https://www.gps.gov/systems/gps/space) (figure 5.11). This arrangement ensures that at any given time a GPS receiver can view at least four satellites from any point on earth.

GPS satellites orbit the earth at an altitude of approximately 20,200 kilometers (12,552 miles). Each satellite orbits the earth twice a day. GPS satellites broadcast radio signals to earth from their orbit. These signals contain information about the satellite location and status and the precise time at which the signal was transmitted from the satellite. GPS satellites are equipped with an atomic clock on board to measure precise time. To maintain coverage, the US Space Force normally flies 31 operational GPS satellites in the eventuality of a GPS satellite being serviced. In June 2011, the GPS satellite constellation was expanded to reposition 6 satellites and include 27 satellites in the constellation. Thus, the GPS system now consists of a constellation of 27 satellites orbiting the earth.

The control segment of a GPS system consists of a global network of ground stations that track and monitor GPS satellites and their transmissions. The ground stations, also called control stations, are located all over the world, with the master control station located at Schriever Space Force Base in Colorado and an

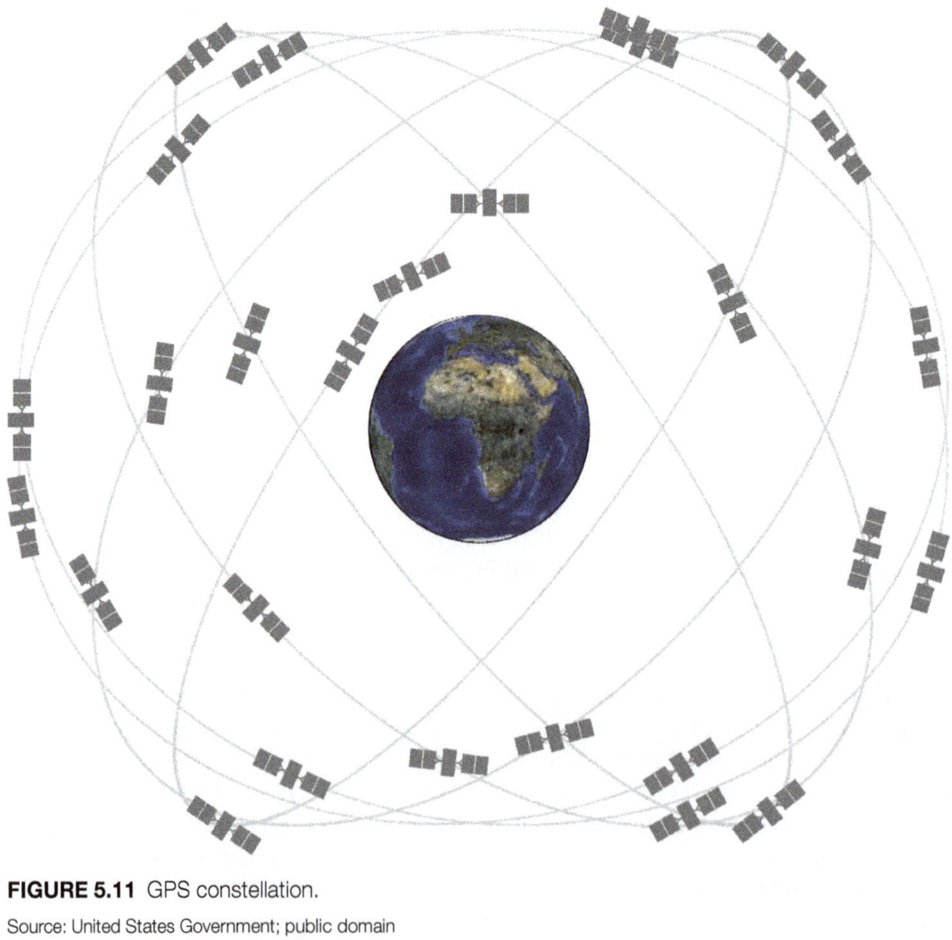

FIGURE 5.11 GPS constellation.

Source: United States Government; public domain

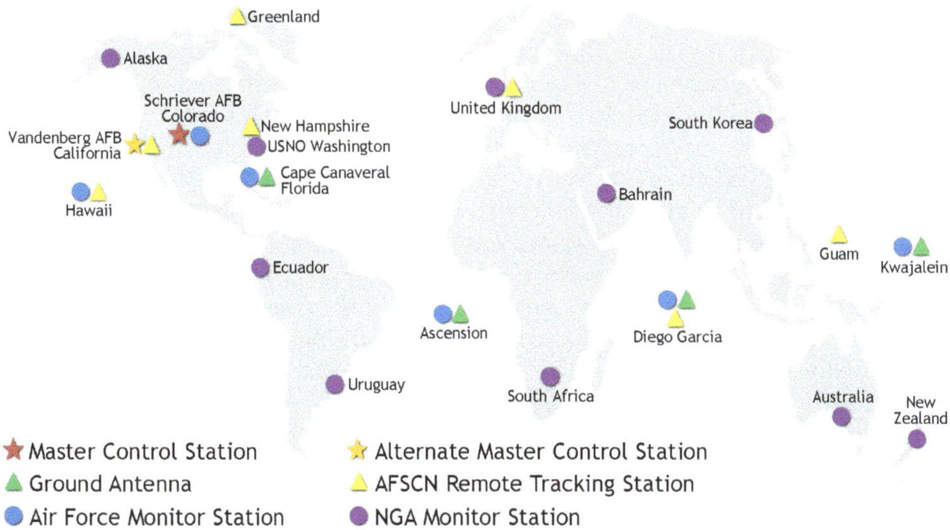

FIGURE 5.12 GPS control segment.

Source: United States Government; public domain

alternate master control station located at Vandenberg Space Force Base in California (figure 5.12). The control stations collect data transmitted from GPS satellites and broadcast the data to the master control station. Corrections, new data, and updated orbit information are constantly transmitted to the GPS satellite constellation from the control stations so that the satellites broadcast accurate data.

The user segment of a GPS system consists of a GPS receiver device that receives signals from GPS satellites and uses the information to calculate its three-dimensional position on earth. All GPS devices available on the market, whether handheld or stored in a watch, car, or smartphone, are GPS receivers. The core components of a GPS receiver device consist of an antenna, a processing unit, and a clock. The antenna is tuned to the frequencies transmitted by GPS satellites. GPS satellites transmit signals on two frequencies: L1 frequency broadcast at 1575.42 MHz and L2 frequency broadcast at 1227.60 MHz. Another important component of a GPS receiver device is the number of channels in the receiver, which determines from how many satellites it can receive signals. All GPSs also have some sort of a display for user interaction, showing your location information. Most GPS receiver devices are also accompanied by a map display so that location information is displayed on a street map or a remotely sensed image.

The accuracy of information obtained by a GPS receiver depends on several factors. A GPS receiver with a higher number of channels can pick up signals from more GPS satellites. Similarly, a single-frequency receiver will only receive L1 frequency, whereas a dual-frequency receiver will be able receive both L1 and L2 frequencies. Storage capacity and the receiver's firmware also impact data accuracy.

All three segments of GPS, discussed above, work together to determine location on earth. GPS satellites transmit signals that are received by the control stations. The signals contain important information about the satellite's orbit, status, and location. Data containing information about GPS satellite constellation and the status of each satellite in the constellation is known as almanac. Data containing information about the GPS satellite's orbital position is known as ephemeris. Almanac data helps a GPS receiver to locate satellites, and ephemeris data helps the receiver to deliver precise location information.

The signals from GPS satellites are sent in the form of codes. Signals are transmitted either in C/A code or P code. **C/A code** is a coarse acquisition code transmitted at L1 frequency, while **P code** is a precise code transmitted on L1 and L2 frequencies. All civilian GPS receivers receive C/A code, whereas military-grade GPS receivers are required to receive information transmitted in P code.

When a GPS receiver is activated, it locates the nearest satellites and receives signals transmitted from each visible satellite. A GPS receiver uses a mathematical process called **trilateration** (figure 5.13) to convert the information it receives from satellite signals into position information. Trilateration means pinpointing positions from three distances on a two-dimensional map. In the case of a GPS receiver, this position is being located on the earth's three-dimensional surface with respect to GPS satellites orbiting the earth by calculating spherical distance. This process is called *three-dimensional trilateration* and is used by GPS receivers to derive position information on the earth's three-dimensional surface.

In addition to the satellite information transmitted in codes, the time when a signal was transmitted from a satellite and how long it takes for the signal to reach the GPS receiver is important in computing location information. GPS satellites are mounted with highly precise atomic clocks, whereas GPS receivers are equipped with a clock that is not as precise. Due to the difference in precision between the two clocks, an error in timing occurs causing inaccurate distance calculation by a GPS receiver. In order to correct this clock error, GPS receivers collect signal information from a fourth satellite. A GPS receiver must therefore have the location of at least

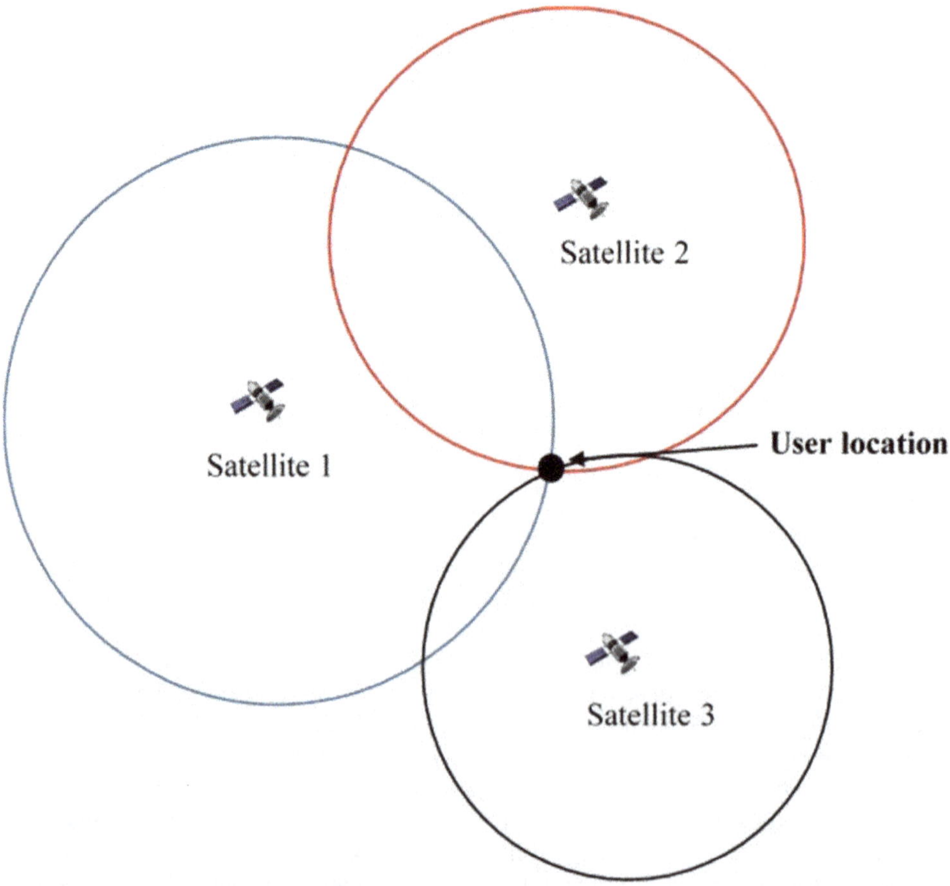

FIGURE 5.13 GPS trilateration process.

four satellites to pinpoint location. By analyzing radio signals transmitted from at least four satellites and calculating the distance between the GPS receiver and each of the GPS satellites, coupled with the time it takes for transmitted signals from the satellites to reach the GPS receiver, a GPS receiver is able to pinpoint location on earth. If a GPS receiver cannot locate at least four satellites, it will be unable to pinpoint its location. If it is able to locate more than four satellites, the accuracy of the three-dimensional trilateration process increases. GPS receivers provide position information in WGS84 datum.

GPS Data Accuracy and Augmentation Systems

An obvious question about location information derived from a GPS concerns its accuracy. Ultimately how accurate is your GPS and the information it provides? GPS accuracy is impacted by a number of factors. Signals transmitted by GPS satellites are fairly accurate. Inaccuracies in reading occur due to factors that interfere with, impede, or delay the transmission and reception of signals by receiver units. Most commercial GPS receivers are accurate within 10 to 15 meters. GPS-enabled smartphones are also fairly accurate within 5 meters. Nonetheless, in order to provide accurate information, GPS receivers need a clear view of the sky for appropriate signal reception. Therefore any kind of hindrance such as tall buildings, bridges, thick forests, or canopy cover that interferes with signal reception causes errors in position calculation.

Another cause of error in position calculation is an effect called **multipath**. This happens when signals, instead of following a straight path from a satellite to the receiver, are reflected off buildings or walls before they reach a GPS receiver. This causes a delay in signal reception by the receiver, introducing an error in position calculation. Errors also occur due to issues with satellites. If there is an error in satellite position, it will transmit incorrect orbital information, causing an error in position calculation by the receiver. Such satellite errors must be fixed by the control station. Poor arrangement of satellites can also be a cause of error.

An important source of error that existed until 2000 was a feature called *selective availability* (SA). This feature was introduced by the US government to deliberately make civilian GPS less accurate globally. By degrading the C/A signal transmitted from GPS satellites, civilian GPS was only able to achieve accuracy within 100 meters. In May 2000, the US government under the Clinton administration ended its use of selective availability.

In order to improve the accuracy of GPS receivers, different augmentation systems have been developed. One such system is differential GPS. **Differential GPS (DGPS)** is based on the use of two GPS receivers. One receiver is positioned at a location that receives satellite signals and its x, y, and z coordinates are known. This is the base, or reference, station receiver. The second GPS receiver is a mobile, or roaming, receiver. Both receive signals transmitted from at least four visible satellites. Because the precise location of the base station receiver is known, it can calculate corrections to any errors that cause miscalculation of position in a roaming receiver. The base station receiver then transmits this correction to any roaming receiver capable of receiving this correction, called *differential correction*. The roaming GPS receiver receives the correction information to achieve measurement accuracies within 5 meters. Roaming GPS receivers capable of receiving differential correction thus receive satellite signals from at least four visible satellites plus an additional differential correction from a base station receiver. Base stations for differential GPS are located all over the world. Global Differential GPS (GDGPS) is a global augmentation system that employs a network of global base station receivers.

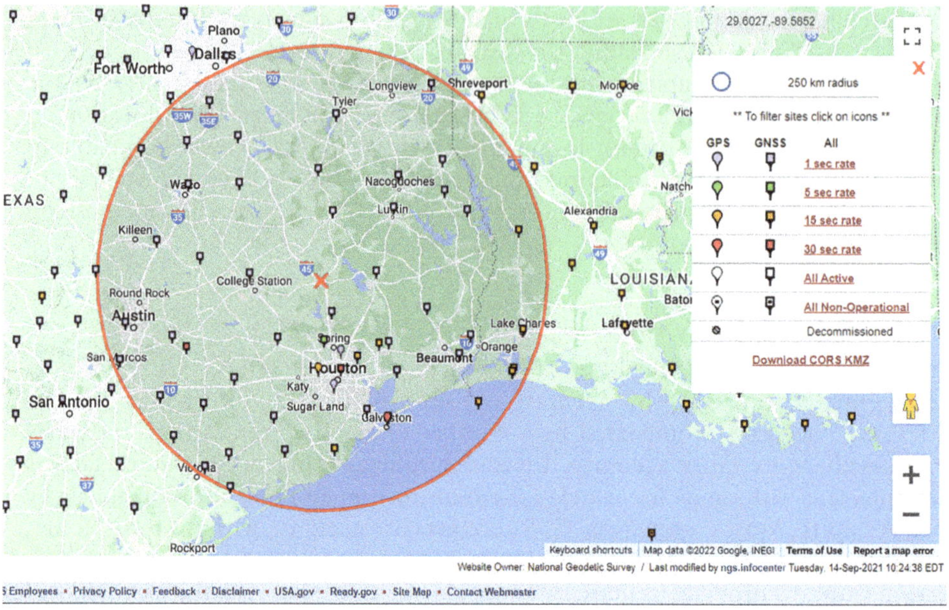

FIGURE 5.14 NOAA CORS Network (NCN) map.

Source: National Geodetic Survey; public domain

Another augmentation system is the **Continuously Operating Reference Stations (CORS)** program operated by the National Geodetic Survey and managed by the National Oceanic and Atmospheric Administration and the National Geodetic Survey. CORS base stations are located throughout the United States and Canada. The **NOAA CORS Network (NCN)** (figure 5.14) provides data to improve the accuracy of position measurements throughout the United States. CORS collects GPS data from approximately 200 private, public, and academic organizations and almost 2,000 GPS tracking stations and distributes the data to improve the accuracy of position measurements relative to the National Spatial Reference System. Any GPS receiver within 200 kilometers of a CORS station can receive corrections to improve its measurements. CORS stations are located in a way to receive unobstructed signals from GPS satellites at all times and are equipped with very high-quality GPS receivers.

Another augmentation system to improve accuracy of position measurements is based on corrections transmitted from a satellite and is called a **satellite-based augmentation system**. In a satellite-based augmentation system a correction is calculated and transmitted from a satellite and received by a receiver GPS. Thus, in this case the receiver GPS receives signals from four visible satellites plus a correction transmitted from an additional satellite. An example of a satellite-based augmentation system is the **Wide Area Augmentation System (WAAS)**. This system was developed by the Federal Aviation Administration specifically for civil aviation purposes. The system operates through a series of Wide Area Reference Stations (WRS) located throughout the United States. The locations of these WRS are precisely surveyed, allowing rapid detection of any errors in GPS signals received at the WRS location. GPS information received at WRS locations is transmitted to a WAAS Master Station (WMS). The WMS generates a correction to remove errors in GPS signals. This correction is transmitted by the WMS to a geostationary WAAS satellite that transmits the correction to receiver GPS on earth. Receiver GPS that have the capability to receive such corrections transmitted by a WAAS satellite can use the information to improve their position measurements.

Remote Sensing Data Collection

Another important and widely used source of geospatial data is remote sensing. Remote sensing is a process where data is collected by sensing the area of interest remotely. The instrument used to collect data is not in direct contact with the area of interest, rather it is contacting it remotely. Take the example of capturing a picture of a park or your friends. You take the picture using a digital camera or your smartphone. Once you have clicked, the picture will be stored as a file in a designated folder either in the device or in a memory card. Taking a picture of a park or your friends is a form of remote sensing. You hold the camera, whether a digital camera or your smartphone, at some distance from the park or your friends. The camera then senses the park or your group of friends from that distance while you click and take a photograph. The remote sensing device here is the camera that senses your "area of interest" remotely.

You may be surprised to know that when you use your eyes to read a book or look at something, you are remotely sensing the book or whatever it is you are looking at. This is also a process of remote sensing performed using your eyes. When you read a book, you don't place your eyes on the book. You read the pages of the book from a distance, "remotely." Your eyes, your smartphone, or your digital camera are performing remote sensing as they "acquire" information (in case of reading) or data (in case of photographs) from a distance without actually contacting the object. Remote sensing is thus defined as the process of measuring, recording, or collecting information about the physical characteristics of an object or a phenomenon by means of a device that is not in physical contact with the object or phenomenon.

Acquiring geospatial data by means of remote sensing involves the use of specialized instruments. Earth-orbiting satellites and airplanes equipped with special cameras and scanners remotely capture photographs of large areas of the earth. Similarly, images of ocean floors can be generated by means of remote sensing. This gives us access to and allows us to visualize physical characteristics of much larger and at times inaccessible areas of the earth's surface than is possible manually. However, acquiring geospatial data by means of remote sensing involves more than simply capturing a photograph remotely. A lot more goes into generating remotely sensed data. The process of remote sensing and the different devices utilized to generate geospatial data remotely is a topic for an entire book (see Sabins and Ellis 2020).

Satellite images and aerial photographs have become commonplace. These are geospatial data that were obtained by remote sensing. One can view a satellite image or an aerial photograph simply by opening a smartphone's mapping application and turning on the option for satellite image (figure 5.15a). You can search for an address and it will display on a satellite image. Alternatively, you can open Google Earth and look up your neighborhood or even your street or your house (figure 5.15b).

So how is remotely sensed data generated? What do you actually visualize when you see a remotely sensed data? How are the images in figures 5.15a and 5.15b able to give you such a realistic view of the world? Yes, we know a camera captured the image that you see on your smartphone or on Google Earth. But how is the data that was captured by a camera able to generate images that look like the ones in figures 5.15a and 5.15b that we have all become so familiar with?

What you are actually seeing is light energy that was reflected back from different objects and captured by a camera. What the camera captures is electromagnetic energy reflected off the earth's surface. A sensor measures, records, and stores information about that energy. With remotely sensed geospatial data, the camera is

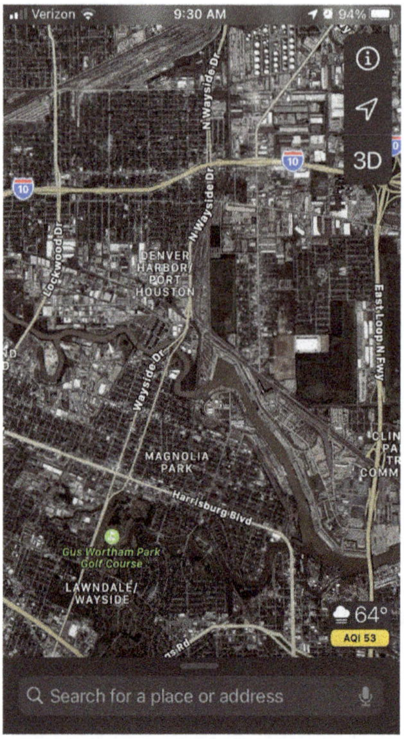

FIGURE 5.15A Remotely sensed geospatial data viewed on an iPhone's mapping application.

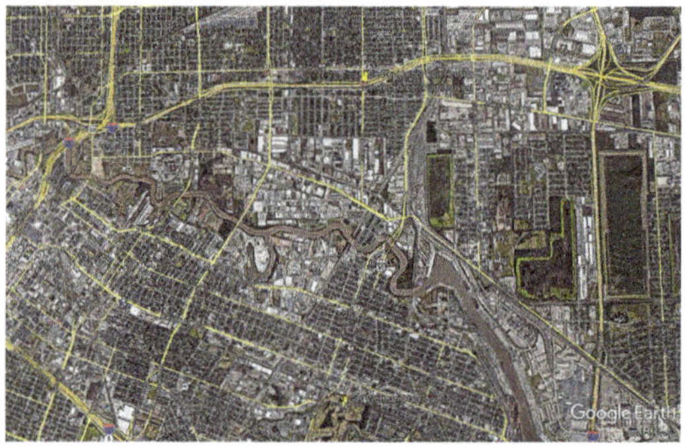

FIGURE 5.15B Remotely sensed geospatial data viewed on Google Earth.

Source: Google Earth

at a considerable distance from the earth's surface and is mounted on an aircraft or satellite. So let's revisit what remote sensing is. Remote sensing is the process of collecting, measuring, and recording information pertaining to electromagnetic energy reflected off the earth's surface using an instrument mounted on an aircraft or satellite. Some remotely sensed data are also generated by measuring forms of energy other than light such as sound energy or thermal energy. Geospatial data is sensed and captured remotely in two main ways. One is aerial photography, and the other is satellite imaging. Satellites collect data from hundreds of miles above the earth's surface and are able to cover a large area, while aerial vehicles capture data from an altitude that ranges from a few thousand to over 60,000 feet.

Aerial Photography

Aerial photography is the process of taking a photograph of the earth's surface by means of a camera mounted on an aircraft. Aerial photography has a very long history. The practice of taking photographs of the earth's surface from airborne vehicles has existed for over a century. Early photographs of the earth were captured from hot-air balloons in the late 1800s. The first aerial photograph was captured in 1858 by Gaspard-Félix Tournachon (Waxman 2018). The photograph, which was captured from a hot-air balloon of an area outside Paris, no longer exists. The oldest evidence of an aerial photograph is that of Boston taken by James Wallace Black from a hot-air balloon 2,000 feet above ground in 1860 (figure 5.16).

FIGURE 5.16 An aerial photograph of Boston taken by James Wallace Black from a hot-air balloon in 1860.

Source: The Metropolitan Museum of Art; public domain

Thereafter there is evidence of aerial photographs captured by means of cameras attached to kites and pigeons in the 1900s (Waxman 2018). In 1906 George Lawrence mounted a camera on 17 kites to capture an aerial photograph of the San Francisco Bay from 2,000 feet above ground after the 1906 earthquake (figure 5.17).

We have come a very long way from hot-air balloons, kites, and pigeons, thanks to technological innovation and advancements. In fact, aerial photography by means of a camera mounted on an aircraft was used for cartographic purposes during World War I. We, of course, have accomplished great strides from those days. Advancements in the aviation field and camera and lens technology have particularly influenced the field of aerial photography. Traditionally, aerial photography has been

FIGURE 5.17 Aerial photograph of San Francisco Bay captured by George Lawrence in 1906 by means of a camera mounted on kites.

Source: Library of Congress, Prints and Photographs Division, Washington, DC. Reproduction number LC-USZ62-136183

conducted with cameras mounted in an aircraft flown by a pilot sitting in the cockpit. Today, aerial photography is conducted from cameras mounted on unmanned aerial vehicles.

Unmanned aircraft systems (UAS) are aircraft that are driven from the ground using a remote control. We are all familiar with UAS. We know them as drones. Unmanned aircraft systems provide a rapid and cost-effective solution to generating aerial photographs. Unlike traditional aircraft, UAS are not flown by a pilot sitting in a cockpit. Instead, the pilot is on the ground and controls the UAS by means of a remote control. UAS have been used extensively to generate data and aerial photos for defense purposes. They were used during the Gulf, Afghanistan, and Iraq Wars. The military utilizes UAS for multiple purposes that include recording intelligence and reconnaissance data, surveillance, and also for striking targets. UAS use by the military includes the use of drones such as the MQ-9 Reaper (figure 5.18), the MQ-1B Predator (figure 5.19), and the RQ-4 Global Hawk (figure 5.20).

Unmanned aircraft systems are widely used by the general public today for leisure, research, and commercial purposes. If you look around, you will find that there are many uses of unmanned aircraft other than for defense purposes. For instance, unmanned aerial vehicles have been very useful during disasters to assess damage (figure 5.21), deliver aid, and locate victims. They are also used for agricultural purposes, where drones aid farmers to gather data about crops, pests, and irrigation. Drones are being used to track wildlife for conservation purposes. Use of unmanned aircraft is also prevalent among law enforcement and security. UAS have become a major source of geospatial data. Unmanned aircraft are widely used by government agencies like the USGS, private companies, academic institutions, and research centers to generate geospatial data.

Unmanned aircraft are equipped with sensors, cameras, and other equipment to collect aerial photographs, images, and other geospatial data crucial for the GIS world. For instance, in figure 5.22, USGS pilots Elizabeth Pendleton and Seth Ackerman are flying an unmanned aircraft to collect images that will be used to create 3D surface models.

FIGURE 5.18 An MQ-9 Reaper UAS.

Credit: US Air Force photo / Lt. Col. Leslie Pratt

FIGURE 5.19 An MQ-1B Predator UAS.

Credit: US Air Force photo / courtesy photo

FIGURE 5.20 A RQ-4 Global Hawk UAS.

Credit: US Air Force Photo / courtesy photo

FIGURE 5.21 A pilot for the USGS Woods Hole Coastal and Marine Science Center's Aerial Imaging and Mapping group flies a UAS mission in Hawaii while a river of lava flows in the background.

Public domain. Attribution: Natural Hazards, Coastal and Marine Hazards and Resources Program, Region 1: North Atlantic–Appalachian, Woods Hole Coastal and Marine Science Center

FIGURE 5.22 USGS drone operation at Sandy Neck, Barnstable, Massachusetts.

Types of Aerial Photographs

Aerial photographs can be classified into different categories based on three main criteria: axis of the camera, scale of the image, and type of film used to capture the photograph.

Aerial photos captured with a vertical camera axis are known as **vertical photographs**. Imagine a camera is attached to the bottom of the aircraft and is capturing photographs of terrain directly under the plane. Photographs captured in this manner are vertical photographs. The point on the ground that is directly under the camera is called the **nadir**. Vertical photographs have very little relief and cover a relatively small area.

Aerial photographs captured with a camera tilted at an angle of more than 3 degrees are known as **oblique photographs**. In this case, the camera is mounted in such a way that it is not positioned directly at the nadir but at an angle. Oblique photographs are of two types: low oblique aerial photographs where the tilt of the camera axis is more than 3 degrees and high oblique aerial photographs where the tilt is around 60 degrees. Low oblique aerial photographs do not cover the horizon and cover a relatively small area, whereas high oblique aerial photographs include the horizon and cover a larger area, providing a wide view of the area captured.

In oblique aerial photographs a viewer sees the photograph from a perspective view, so the viewer can visualize other features such as sides of buildings, entrances, or terrain features. Many companies commercially sell high-quality and high-resolution oblique aerial photographs. Microsoft's Bing Maps offers oblique 45-degree-angle aerial photographs of several locations in the United States for free, known as Bird's Eye. Figure 5.23 is an oblique photograph of Rice University shown in Bing Maps Bird's Eye imagery.

Aerial photographs can also be classified as large-scale and small-scale. **Large-scale photographs** are taken in an aircraft flying at a lower elevation, while **small-scale photographs** are captured in an aircraft flying at a higher elevation. Therefore, large-scale aerial photographs cover a smaller area and small-scale aerial photographs cover a larger area.

Aerial photos are also classified into different categories based on the type of film used to capture the image. Different camera films are sensitive to different wavelengths of electromagnetic energy. Remember cameras capture light reflected off the earth's surface. What they actually capture is electromagnetic energy. The color a human eye sees in an image depends on the wavelength of the electromagnetic energy captured by the camera. Electromagnetic radiation spans a wide range of wavelengths known as the *electromagnetic spectrum*. Electromagnetic spectrum is divided into seven regions: radio waves, microwaves, infrared (IR), visible light, ultraviolet (UV), X-rays, and gamma rays (figure 5.24).

Images captured with panchromatic films are called *panchromatic aerial photographs*. Panchromatic films are sensitive to the entire visible region of the electromagnetic spectrum and capture the image in shades of gray ranging from black to white. Therefore, the aerial photos are black and white.

Color photos are captured by a camera that has a film sensitive to three main sections of the visible region of the electromagnetic spectrum: red, green, and blue. A color infrared aerial photo is taken with a camera that has a color infrared film, sensitive to the green and red bands of the visible portion and extending into the infrared region of the electromagnetic spectrum. In a color infrared photo, objects that reflect near infrared wavelength of electromagnetic energy are displayed in red,

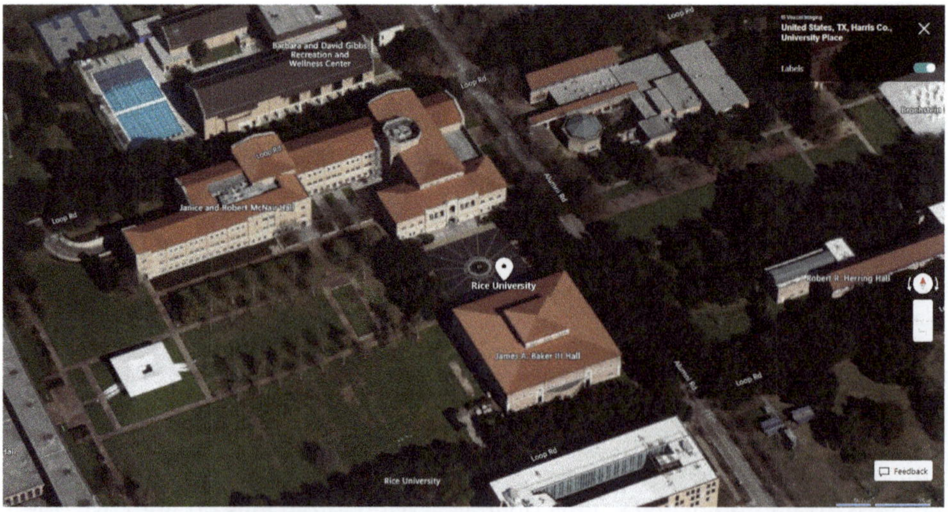

FIGURE 5.23 Oblique aerial photograph of the Rice University campus in Houston, Texas, as shown in Bing Maps Bird's Eye imagery.

Credit: Bing Maps Bird's Eye

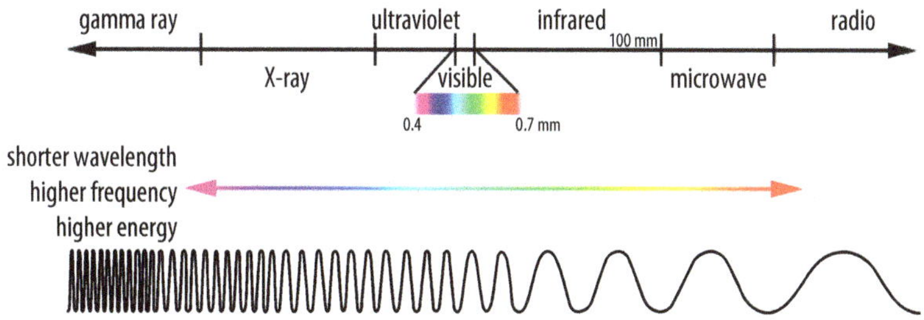

FIGURE 5.24 Regions of electromagnetic spectrum.

Credit: NASA's Imagine the Universe

objects that reflect red wavelength of electromagnetic energy are displayed in green, and objects that reflect green wavelength of electromagnetic energy are displayed in blue. Color infrared photos are also called "false color" photos because they show an image in colors not normally visible to the human eye.

Color infrared photos are very useful for studying natural resources, particularly vegetation. Figure 5.25 is a color infrared aerial photo of Horry County located on the North Carolina–South Carolina border. Healthy vegetation appears red to intense red in color. In any color infrared photo, as the vitality of vegetation decreases, it appears in the photo in lighter shades of red and pink and in different shades of green. Water appears in different shades of blue, ranging from almost black, indicating only water with no deposits, to lighter shades of blue as the amount of sediment deposit increases. Built-up area and bare soil appear in light tones as white or cyan. The colors that you see in the color infrared aerial photo are indicative of the wavelength captured by the camera.

An aerial photograph cannot be used as a map, no matter what type of an aerial photo you have. Remember it is a photo taken with a camera. It may show all the features, but the photo does not have uniform scale throughout. When an aircraft mounted with a camera flies over a terrain, some features will be closer to the camera, such as tall buildings, mountains, and towers, and will have a larger scale, while other features will be farther away from the camera and will have a smaller scale. This causes an effect known as relief displacement in aerial photos. Relief displacement is a shift in the position of a feature in an image due to its height. As a result, tall features in the photo appear to be leaning away. In order to rectify this issue, aerial photos go through a process known as orthorectification. This process removes the effect of terrain and relief displacement from the aerial photo, producing an orthophoto. Orthophotos as a result have a uniform scale throughout the photo. Orthophotos are used to produce DOQs, covered earlier in this chapter.

Satellite Imaging

Another excellent source of geospatial data derived through remote sensing are satellite images. Before we start our discussion on different satellite datasets, let's revisit how the process of remote sensing happens. As already discussed, what the process of remote sensing really captures is light reflected off the earth's surface. So how does this actually happen? Solar energy passes through the atmosphere and reaches the earth's surface. Here it interacts with objects on the surface, and some of this energy is reflected back. The reflected electromagnetic energy travels in waves into space, passes through the atmosphere, and is captured by either an aerial vehicle or an earth-orbiting satellite (figure 5.26).

FIGURE 5.25 Color infrared aerial photo of Horry County, South Carolina, acquired in 2007.
Source: USGS, Earth Explorer

What is really captured by an aerial vehicle or an earth-orbiting satellite is electromagnetic energy reflected off an object. This process depends on two things: the type of lens or film used to capture electromagnetic energy and the wavelength of electromagnetic energy reflected off the earth's surface. Wavelength is an important parameter used to measure the properties of electromagnetic energy. Wavelength of electromagnetic energy covers a broad range, called the **electromagnetic spectrum** (figure 5.24). You will be surprised to know that we actually encounter electromagnetic energy in our everyday lives. For instance, our microwave oven, remote controls, radio, and night vision goggles all use electromagnetic energy. Figure 5.27 shows different portions of the electromagnetic spectrum that we encounter in our daily lives.

We may not see the reflected electromagnetic energy, because the human eye is not capable of seeing electromagnetic energy at all wavelengths. The human eye is sensitive to reflected light that has a wavelength ranging between 0.4 and 0.7 micrometer. The portion of the electromagnetic spectrum that covers those wavelengths is called the **visible light spectrum**. The visible portion of the spectrum is composed of different colors and can be divided accordingly from the shorter wavelength to the longer wavelength into blue, green, and red portions of the spectrum.

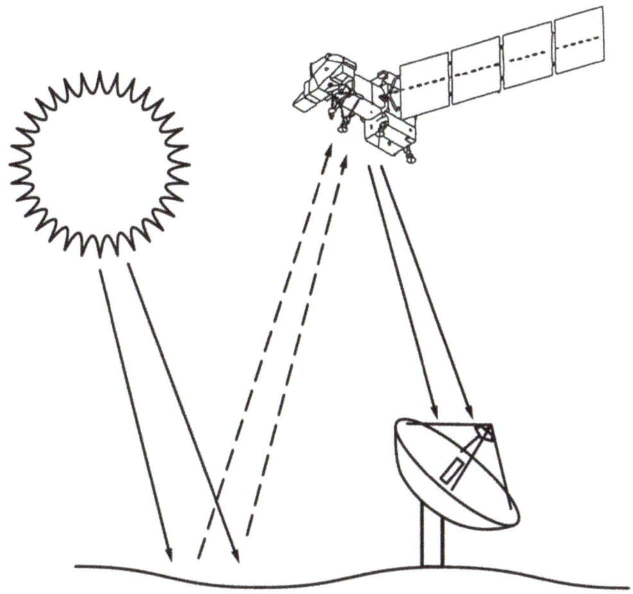

FIGURE 5.26 Process of remote sensing.

Source: USGS

The blue portion lies between 0.4 and 0.5 micrometer, green lies between 0.5 and 0.6 micrometer, and red lies between 0.6 and 0.7 micrometer. So when you see something that is green in color, your eye is actually sensing electromagnetic energy reflected from that object and the wavelength of the energy lies between 0.5 and 0.6 micrometer. Each portion of the electromagnetic spectrum that represents electromagnetic energy of a certain wavelength is called a band. So wavelength between 0.4 and 0.5 micrometer is called blue band, between 0.5 and 0.6 micrometer is called green band, and between 0.6 and 0.7 micrometer is called red band.

While the human eye can only see reflected electromagnetic energy that lies between 0.4 and 0.7 micrometer, remote sensing instruments can sense energy beyond those bands of the spectrum. The infrared portion of the electromagnetic spectrum covers wavelength between 0.7 micrometer and 100 micrometers. In the infrared portion, band covering 0.7 to 1.3 micrometers is called *near infrared band*, band covering 1.3 to 3.0 micrometers is called *shortwave infrared band*, and 3.0 to 14.0 micrometers is called *thermal infrared band*. These three bands are extremely useful in examining different aspects of the earth's surface using satellite imagery. For instance, the thermal infrared band is used in land surface temperature and urban heat island studies.

Each of these bands may not be visible to the human eye, but they can be captured by sensors on either an aerial vehicle or a satellite. Different sensors can capture and measure different bands of the electromagnetic spectrum. The sensor converts the measurement into a series of pixels. Each pixel is assigned a value that corresponds to the amount of reflected energy measured by the sensor at that point on the ground. This value is called brightness value. This measurement is made for each band of the electromagnetic spectrum that the sensor senses and captures.

Remote sensing devices are capable of capturing multiple bands of the electromagnetic spectrum concurrently. Imagery produced by sensors that can sense and capture multiple bands of the electromagnetic spectrum is called multispectral imagery. The sensors in this case can sense anywhere between 3 and 10 bands.

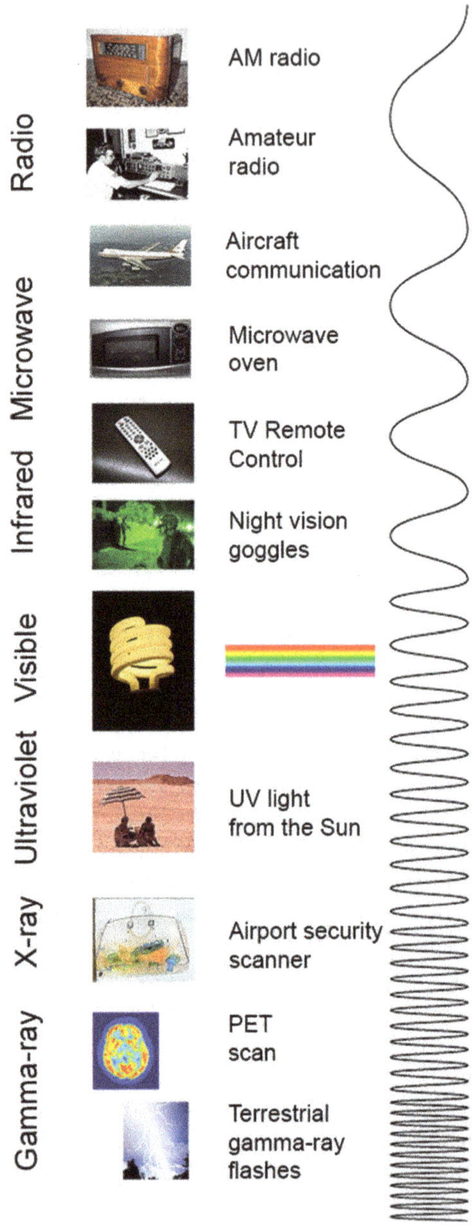

FIGURE 5.27 Different portions of the electromagnetic spectrum encountered in daily life.

Credit: NASA's Imagine the Universe

Imagery produced by sensors that can sense and capture over 200 bands is called hyperspectral imagery.

Each band of a multispectral imagery can be displayed one band at a time or as a combination of three bands at a time. The former case, where only one band is displayed, will produce a grayscale image, whereas in the latter case, where three bands are displayed at a time, a color composite image will be produced. Therefore, to display a color composite imagery, multiple bands of the electromagnetic spectrum must be captured and displayed simultaneously or else it will be displayed in grayscale. When displaying a color composite imagery on a computer screen, three primary colors of light—red, green, and blue—are used. Display devices such as television screens, computer monitors, and laptop screens are equipped with a special device to draw color pixels by using a combination of the three primary colors. When displaying color composite images, display devices associate each primary color to a band of an image. This process produces a color image where each pixel's color is determined by a combination of the primary colors of different brightness values. Such an image is referred to as a color composite image.

If the red band is assigned to the red channel of the display device, the green band is assigned to the green channel of the display device, and the blue band is assigned to the blue channel of the display device, the imagery is called a **true color composite**. Features in a true color composite image will look exactly the way they appear on the earth's surface, as the human eye would see them. An imagery produced by assigning the red band to either the green or blue channel of the display device, the green band to either the red or blue channel of the display device, and the blue band to either the red or green channel of the display device is called a **false color composite**. A panchromatic image is produced when the sensor combines the red, green, and blue bands of the visible spectrum into a single band. The result is a grayscale image. Figure 5.28 shows a false color composite, a panchromatic image, and a standard color composite of an urban area in India.

Remember that satellite images are pictures captured by a camera mounted on an earth-orbiting satellite that visits the same location multiple times. The data generated thus is a raster data model, presented as a collection of pixels (chapter 3). Satellite datasets therefore have two kinds of resolutions, spatial resolution and temporal resolution. **Spatial resolution** is the area on the ground covered by one pixel. **Temporal resolution** is the time is takes for the satellite to return to the same location to capture an image of the same spot on the earth's surface.

Earth-Orbiting Satellites

Today we have many satellites orbiting around the earth and capturing images of the earth's surface. Satellites are launched by government-based satellite programs as well as by commercial satellite programs. For instance, SPOT is a French satellite program initiated by the French space agency CNES in collaboration with Sweden and Belgium and operated by the SPOT Image Corporation. Landsat is a joint NASA and USGS satellite program. Similarly, Sentinel is a satellite program operated by the European Space Agency and Copernicus, the European Commission's Earth Observation program.

Many commercial satellite programs have become key to our day-to-day applications. Digital Globe, a US-based company, is one of the largest producers of commercial satellite imageries today. It has launched and acquired several earth-orbiting satellites, namely Quickbird and WorldView 1, 2, 3, and 4. The high-resolution

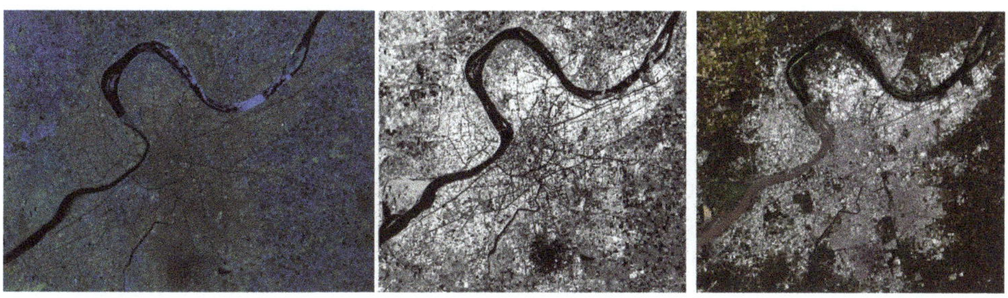

FIGURE 5.28 Three images of an urban area in western India. The first is a false color composite, the second is a panchromatic image, and the third is a standard color composite.

Credit: Mukherjee and Singh 2020

imagery that we are all very familiar with through Google Earth and Google Maps is provided by Digital Globe.

The first satellite to orbit the earth was Sputnik, launched by the Union of Soviet Socialist Republics (USSR) in 1957. The United States launched its first satellite, Explorer I, in 1958. Today, China, Russia, India, and many European countries have launched their own satellites that orbit the earth. Different satellites have different capabilities and accordingly are able to gather different types of information. This depends on a number of factors.

All satellites orbit the earth on a specific path. Some satellites follow an orbit where the satellite rotates at the same speed as the earth. These satellites are called **geostationary satellites** because they follow a geostationary orbit. Geostationary satellites take 24 hours to complete one orbit and will be at the same location at the same time every day. Another type of an orbital path many satellites take is called **near-polar orbit**. These satellites move close to the North and South Poles while completing multiple orbits in a day. Some satellites also follow a **sun-synchronous orbit** where the satellite covers an area on the earth's surface at the same local time.

Sensor characteristics on board the earth-orbiting satellites are very important. A key characteristic of any sensor is which band the sensor can capture and measure. A **panchromatic sensor** measures the visible band and combines it into a single band. This produces a grayscale panchromatic image. The French satellite SPOT is equipped with sensors that capture different bands of the electromagnetic spectrum. For instance, SPOT 6/7 is equipped with a panchromatic sensor. A **multispectral sensor** measures several bands of the electromagnetic spectrum. For instance, Landsat MSS and Landsat Thematic Mapper launched by NASA and sensors on board the commercial French satellite SPOT capture and measure the visible and infrared portions of the electromagnetic spectrum, generating multispectral images. Some satellite sensors are capable of measuring hundreds of bands of the electromagnetic spectrum and are called **hyperspectral sensors**. They produce hyperspectral imagery. Hyperion launched by NASA, HISUI by Japan, and PRISMA by Italy are all hyperspectral imaging missions. There are many such satellite missions operated by different countries, each with a particular goal and each with its own sensors to measure different bands of the electromagnetic spectrum. Table 5.1 lists a few key global satellite programs.

Geocoding

A large amount of geospatial data already exists, created by different government agencies. Data is collected using GPS devices or captured by cameras and sensors mounted on an aircraft or an earth-orbiting satellite. A great amount of data also exists in analog form as hard-copy paper maps. But what if you want to create your own data? Or create data from a paper map or a satellite image or an aerial photograph? There are a few important GIS techniques that can help you achieve those goals. Geocoding is one such process.

Geocoding is the process of converting spatial information into a digital form. The spatial information can be extracted and converted into a digital form using different data sources such a paper map, table of addresses, aerial photo, or satellite image. Remember that you are creating geospatial data through this process. So when you are converting spatial information in a digital form, you are extracting the map information as well as the attribute information. At times, the attribute information may be created after the conversion process is complete. The type of geospatial data

TABLE 5.1 Key Global Government and Commercial Satellite Programs

Name	Country/Owner	Orbit	Sensor	Bands	Purpose	Resolution
Landsat 4 and 5 (decommissioned)	USA NASA	Polar, sun-synchronous	Multispectral Scanner (MSS), Thematic Mapper (TM)	MSS: 4 bands (red, green + 2 bands in near infrared) TM: 7 bands (3 visible bands + 1 near infrared band + 2 shortwave infrared bands + 1 thermal band)	Earth observation	Temporal resolution: 16 days Spatial resolution: MSS, 68 m × 83 m; TM, 30 m
Landsat 7	USA NASA/USGS	Sun-synchronous	ETM+ (Enhanced Thematic Mapper Plus)	ETM: 8 bands (3 visible bands + 1 near infrared band + 2 shortwave infrared bands + 1 thermal band + 1 panchromatic band)	Earth observation	Temporal resolution: 16 days Spatial resolution: ETM, thermal band, 60 m; panchromatic band, 15 m
Landsat 8	USA NASA/USGS	Sun-synchronous	Operational Land Imager (OLI), Thermal Infrared Sensor (TIRS)	OLI: visible and infrared bands (30 m resolution) + panchromatic band (15 m resolution) + 2 new coastal/aerosol and cirrus bands TIRS: 2 different thermal bands	Earth observation	Spatial resolution: OLI, 30 m; panchromatic band, 15 m; two new bands, 30 m
WorldView 1	USA Satellite Imaging Corporation	Sun-synchronous	Panchromatic	Panchromatic	Commercial, earth observation	0.31 m
WorldView 2	USA Satellite Imaging Corporation	Sun-synchronous	Multispectral sensor	1 panchromatic band + 8 multispectral bands (red, green, blue, near infrared + 4 new bands: coastal, yellow, red edge, and near infrared 2)	Commercial, earth observation	Temporal resolution: 1.1 days Spatial resolution: panchromatic, 0.46 m; multispectral, 1.8 m

(continued)

TABLE 5.1 Continued

Name	Country/Owner	Orbit	Sensor	Bands	Purpose	Resolution
WorldView 3	USA Satellite Imaging Corporation	Sun-synchronous	Super spectral	1 panchromatic band + 8 multispectral bands (red, red edge, coastal, blue, green, yellow, near IR1, near IR2) + 8 shortwave infrared bands + 12 CAVIS bands (cloud, aerosol, vapor, ice, snow)	Commercial, earth observation	Temporal resolution: less than 1 day Spatial resolution: panchromatic, 31 m; multispectral, 1.24 m; SWIR, 3.7 m; CAVIS, 30 m
Sentinel 2A, 2B	European Space Agency and European Copernicus Programme	Sun-synchronous	Multispectral	13 spectral bands (visible, near infrared, shortwave infrared, Cirrus + vegetation detection and water vapor bands)	Earth observation	Spatial resolution: 4 bands at 10 m; 6 bands at 20 m; 3 bands at 60 m
Sentinel 3	European Space Agency and European Copernicus Programme	Near-polar, sun-synchronous	OLCI (Ocean and Land Cover instrument), SLSTR (Sea and Land Surface Radiometer), MWR (microwave radiometer), SAR (synthetic aperture radar)	21 spectral bands from 400 to 1,200 nm	Ocean surface topography, land and sea surface temperature	Spatial resolution: 300 m
Sentinel 5	European Space Agency and European Copernicus Programme	Polar, sun-synchronous	High-resolution spectrometer system	7 spectral bands (UV-1 at 270 to 300 nm, UV-2 at 300 to 370 nm, VIS at 370 to 500 nm, NIR-1 at 685 to 710 nm, NIR-2 at 755 to 773 nm, SWIR-1 at 1,590 to 1675 nm, and SWIR-3 at 2,305 to 2,385 nm)	Air quality measurements, stratospheric ozone monitoring, solar radiation measurements, climate monitoring	Temporal resolution: 29 days Spatial resolution: below 8 km for wavelengths above 300 nm; below 50 km for wavelengths below 300 nm
SPOT 6, 7	French Commercial	Sun-synchronous	Multispectral	4 bands (blue, green, red, near infrared)	Earth observation	Spatial resolution: panchromatic, 1.5 m; multispectral, 6.0 m

created after the conversion will depend on the conversion process used. If information is extracted as a series of points or lines as in the case of digitizing, the geospatial data will be a vector data. On the other hand, if data is extracted in a digital form by scanning a paper map, the geospatial data will be a raster data format because here data is extracted from a paper map as a group of pixels.

Digitizing

Digitizing is a common technique used to create vector data. It is also an excellent form of data creation to capture data from hard-copy paper maps or aerial photos. **Digitizing** is the process of capturing map data by tracing over the top of a map or image to model the features represented on the image or map in a GIS. For instance, if you open an aerial photograph of your neighborhood in a GIS application, you can use your computer's mouse and draw lines to sketch neighborhood streets; draw polygons to sketch the outline of parks, tennis courts, and houses; and save it in your GIS application. What you have done is create a vector representation of your neighborhood from an aerial photo. You have created a line data representing streets, a polygon data representing parks, another polygon data representing tennis courts, and a third polygon data representing houses. So you have created four different vector datasets to model your neighborhood in a GIS from the aerial photo. Let's say that you have eight houses on the street that you have sketched. Your polygon data representing houses will store all eight polygons as a single GIS data representing houses. When you digitize in this way, data is captured in a series of points.

Traditionally, digitizing was performed on a digitizing tablet using an electronic handheld cursor or an electronic pen. Today digitizing is performed using a process called "on-screen digitizing" or "heads-up digitizing." The process is the same where you sketch or trace outlines on the top of a map or image. The only difference is that in on-screen digitizing, an image, either an aerial photo or a satellite image, is displayed on the computer screen and the user sketches outlines of features on top of the image using a mouse. You must first create a new data layer. The new data will be a point, line, or polygon depending on the feature that you will be sketching. For instance, in the example above, you would first create a new line data in one of the GIS vector data formats to sketch the streets in your neighborhood. Similarly, you would create three new polygon data to sketch the outlines of houses, parks, and tennis courts in your neighborhood. The new data will be empty when you first create it. It will store the sketches or tracings of the outlines that you will perform on top of an aerial photo or satellite image (figure 5.29).

FIGURE 5.29 On-screen digitizing to create polygon data representing houses. The outline boundary of houses on the aerial photo is traced to create the pink polygons in the second image.

Credit: Aerial imagery: Esri and its users and partners. ArcMap interface courtesy of Esri, Inc.

In order for a GIS to understand how the data that were created by digitizing real-world features on an aerial photo or a satellite image model, it would require information about the spatial relationship between the geospatial datasets. Are the polygons adjacent to each other, do the streets intersect, is the park adjacent to the tennis courts? A GIS will need this information in order for the geospatial data to model their real-world behavior. This information is derived from a topological data structure covered in chapter 3.

Address Matching

Another common form of geocoding is address matching. We are all very familiar with the process of address matching, thanks to our smartphones and the everyday use of mapping applications on our smartphones to locate addresses. How do you find an address? You type in a string of letters and numbers, and the map application somehow translates that information into geospatial data. This process of converting a combination of letters and numbers into a form of information that can be located on a map is called **address matching**. This may seem like a process that occurs instantaneously. However, multiple components and steps are involved in this process.

Address matching requires three important elements: input data, reference data, and an address locator. The input data is the list of addresses that must be located on a map. Reference data is a geocoded database, typically a road network data against which the input data is matched. One of the most widely used reference data are the US Census Bureau's TIGER/Line files. TIGER/Line files contain detailed information for each street segment in the United States such as street name, street type, street direction, suffix, address ranges on the left and right sides of the street, and zip code to the left of the street and zip code to the right of the street. The third element is the address locator.

Before geocoding can be conducted, an address locator must be created. An address locator is created by a geocoding engine embedded in a GIS. The geocoding engine constructs the address locator by examining the addresses. During this stage, two important processes take place: parsing and address standardization. **Parsing** is a process of breaking up the address into individual components. This enables the address to be matched with the reference database. For instance, consider the address 4848 N Maricopa Ave, Tomball, TX 77377. The address consists of the following components:

Street number = 4848
Direction = N
Street name = Maricopa
Street type = Ave
City = Tomball
State = TX
Zip code = 77377

The next step is for the geocoding process to understand different items in the address that mean the same thing. For example, if you were using your smartphone's map application you would type in "4848 Maricopa" or "4848 N Maricopa Avenue" or "4848 Maricopa Ave." All of these mean the same thing to us, but even a GIS needs to understand that these different things mean the same. This is done through a process of **address standardization** where the input addresses are standardized in a consistent format so that it can be read by the address locator to match the appropriate attributes in the reference data no matter how it is represented in the input data.

During address standardization, street types such as "Ave," "AVE," and "Avenue" are all given the same code so that whichever way "avenue" is recorded in the input address, it will match the reference data.

After parsing and address standardization, the address matching process takes place. The address locator takes individual components of the input data and matches it against the reference data. It will find the street segment in the reference data that matches individual components of the input address and place a point matching the street number on the corresponding street segment at an approximate location. As you can imagine, this process of plotting addresses on a map is an approximate process. If a street has address numbers ranging from 42 to 52, an input address number of 46 will be placed closer to the start of the street, whereas an address number of 50 will be placed closer to the end of the street. This method of plotting a point at an approximate location along a line segment is called linear interpolation.

There are many sources of error in plotting addresses in this manner. The address locator can only match the input address if the address is in the reference data. If the reference data is not updated, the address locator will be unable to locate addresses that came into existence after the reference data was created. The geocoding process will be as accurate as its reference data. If the reference data contains incorrect information or is missing information, the geocoding process will not be able to accurately match input addresses. Another important factor is positional accuracy. How close are the geocoded addresses matched by the geocoding process to their real-world locations? In reality, geocoded points must be as close as possible to their real-world locations. The accuracy of the geocoded points will impact the accuracy of other analyses that are conducted using those point locations.

Questions and Activities

1. What are the important characteristics of geospatial data used in any GIS project that sets them apart from other types of data?
2. Pick any five examples of maps that you use in your everyday life. Are the maps analog or digital? What kinds of information do you derive from those maps?
3. Next we will use USGS's Earth Explorer to examine geospatial data. Open USGS Earth Explorer at https://earthexplorer.usgs.gov. Select the "Data Sets" menu to review a list of datasets available through the Earth Explorer (figure 5.30).

   Before you search for data, you will define the area you want to focus on. This way we limit the number of results and only focus on an area of interest. Here we will focus on Texas. Select the "Search Criteria" menu on top. You can specify your search criteria in different ways.

   The first way to search for data is to select a state from the "State" drop-down menu and select from a list of data in the "Feature Type" drop-down menu. Let's select *Texas* for the state and *Academy College High School University* for the feature type. Hit the "Show" button. A list of schools, colleges, and universities in Texas will be displayed with their latitude and longitudinal coordinates (figure 5.31). You can display the location on the map to the right by selecting a feature from the list. (Take note of the message below the map indicating that the provided maps are not for purchase or download, only for reference purposes.)
4. Next we will search for data by using the predefined area option. Select the "Predefined Area" menu. Next select the "Add Shape" button. In the "Add Shape" window, select *Texas* for state, *County* for area type, and *Galveston County* for area. Hit the "Add" button. You will see that a polygon shape outlining Galveston County is added to your map (figure 5.32).

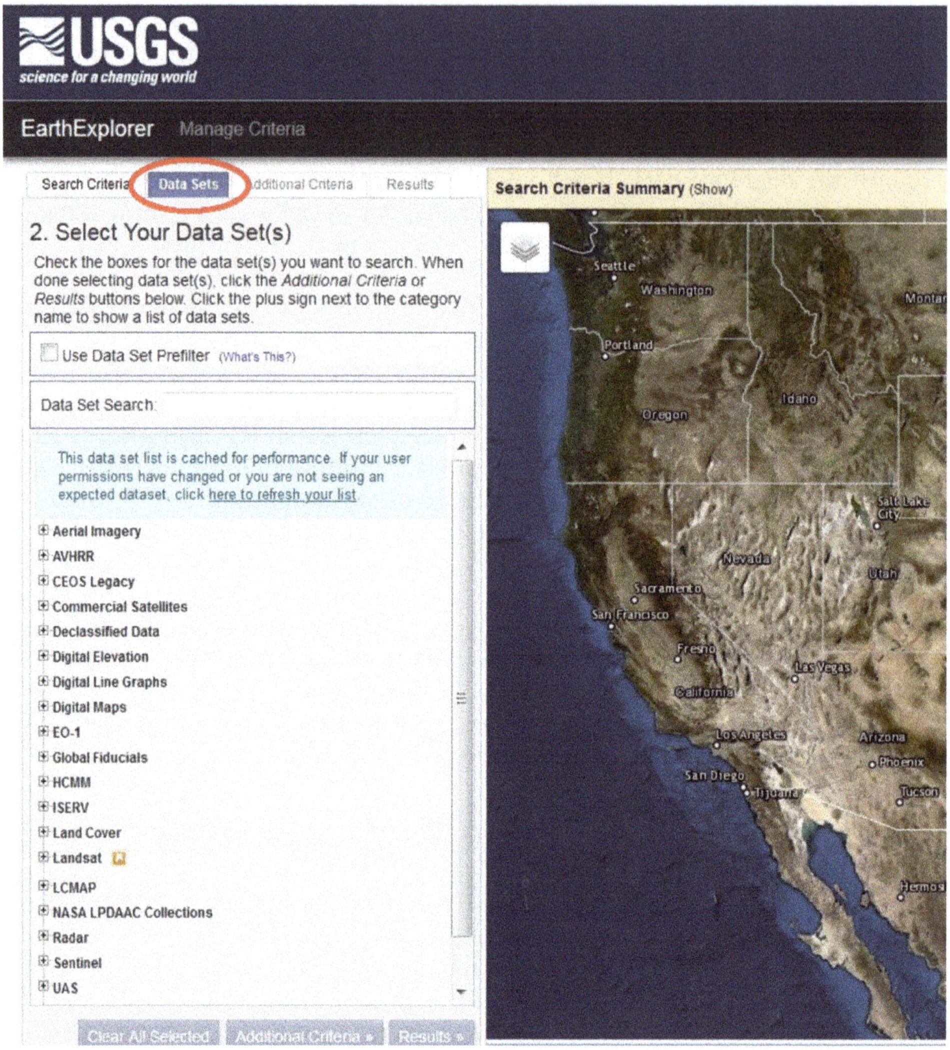

FIGURE 5.30

Now we will search for aerial imagery of this county. Select the "Data Sets" button at the bottom (figure 5.33).

Click the plus sign for *Aerial Imagery*. We will examine DOQs of Galveston County. Check the box for DOQ and hit the "Results" button at the bottom. A list of DOQs covering Galveston County will be displayed. You can click on the footprint icon to view which part of the county the DOQ represents on the map to your right. Let's examine the map name *FLAKE*. Click on the "Show Footprint" button to see which area of Galveston County this DOQ represents. Click on the "Show Metadata" button to examine the metadata (figure 5.34).

What is the source of the aerial photo? When was the photo acquired? Click on the image to open it in a new window. In what format does the photo open? What can you tell about the features that you see in the photo? Write a one-paragraph description of the features that you see.

Compare the aerial photo with the orthophoto imagery used as a base layer on the Earth Explorer. Do you see any changes in the features?

Explore DOQs of other areas of your choice.

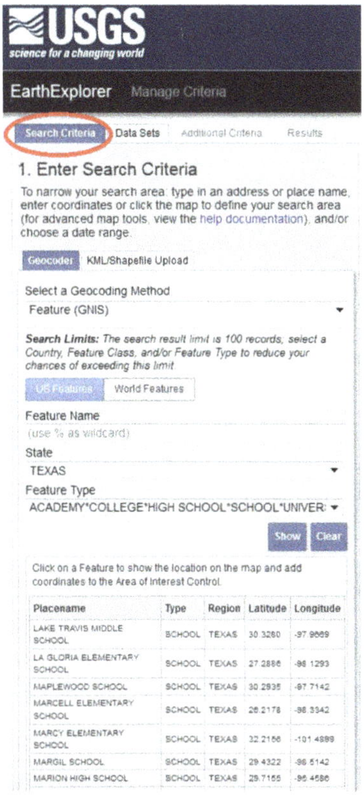

FIGURE 5.31

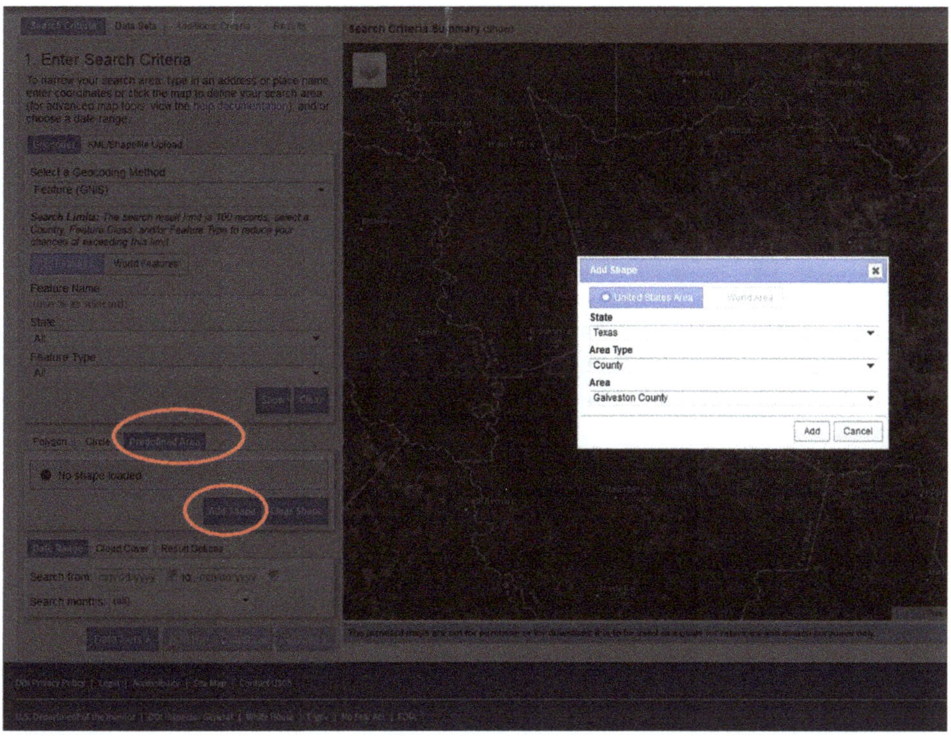

FIGURE 5.32

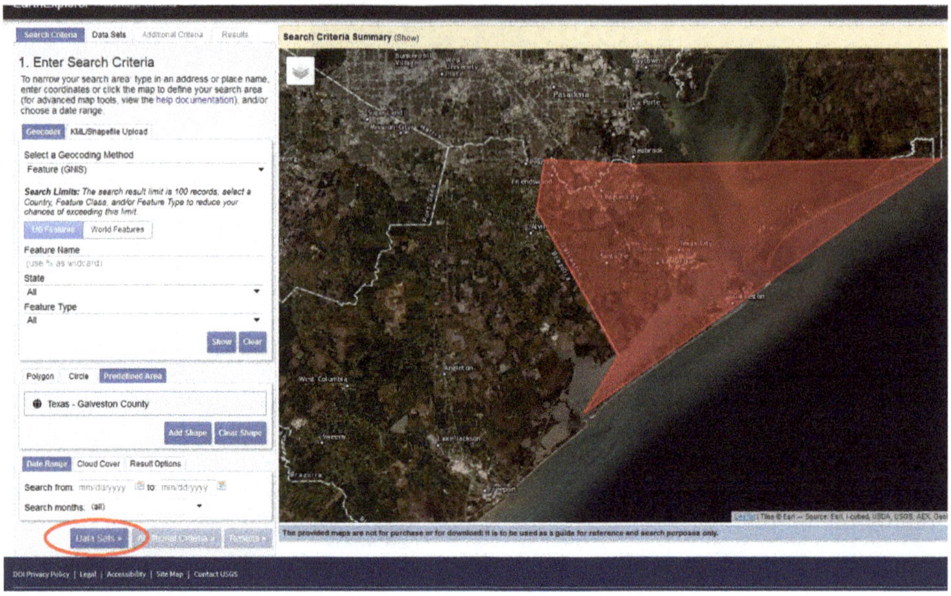

FIGURE 5.33

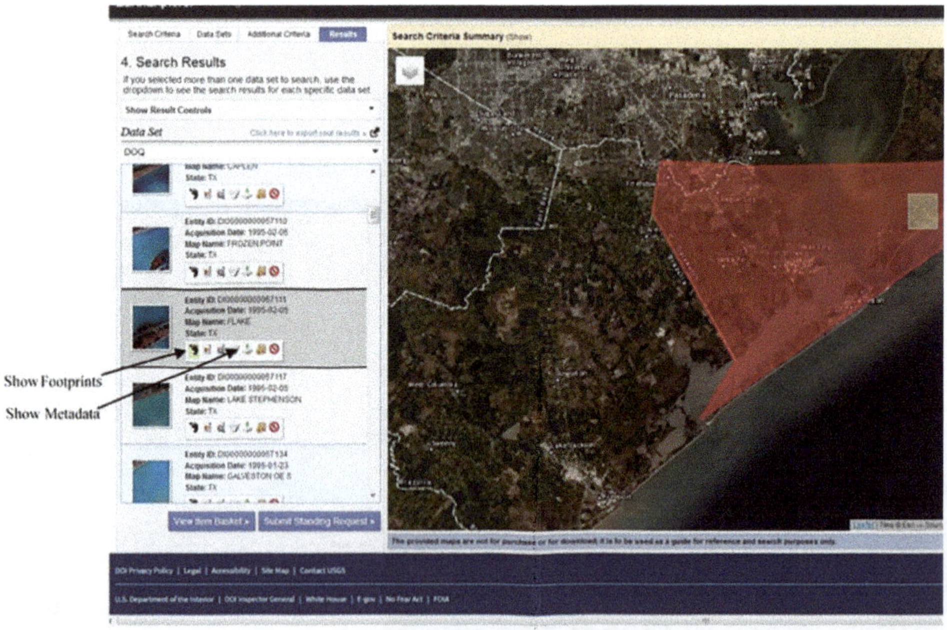

FIGURE 5.34

5. Next we will examine aerial imagery of Galveston County available from NAIP. Return to the "Data Sets" tab and uncheck the box for DOQ. Check the box for NAIP. Make sure that a polygon shape outlining Galveston County is added to your map. If not, go back and add it. Before you search for NAIP Aerial Imagery, click on the "Collection Info" button (blue circle with the letter *i*). A new window will open with information about the NAIP program.

Write a short description about the NAIP program. What is it? What kind of data is available through the program? And what are the specifications of the geospatial data?

6. Now click on the "Results" button. From the list of images, scroll down and select the image with Entity ID *M_2909435_NE_15_060_20181204*. It is an aerial photo of the same area as in the DOQ you examined in question number 5. What can you say about the datum, resolution, and type of sensor used to procure this imagery?
7. Next you will examine Landsat data of Galveston County. Uncheck the box for NAIP. Click the plus sign in front of Landsat. We will examine Landsat 8 OLI/TIRS image. Check the box in front of Landsat, then expand Landsat Collection 2 Level-2. Check the box for Landsat 8-9 OLI/TIRS C2 L2. Then click the "Results" button. A list of images will be displayed. You will find multiple images covering the same area. Do all of the images have the same amount of clarity? Why or why not? Examine the metadata of one of the images. When was the satellite image captured? Can you tell what time the image was captured? What is the datum for the satellite image?
8. What is geocoding? What are the different types of techniques or processes used in geocoding? What impacts do these processes have on the data created through geocoding?
9. Next we will examine geocoding. There are many online resources that will perform address matching interactively for you, such as United States Census Geocoder, BatchGeo, and Geocode Earth. We will use one such resource available through Texas A&M University, Texas A&M Geocoder.

   Go to https://geoservices.tamu.edu/Services/Geocode. Hit the "Interactive Geocoding" button. Select the drop-down menu for "Census Year(s)." We will select *TwoThousandTen*. Now enter an address familiar to you such as your home address or work address or a local spot that you are familiar with and hit the "Geocode" button. Examine the results. See how the address that you put in was divided into individual components as a parsed address.

   The geocoding service available through Texas A&M Geoservices lets you display geocoded address on a Google map, Bing map, and Yahoo map. Click on "Google Maps" to view the geocoded address on Google Maps. Does the geocoded point appear exactly where its real-world location is? What can you say about the positional accuracy of the geocoded point?

   Now select Bing Maps. Is a Bird's Eye imagery available for your location? If it is, select "Bird's eye" and examine the location.

Key Terms

address matching (p. 130)
address standardization (p. 130)
aerial photography (p. 114)
almanac (p. 109)
band (p. 123)
brightness value (p. 123)
C/A code (p. 110)
color composite image (p. 124)
Continuously Operating Reference Stations (CORS) (p. 112)
control segment (p. 108)
differential GPS (DGPS) (p. 111)
digital orthophoto quadrangle (DOQ) (p. 100)
digitizing (p. 129)
Earth Explorer (p. 102)
electromagnetic spectrum (p. 122)
ephemeris (p. 109)
false color composite (p. 125)
geocoding (p. 97)
geostationary satellites (p. 126)

global navigation satellite system (GNSS) (p. 107)
Global Positioning System (GPS) (p. 107)
Global Visualization Viewer (GloVis) (p. 101)
Historical Topographic Map Collection (HTMC) (p. 99)
hyperspectral imagery (p. 124)
hyperspectral sensor (p. 126)
large-scale photograph (p. 120)
linear interpolation (p. 131)
metadata (p. 102)
multipath (p. 111)
multispectral imagery (p. 123)
multispectral sensor (p. 126)
nadir (p. 119)
National Agriculture Imagery Program (NAIP) (p. 99)
The National Map (p. 98)
near-polar orbit (p. 126)
NOAA CORS Network (NCN) (p. 112)
oblique photographs (p. 119)
orthophoto (p. 99)
orthorectification (p. 121)
P code (p. 110)
panchromatic sensor (p. 126)
parsing (p. 130)
relief displacement (p. 121)
remote sensing (p. 113)
satellite-based augmentation system (p. 112)
small-scale photograph (p. 120)
space segment (p. 108)
spatial resolution (p. 125)
sun-synchronous orbit (p. 126)
temporal resolution (p. 125)
Topologically Integrated Geographic Encoding Referencing (TIGER) (p. 103)
trilateration (p. 110)
true color composite (p. 125)
United States Geological Survey (USGS) (p. 97)
unmanned aircraft systems (UAS) (p. 116)
user segment (p. 109)
vertical photographs (p. 119)
visible light spectrum (p. 122)
Wide Area Augmentation System (WAAS) (p. 112)

References

ArcGIS Living Atlas of the World. n.d. Accessed January 15, 2021. https://livingatlas.arcgis.com/en/home.

Jensen, John, and Ryan Jensen. 2013. *Introductory Geographic Information Systems.* Boston: Pearson.

Sabins, Floyd, and James Ellis. 2020. *Remote Sensing: Principles, Interpretation, and Applications.* Long Grove, IL: Waveland Press.

Shellito, Bradley. 2018. *Introduction to Geospatial Technologies.* New York: W. H. Freeman– Macmillan Learning.

Union of Concerned Scientists. 2005. "UCS Satellite Database." December 8, 2005; updated May 1, 2023. https://www.ucsusa.org/resources/satellite-database.

US Census Bureau. n.d. "Explore Census Data." Accessed January 12, 2021. https://data.census.gov/cedsci.

USGS. n.d. "Earth Explorer." Accessed January 12, 2021. https://earthexplorer.usgs.gov.

———. n.d. "The National Map." Accessed January 12, 2021. https://www.usgs.gov/core-science-systems/national-geospatial-program/national-map.

———. n.d. "Tracking Change Over Time—Understanding Remote Sensing." Accessed January 12, 2021. https://pubs.usgs.gov/gip/133/pdf/RemSens-Student_web.pdf.

Waxman, Olivia. 2018. "See the History of Aerial Photography in Pictures." *Time*, May 31, 2018. https://time.com/longform/aerial-photography-drones-history.

GIS and Cartography 6

Learning Objectives

- Explain what cartography is and the important role of mapmaking in GIS
- Describe various cartographic elements of a map
- Explain and classify different types of GIS maps
- Use mapping tools available in a GIS software to create a map layout
- Discuss different classification methods available in a GIS software
- Use data classification methods in GIS to create a map layout
- Explain the use of imagery in GIS maps

Why Cartography?

There is no doubt that GIS is interdisciplinary in nature. As we have seen so far in the text, GIS has roots in many disciplines such as information science, database management systems, surveying, and mathematics. One particular discipline that has made an important contribution to the field of GIS is cartography, the science of maps. Maps play a very big role in GIS. No matter what kind of GIS data you work with or GIS analysis you perform, ultimately the results of your analysis or the GIS information that you wish to convey must be communicated through some sort of a medium. And that medium is a map.

GIS users communicate with their audience through the language of maps. One of the unique characteristics of geospatial data is that it can be mapped on a GIS platform. In fact, in the current era of digital technology, geospatial data are increasingly visualized on digital maps using GIS. GIS has become a key tool to produce many important maps. However, creating maps in a GIS is not a straightforward process of printing out the results of your analysis or your data and putting a border around it. Many design considerations must be taken into account. In fact, there is an entire science behind mapmaking. Cartography is the science of designing and creating maps to represent features on the earth's surface. Maps have been around for centuries. With advances in technology, the art and science of creating maps have evolved over time accompanied by more-complex design considerations.

A map is a representation of data that conveys specific information to its audience. Maps are created for a variety of purposes. A map may be created for use by the mapmaker or may be created for a specific audience. At times you may create a map to visualize your data and conduct a preliminary analysis to examine if there are any existing patterns in your data. Some maps are created with a specific objective. For instance, maps are created for locating specific features such as hospitals, state parks, restaurants, hiking and biking trails, or gas stations. Maps are also an important tool for navigation purposes such as subway maps, highway maps, or street maps. Some maps show very specific data in great detail such as soil maps, geology maps, utility network maps, or land use maps. Thus every map has a specific goal and has been designed accordingly. Maps are designed and created following a process that includes certain steps starting from envisioning the objective of the map and concluding

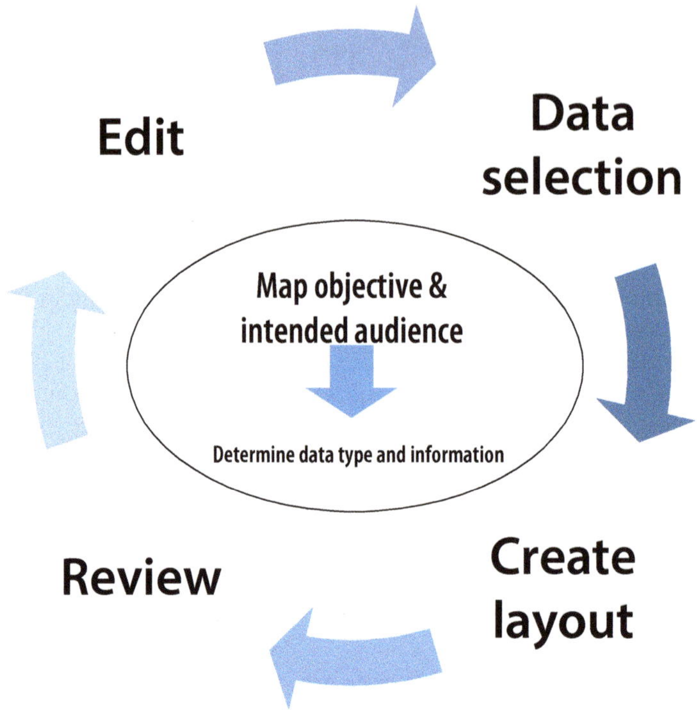

FIGURE 6.1 The cartographic design process.

with the final creation of a map. This process is known as the cartographic design process (figure 6.1).

The first step in this process is to determine the overall goals of your mapping project and its intended audience. This is an important step and will shape the subsequent steps involved in the cartographic design process. For instance, if the map is geared towards those accessing a hiking trail, the map must include important information related to hiking such as location of entrances to the trail, parking, length of the trail, and elevation. It is also crucial that the locations are accurately mapped and the trails stand out on the map. If the map is created to show the location of hospitals or restaurants in an area, the map must highlight accurate locations of hospitals or restaurants using appropriate symbols, along with other important features such as roads and highways and neighborhood locations.

Once the objective and intended audience are established, the next step is to assemble appropriate datasets to create the map. During this step a mapmaker must also consider what kind of data will be used to represent the features on the map. For instance, how will you present trails on your map versus locations of hospitals or restaurants or neighborhoods? Trails will be represented as lines, locations of hospitals or restaurants will be represented as points, and neighborhoods will be represented as polygons. During this step a mapmaker must also consider the amount of information that will be included in the map in order to effectively communicate the information to the audience. This step is followed by the actual map design and creation process using cartographic principles during which a mapmaker creates a map layout, reviews the map elements to determine if the map effectively communicates information to the intended audience, and edits the layout to improve the map.

There are two other important considerations a mapmaker must bear in mind: the medium used to convey a map to its intended audience and map scale. Maps were traditionally designed and created on paper, providing the audience a static perspective to information. But in today's digital era, digital mapping has become a common cartographic practice, providing the audience a dynamic perspective to information. Maps are being viewed on smartphones, tablets, and computer and television screens. Digital maps have very different design requirements and present unique challenges to mapmakers. Maps have become interactive. They include different types of data derived from many different sources and provide a variety of options to users to perform directly on the map. User interface and user interaction hence require special consideration during map design.

Map scale is also an important factor. The amount of information that will be conveyed on a map and how that information will be conveyed will depend on map scale. A large-scale map will show a small area in detail, while a small-scale map will show a large area with less details. Therefore the choice of features, symbols to represent the features, and the cartographic generalization that will occur will be influenced by the scale at which a map is created and will be an important consideration.

Ultimately, the goal is to create a product at the end of the cartographic design process that effectively communicates the information to its intended audience. This product, a cartographic map, has certain basic characteristics. Maps are a representation of reality, of the features, patterns, events, activities, and phenomena that occur in the geographic space around us. Maps represent the world in different ways. However, ultimately every type of map represents two things: feature locations and feature attributes. All maps represent those two things with some amount of generalization (covered in chapter 2). The generalization will depend on a mapmaker's decision to include select features to display while excluding some other features that are insignificant or not relevant to the objective of the map. Thus all maps include a generalized representation of geographic space.

Maps represent features and phenomena from geographic space that occur on a three-dimensional earth surface onto a flat, two-dimensional space and therefore involve projections and coordinate systems (covered in chapter 2). The choice of a projection system used to create a map must conform to the objective of the map. For instance, a map that requires accurate area calculation must use a projection that preserves area. All maps represent the geographic space around us at a reduced scale and hence involve scale conversion (covered in chapter 2). And finally, all maps present a generalized view of the geographic space around us by representing features, patterns, activities, events, or phenomena by using a variety of symbols. Symbols used to represent features will be influenced by the type of information, characteristics of data used to represent the information (covered in chapter 4), and map scale. It is important that symbols chosen by a mapmaker to represent features or phenomena effectively communicate spatial information.

Cartographic Elements of a Map

Ultimately, as a GIS user you are communicating your information to the audience through a story that you are telling via a map. In order to weave this story together, a mapmaker must make use of spatial as well as nonspatial information to design and create a map. In addition to geospatial information, all maps must therefore include other types of information that are not necessarily spatial in nature. This information incorporates some of the basic elements of cartographic map design such as the north arrow, scale bar, title, and legend.

As we know, all maps are created at a reduced scale. Therefore, anyone reading a map must be able to figure out fairly easily at what scale the map was created. This information is included in a map by using a *scale bar*. Scale information can be included in two ways, either by simply including a text to denote the representative fraction or by using a graphic scale bar (covered in chapter 2). Another important map element is the north arrow. A north arrow helps to orient the map and understand cardinal directions on the map. No matter how a map is oriented, the north arrow must point in the direction that is true north. GIS softwares have options for including different types of north arrows in a map. A north arrow can be a simple line with the letter *N* on the top or it can be a complex design of a compass with all four cardinal directions shown on it. Another element that all maps include is legend. A map legend is a guide for a map reader to understand what the symbols and colors shown on the map represent. GIS softwares provide their users different options to design and create a map legend. A good legend design is part of a mapmaker's effective storytelling strategy.

An important element of map design is typography. Typography is the art of lettering or type. If you look at any map, you will see that maps involve the use of a lot of letters. Mapmakers communicate with their audience through the use of different cartographic elements in a map. Several of those cartographic elements such as legend, map title, labels, and text use letters. It is thus an important component of user interface.

Elements of lettering that particularly impact map design are font style, size, structure, and color. A mapmaker's choice of font type, size, and color will influence the appearance of the final map layout. A variety of different fonts are available for all elements in a map. GIS programs provide users several different types of fonts to choose from. Fonts can be as simple or as ornate as a mapmaker chooses. Ultimately, one must focus on the final layout and how it looks. Just because you have so many different fonts available does not mean you use several different types of fonts on your map, because such a map will look cluttered and will be difficult to read.

Similarly, the font size you choose will also impact how your map looks—is it easy to read or is it cluttered and difficult to read? Font size must match the importance of information portrayed on a map. Ideally, information that must grab a reader's attention first is portrayed in a larger font compared to other text on the map. For instance, the font size of the map title will be larger than the font size of labels and legends on a map. It is good to use varying font sizes on a map to convey the importance of information being portrayed. However, one must maintain a balance between the number of different font sizes to incorporate and the clarity of the map to a reader.

The same holds true for font color and structure. All maps include labels to identify features represented on the map. GIS programs label map features using an automatic placement tool incorporated into the program. By default the tool aligns the labels in an upright position, which may work most of the time. However, upright alignment may not work for topographic features that are aligned in a certain way depending on the shape of the feature. In such cases, GIS programs have the option to select individual labels and move them to place the label interactively along the general shape of the feature. Ultimately the goal is to use font type, size, color, and placement to create an overall layout that grabs a reader's attention and is easy to read.

A map must also include metadata information. Metadata is data about data and includes valuable information about who, what, when, where, and how the data or map was created. Metadata information is a critical part of public geospatial data and map use. At a minimum a map must include metadata information describing the data source used to create the map, when each of the datasets included on the map were obtained, and the agency and department responsible for data and map creation.

In a GIS a map is created by assembling all of the cartographic elements to create a map layout. GIS softwares like ArcGIS create a default map layout using preloaded templates for users. The default layout puts together the geospatial information and cartographic map elements such as scale bar, north arrow, title, and legend and places them on a blank map frame. A user can now design the map layout the way they want by carefully selecting each of the cartographic map elements, editing them to resize the elements or change alignment, and rearranging them to place the map elements at an appropriate location on the map frame.

The goal is to create a well-balanced map layout. This is done by carefully selecting where and how each of the map elements will be placed. You want to create a map that grabs a reader's attention and is easy to read. You don't want to overload the map with too much information or too many unnecessary map elements and create a cluttered layout that would detract a reader's attention. Empty spaces or blank areas on the map must be used carefully. You don't want to place all of the map elements in one area, nor should you place all of the elements very close to each other. Such a layout would look crowded. Size of the map elements also matters. All map elements should be of uniform size to match the importance of information they are conveying and must be placed in a balanced proportion to one another.

The geospatial information is the most important element of a map and must be placed in such a way that it grabs the reader's attention. This is done by maintaining figure-ground relationship (Buckley and Field 2011). The figure is the most important content of your map and must grab the reader's attention. The ground is the background information that includes ancillary data to support figure information, defining the figure-ground relationship. Map elements like title, north arrow, scale bar, legend, and metadata information will reside in the background. Figure 6.2 shows an example of a well-balanced map that has maintained figure-ground relationship by placing the geospatial information and the map elements in proportion to one another.

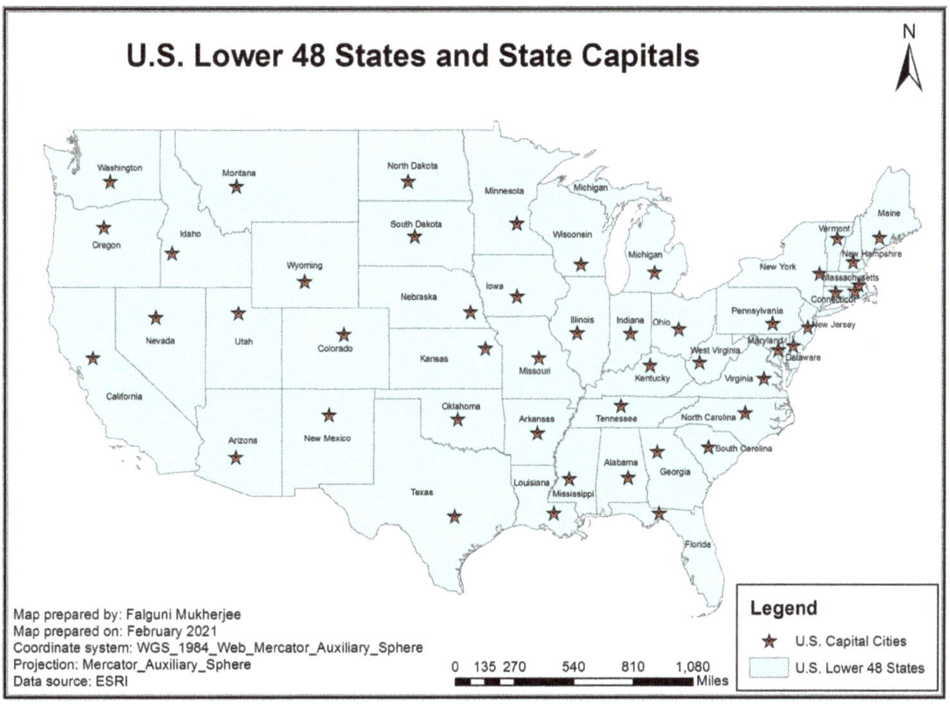

FIGURE 6.2 Example of a well-balanced map.

ArcMap interface courtesy of Esri, Inc. Data sources: ArcGIS Living Atlas of the World. Esri, US Census Bureau, TomTom, U.S. Department of Commerce (DOC), National Oceanic and Atmospheric Administration (NOAA), National Ocean Service (NOS), National Geodetic Survey (NGS)

GIS Maps

All maps, whether paper maps or digital maps, fall into two broad categories: reference maps and thematic maps. **Reference maps** are used to show location information. They have been around for a long time. In fact, the oldest known maps belong to the reference map category. Reference maps are very useful for providing geographic reference of locations of features. On a reference map no one particular feature is emphasized over another feature. All features on a reference map are shown with equal importance. A map of state capitals (figure 6.3), a map of streets and highways, a map of hospitals, a map of all tennis courts in your area, and a map of hiking trails in your area are all examples of reference maps. Topographic maps that show terrain information are also examples of reference maps.

Unlike reference maps, **thematic maps** emphasize a particular theme. Thematic maps emphasize geographical distribution of a specific theme, and that particular theme is given prominence in the map. A map displaying population distribution is an example of a thematic map, where population distribution is the theme of the map. Another example would be a map displaying terrestrial ecosystems (figure 6.4). Similarly, a map of average annual rainfall, a map displaying the different climate zones in a region, and a map showing the distribution of different vegetation species are all examples of thematic maps. Thematic maps show contextual information in the background to provide a map reader geographical context to the theme displayed on the map. Thematic maps may show themes that are abstract concepts and do not actually physically exist on the ground. Such thematic maps that display geographic distribution of an abstract concept are based on some sort of a mathematical, statistical, or theoretical calculation, such as population density.

Different types of maps can be prepared using the mapping tools available in a GIS. You will be able to create most of the basic types of maps using a GIS program. Two important factors that influence the type of map you can create using a GIS are the spatial dimension of the feature and the measurement level of data represented on the map. Vector data in GIS is represented as points, lines, or polygons (covered in

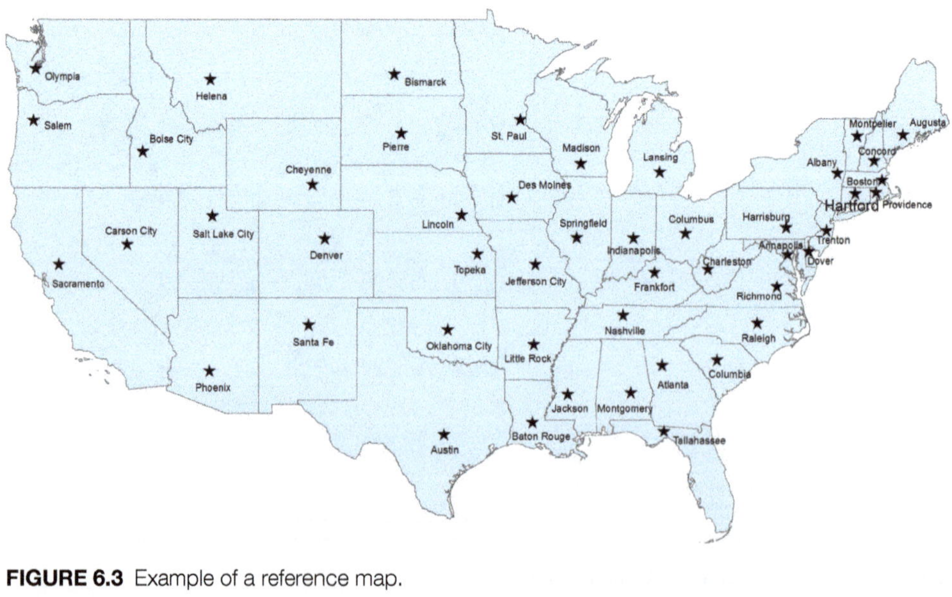

FIGURE 6.3 Example of a reference map.

ArcMap interface courtesy of Esri, Inc. Data sources: ArcGIS Living Atlas of the World. Esri, US Census Bureau, TomTom, U.S. Department of Commerce (DOC), National Oceanic and Atmospheric Administration (NOAA), National Ocean Service (NOS), National Geodetic Survey (NGS)

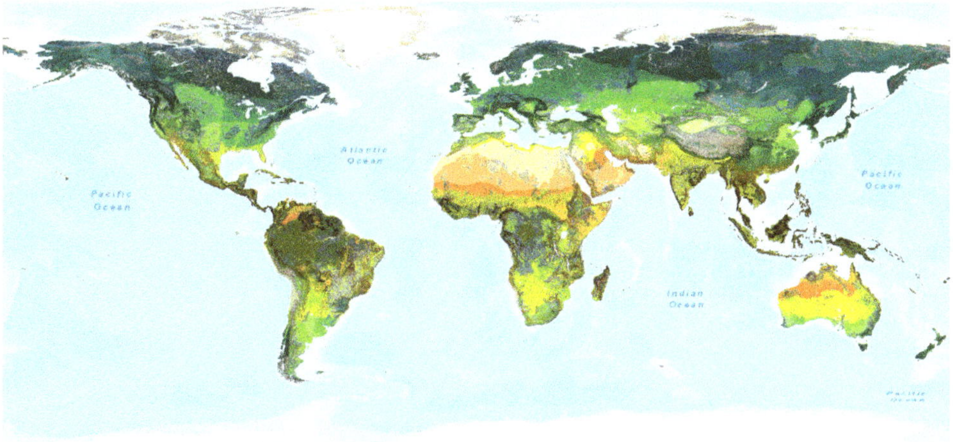

FIGURE 6.4 Example of a thematic map.

Data source: ArcGIS Living Atlas of the World. Attribution: USGS, TNC, Esri. ArcMap interface courtesy of Esri, Inc.

chapter 3) and accordingly you have different types of maps in GIS to represent point data, line data, and polygon data. As covered in chapter 4, data, whether quantitative or qualitative, are mainly of four types: nominal, ordinal, interval, and ratio. Different types of maps can be created in a GIS for quantitative point, line, and polygon features and for qualitative point, line, and polygon features. Every map type has unique characteristics and provides a user different options depending on the map capabilities.

Mapping Point Data

A large amount of geospatial data is collected every day at specific point locations for a variety of different purposes by government agencies, nongovernmental organizations, private corporations, and members of the public. These point data represent a variety of different types of features. The point data can be a measurement for an entire city, county, or country or can be the geographic location of features such as water hydrants, manholes, oil wells, and hospitals. In addition to location information, other types of feature information are also collected for point data that are stored in the attribute table in a GIS. These include different types of information such as population of a county or city, operational status of an oil well, production capacity of an oil well, size of a fire hydrant, or hospital capacity. The attribute information can be used in a GIS to create different types of point maps.

The simplest type of point map is one where point data is shown with simple geometric symbols, as shown in figures 6.2 and 6.3. This type of map is used when the point feature represents either qualitative (nominal) or quantitative (ordinal, interval, or ratio) data. In both figure 6.2 and figure 6.3, state capitals are shown with a star symbol. ArcGIS software uses a circle symbol for all point data by default. Users can change the symbol and choose from different symbol options available within the software (figure 6.5).

Instead of a geometric symbol, point data can also be represented using a picture. Such maps are called picture symbol maps. Here the pictures used to represent the features are a miniature version of the feature itself. However, one must be careful in using pictures for symbolizing point data. Pictures can have unusual shapes and can make it difficult for a map reader to understand the spatial information represented on the map. You don't want your symbols to overlap, creating a cluttered map.

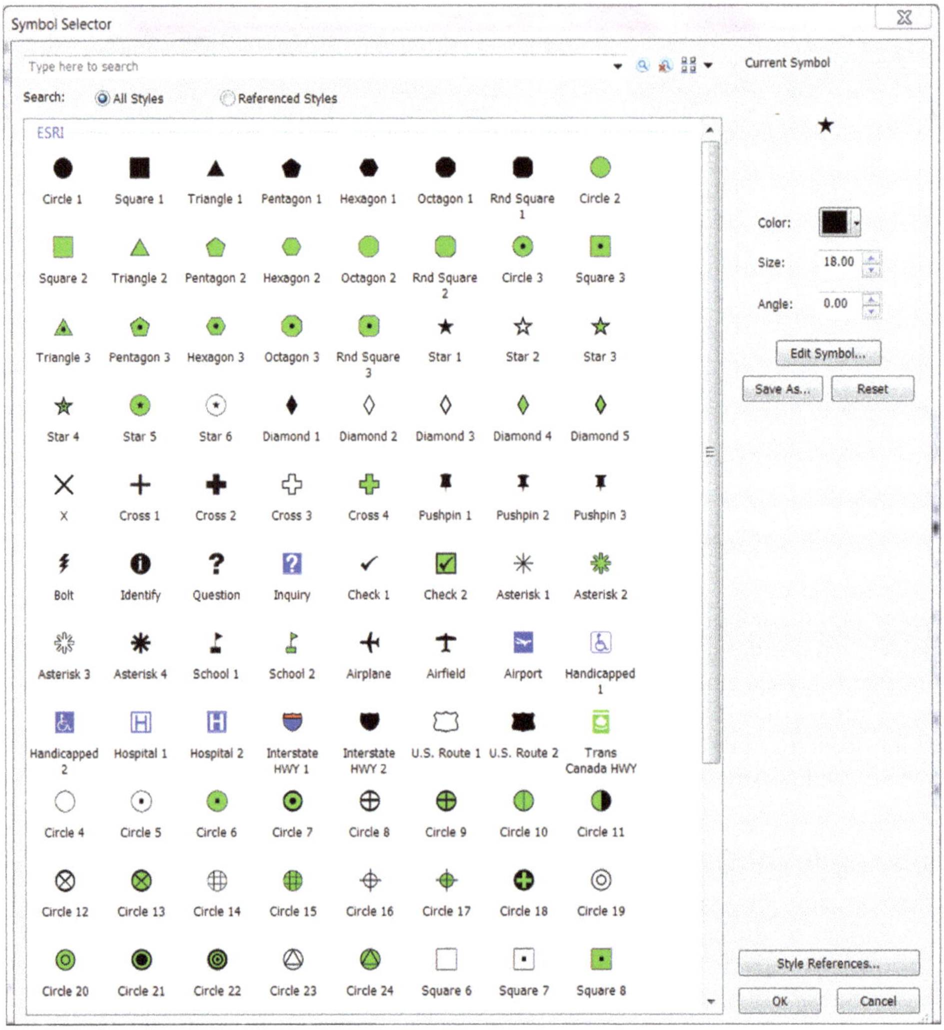

FIGURE 6.5 Symbol selector for point data in ArcGIS software.

ArcMap interface courtesy of Esri, Inc.

Point features that represent quantitative data can be mapped using either a graduated symbols map or a proportional symbols map. A graduated symbols map is created to represent quantitative data at point locations (figure 6.6a). In this type of map, point data is shown with varying-sized symbols to represent ranges of values. Here, the point data may represent either true point measurements or conceptual point measurements. A reader will not be able to tell the exact value at every point location on the map. The value will fall within a certain range representing either numeric values or ordinal level ranking.

Another type of point data map that uses varying-sized symbols is a proportional symbols map. In a proportional symbols map (figure 6.6b), the size of the symbols is scaled in proportion to the magnitude of values at point locations. The point locations can be true points or conceptual points such as population density for US counties. A proportional symbols map uses a specific symbol size for each numeric value rather than a range of values. In a proportional symbols map, readers may have difficulty understanding the data represented on the map if there are too many data points with similar values, making it difficult to distinguish between symbols.

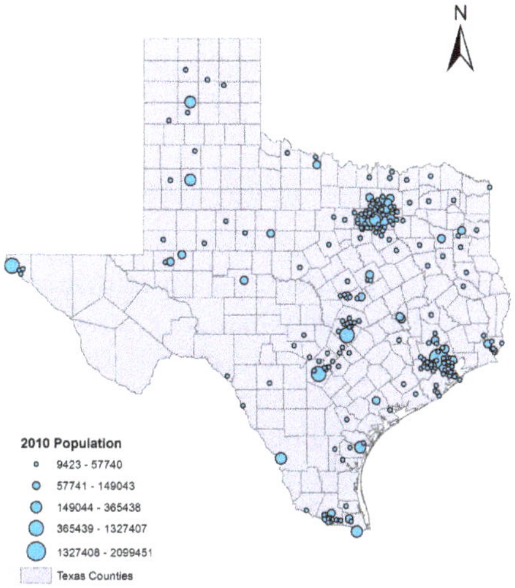

FIGURE 6.6A Graduated symbols map showing 2010 population of major cities in Texas.

Data source: ArcGIS Living Atlas of the World. Attribution: US Census Bureau's American Community Survey (ACS) 2015–2019 5-year estimates, Table(s) B01001. ArcMap interface courtesy of Esri, Inc.

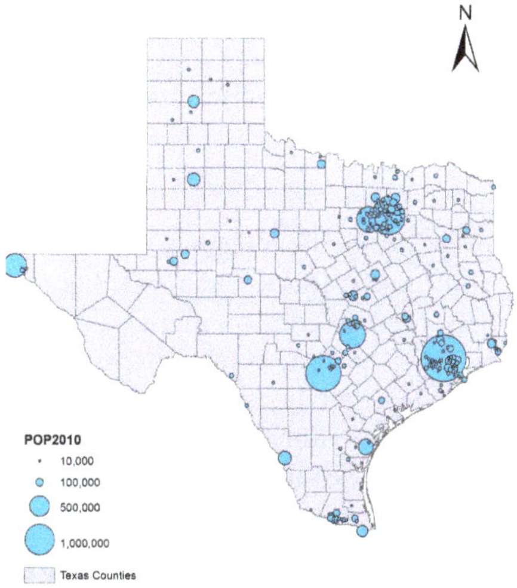

FIGURE 6.6B Proportional symbols map showing 2010 population of major cities in Texas.

Data source: ArcGIS Living Atlas of the World. Attribution: US Census Bureau's American Community Survey (ACS) 2015–2019 5-year estimates, Table(s) B01001. ArcMap interface courtesy of Esri, Inc.

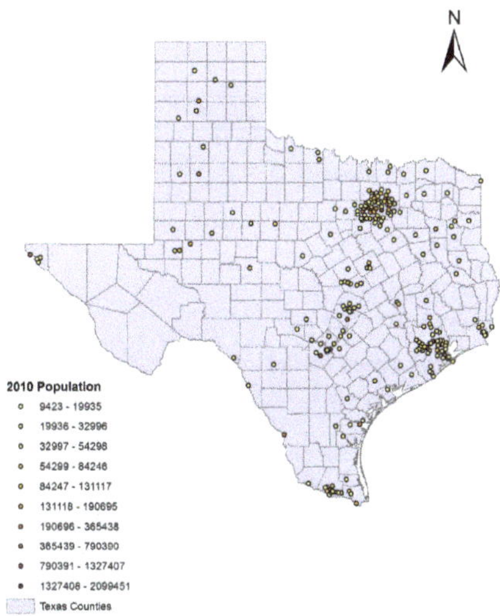

FIGURE 6.6C Graduated colors map showing 2010 population of major cities in Texas.

Data source: ArcGIS Living Atlas of the World. Attribution: US Census Bureau's American Community Survey (ACS) 2015–2019 5-year estimates, Table(s) B01001. ArcMap interface courtesy of Esri, Inc.

Another problem arises if there is a vast difference between data values. In this case, data points with very large values can conceal the other symbols.

Figure 6.6c shows a graduated colors map, where population numbers are distinguished by a range of colors from light to dark. The three maps display the 2010 population of major cities in Texas. All three maps were created using ArcGIS software.

Certain point features are also mapped using standard or standardized symbols. These are symbols that have been designed and are used as a standard in specific types of maps, such as the standardized point symbolization for geologic mapping or topographic mapping developed by the USGS. The use of standardized symbols allows a map user to recognize a feature easily. Figure 6.7 shows standardized point symbols for buildings and related features used by USGS topographic maps (USGS 2009).

Mapping Line Data

If you look at the world around you, you will find many different features that are represented as lines on maps, such as political boundaries, administrative boundaries, parcel boundaries, roads, streams, railroads, bridges, tunnels, pipelines, cable lines, telephone lines, water lines, sewer lines, and contours. Some of these features are truly one-dimensional line features, such as political, administrative, or parcel boundaries, while others such as roads, streams, and bridges are conceived of as lines for mapping purposes. Different types of maps are possible for line features to represent qualitative and quantitative data.

Just like point features, line features can be mapped using standardized symbolization. Look at the standardized symbolization for line features developed for use by USGS topographic maps in figure 6.8. Line features representing qualitative data are mapped using colored symbology or symbology that uses different line patterns or weights. For instance, in figure 6.8 notice how the symbology for boundaries progresses with changes in line weight and pattern. Similarly, notice the use of color and pattern for symbolizing different types of roads in figure 6.8.

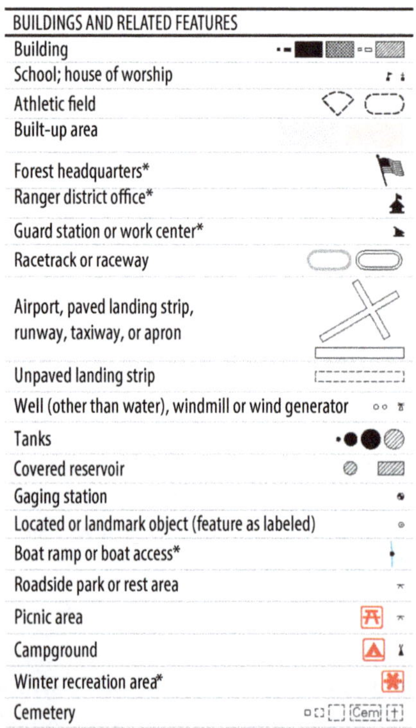

FIGURE 6.7 Standardized point symbols for buildings and related features used by USGS topographic maps.

Source: USGS

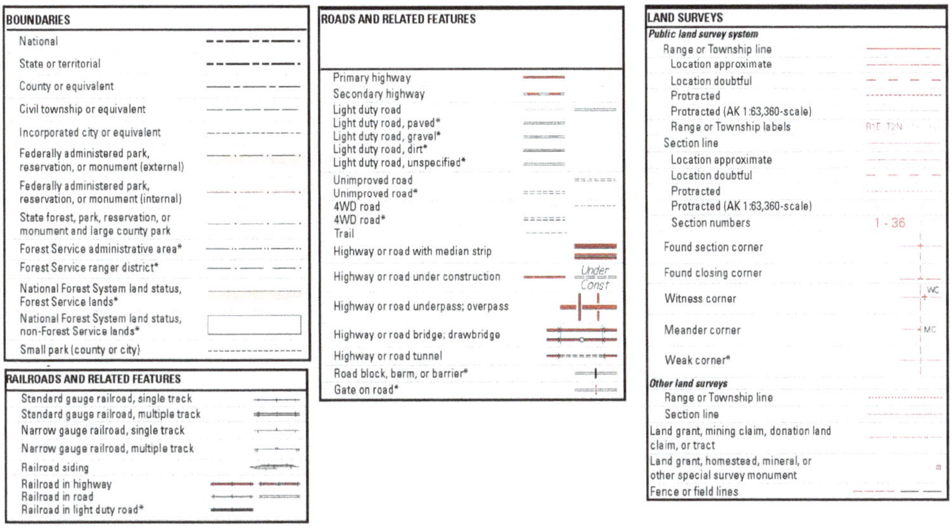

FIGURE 6.8 Standardized line symbols for boundaries, railroads, roads, land surveys, and related features used by USGS topographic maps.

Source: USGS

In a GIS, line features can be symbolized similarly by using different patterns, colors, or line weights that closely follow standardized symbolization (figure 6.9). For instance, the map in figure 6.10 shows the interstate highway, state highways, and arterial roads in Montgomery County, Texas, created using ArcGIS software. If all of the roads in this map were shown with same-colored line with the same thickness, you would end up with a map with dark black spots that would be difficult to read. Instead, the line features are symbolized using different color, pattern, and line weight to differentiate between interstate and state highways and arterial roads represented on the map and in the process makes the map easy for a reader to understand.

Line features representing quantitative data can be mapped using a flow map. A flow map shows changes in magnitude along a line. In a flow map, width of the line representing the line feature is proportionally changed to match the change in magnitude. The line feature may represent changes in stream discharge, air traffic volume, or flow of good and services.

Another type of map used to represent line data is an isarithmic map. An isarithmic map makes use of a specialized cartographic symbolization called an isoline to represent a continuous phenomenon that changes values smoothly, such as elevation, temperature, or pressure. Isolines are specialized lines drawn on a map that connect data points of the same value. Isolines representing a certain type of phenomenon have specific names. For instance, isolines connecting data points of equal temperature are called isotherms, isolines connecting points of equal pressure are called isobars, isolines connecting points of equal elevation above a datum are called topographic contours, and isolines connecting points of equal elevation below a datum are called bathymetric isobaths. Figure 6.11 shows standardized symbolization for topographic contours and bathymetric isobaths used for USGS topographic maps.

Figure 6.12 is a topographic map of Spring, Texas, a suburb of Houston. Contours at five-foot intervals are shown on the map using brown-colored lines, standardized symbols for topographic contours.

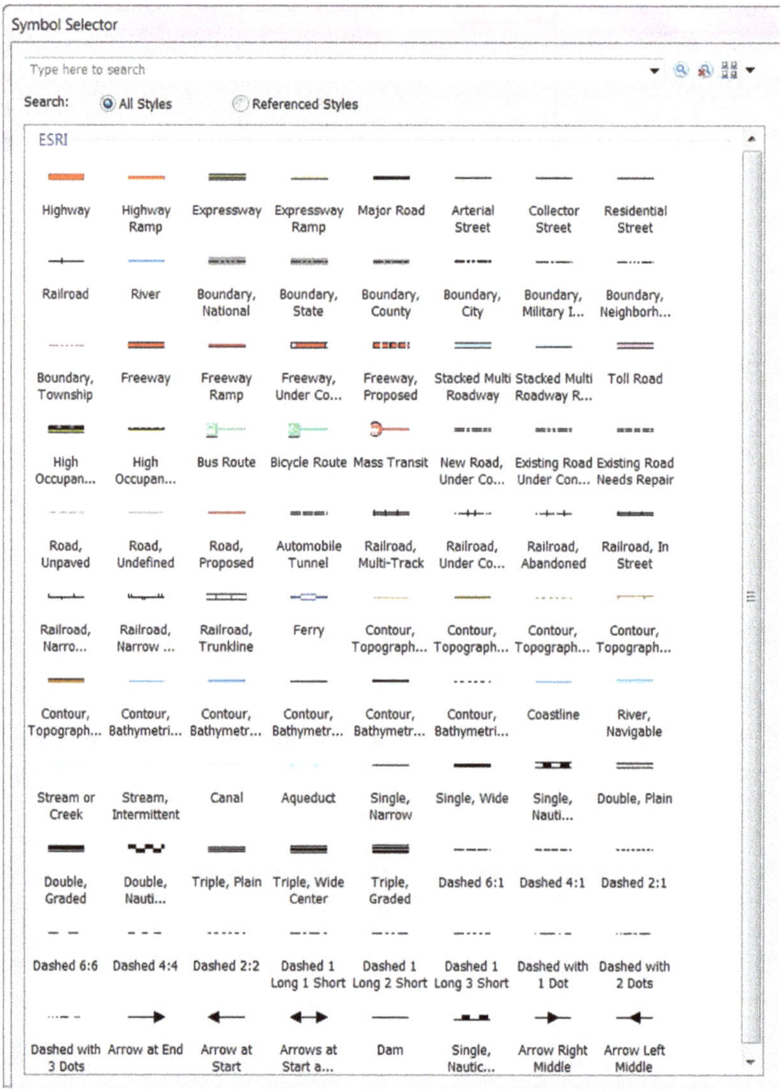

FIGURE 6.9 Symbol selector for line data in ArcGIS software.

ArcMap interface courtesy of Esri, Inc.

Mapping Polygon Data

Many important data are collected and displayed at the polygon level. Consider population, median income, race and ethnicity, land use–land cover, or temperature data collected at the state, county, census tract, or municipality level. Here the data can be mapped in two ways, either as polygon data representing the geographic space it occupies on the earth's surface or as a collection of pixels. The polygon data may be qualitative or quantitative in nature.

Numeric (interval and ratio) polygon data representing a continuous surface that changes values smoothly, such as elevation, precipitation, and temperature, can be mapped using the isarithmic mapping method discussed in the previous section.

Qualitative polygon data such as land use–land cover, soils, and geology can be mapped by displaying the pixels in the case of raster data or polygon area in the case of vector data by using color symbology. The polygon data in this case will represent

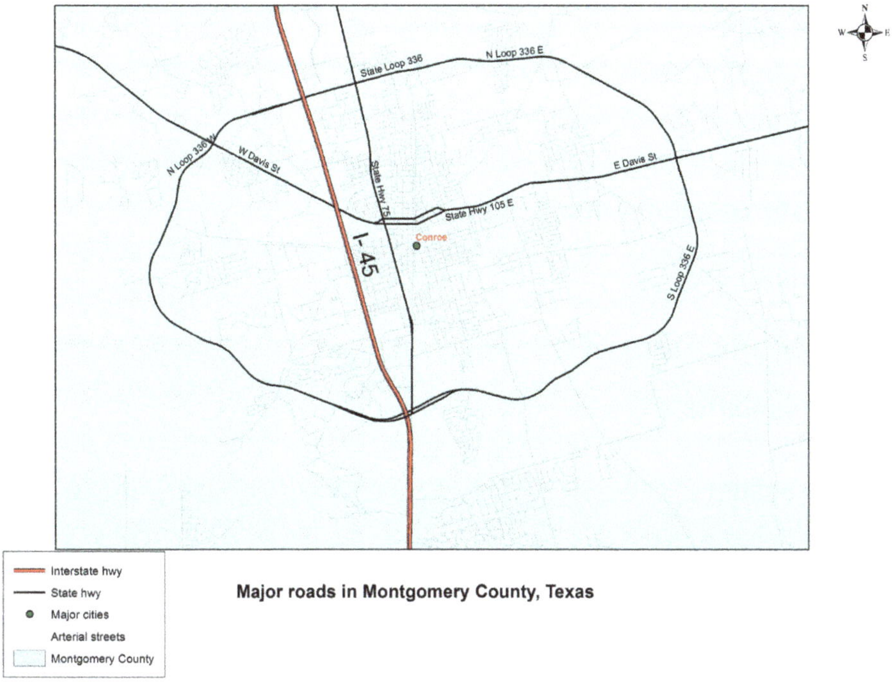

FIGURE 6.10 A map of major roads in Montgomery County, Texas, created in ArcGIS software.

ArcMap interface courtesy of Esri, Inc. Data Sources: United States Census Bureau, 2022, Esri; U.S. Department of Commerce, Census Bureau; U.S. Department of Commerce (DOC), National Oceanic and Atmospheric Administration (NOAA), National Ocean Service (NOS), National Geodetic Survey (NGS)

either nominal or categorical data. Here the different qualitative values are assigned a specific color. The color scheme chosen, however, must be such that the legend classes do not represent any kind of difference in magnitude values. For instance, take the case of the map shown in figure 6.13. This is a land use–land cover map of an urban agglomeration in the western part of India created from a 30 m × 30 m Landsat 8 OLI (Operational Land Imager) / TIRS (Thermal Infrared Sensor) imagery from 2016. The land use–land cover map has four land cover categories. Every pixel in the map is associated with one of the four land cover categories shown in the legend, and each land cover category is represented by a unique color that allows a map reader to distinguish between the four land cover categories.

Also look at the Texas ecoregions map in figure 6.14 derived from Texas Parks and Wildlife. The data here is in polygon format and the map shows the spatial distribution of 11 different timber ecoregions by using different colors to symbolize each ecoregion.

Quantitative polygon data that represent either ordinal, interval, or ratio level data can be mapped using a thematic mapping method called *choropleth mapping*. Choropleth maps are the most commonly created thematic maps. They are ideal for mapping polygon data that are collected for special data collection areas or units where the data values change abruptly at boundaries. The data collection area can be based on political or administrative areas such as state, county, census tract, zip code, school district, or tax zone. Alternatively, the data collection area can be based on a natural resource management zone such as ecoregion or watershed or public health data collection areas. In choropleth mapping, data is categorized into different groups and each data group/category is represented by a unique color depending on its magnitude. Creating a choropleth map in a GIS requires two important considerations: data classification and normalization.

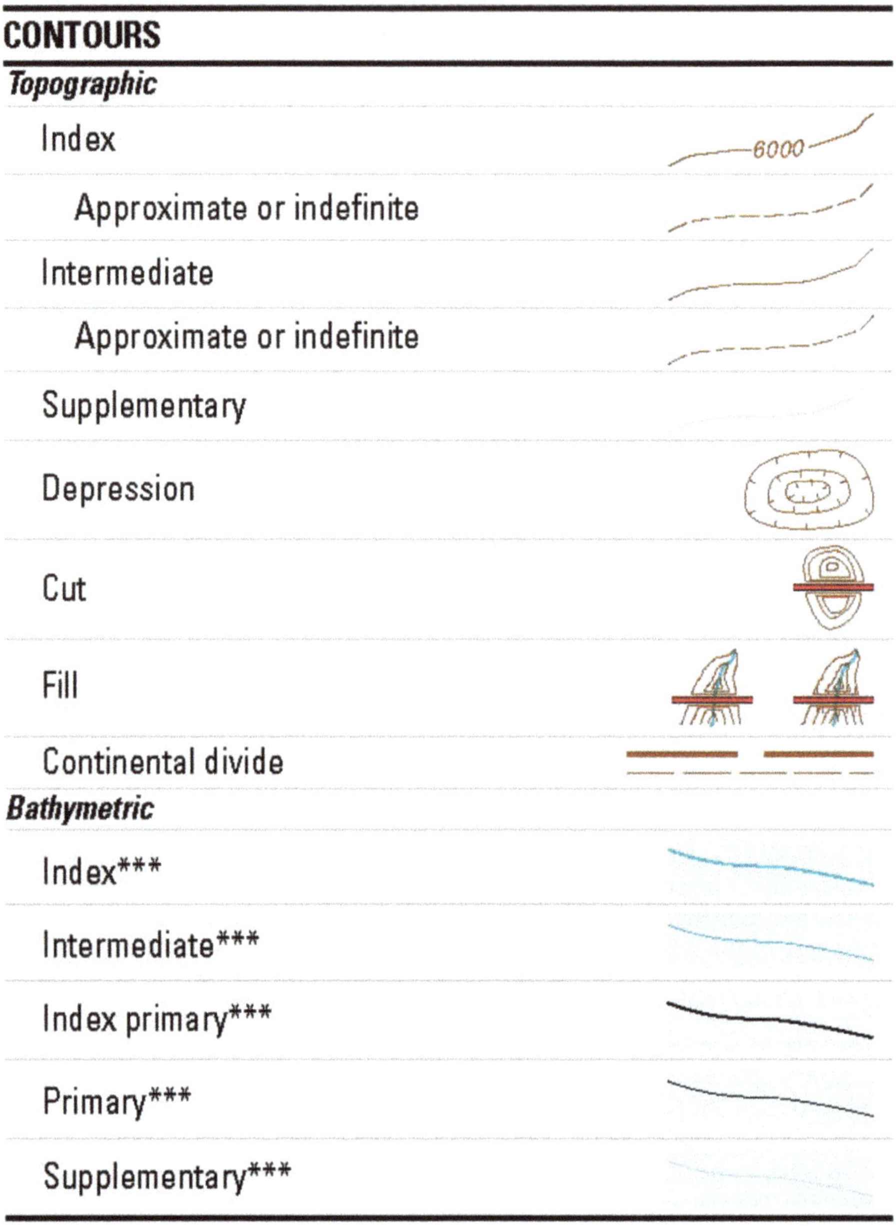

FIGURE 6.11 Standardized symbolization for topographic contours and bathymetric isobaths used for USGS topographic maps.

Source: USGS

The process of dividing data into groups or categories is called data classification. Each group or category is referred to as *class.* GIS gives its users different options to classify data. Different data classification methods will result in different-looking maps because each method will classify your data is a specific manner. Figure 6.15 shows four different maps of percentage of housing units per state that are owner occupied as per US Census Bureau data. Each of the maps representing US states and the District of Columbia was created using a different classification method in ArcGIS. The first three maps have classified the data into four categories.

FIGURE 6.12 7.5-minute quad section of a USGS topographic map of Spring, Texas, a suburb of Houston, downloaded in a GeoPDF file format from the National Map.

Credit: USGS

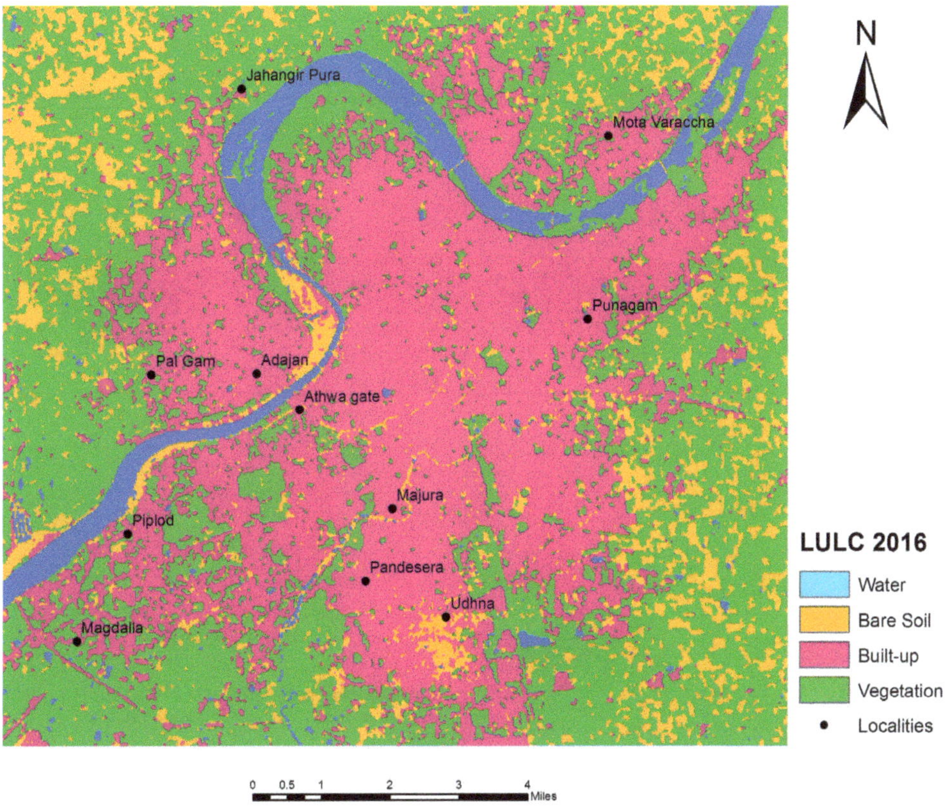

FIGURE 6.13 Symbolizing qualitative nominal polygon data: land use–land cover map of Surat, an urban agglomeration in the western part of India, created from a 2016 Landsat 8 OLI/TIRS imagery.

Map derived from Mukherjee and Singh 2020

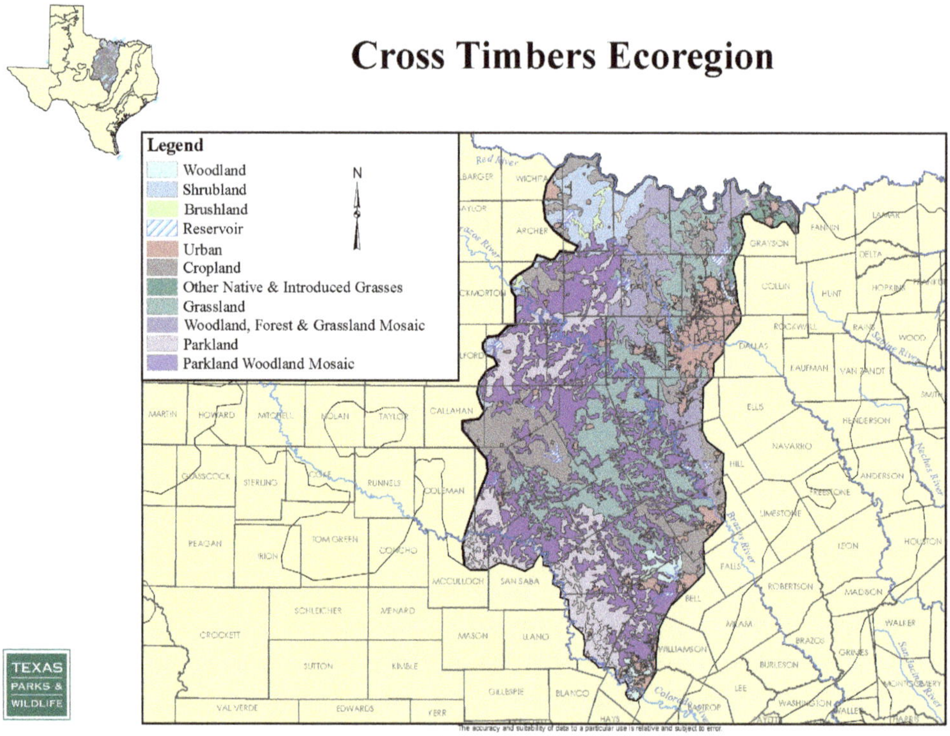

FIGURE 6.14 Symbolizing qualitative polygon data: Cross Timbers ecoregion of Texas.

Source: Texas Parks and Wildlife; public domain

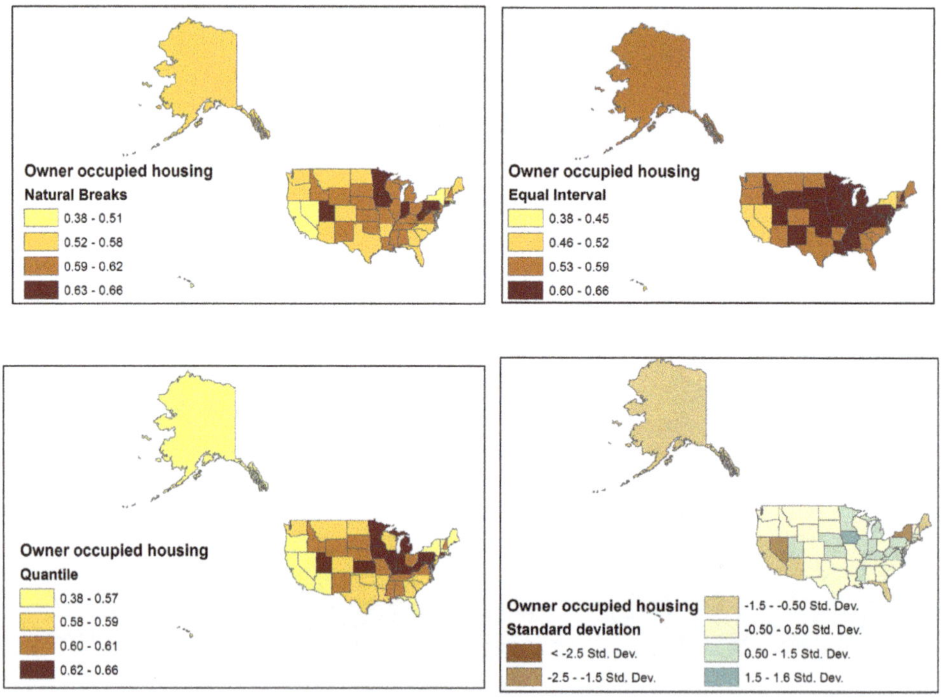

FIGURE 6.15 Examples of choropleth maps displaying percentage of owner-occupied housing in US states created using different classification methods.

Data source: ArcGIS Living Atlas of the World. Attribution: Esri, TomTom, US Department of Commerce, US Census Bureau. ArcMap interface courtesy of Esri, Inc.

The different classification methods available within ArcGIS are natural breaks, equal intervals, quantiles, standard deviation, defined interval, and geometrical interval. Each of these methods classify data in a unique way.

The natural breaks method, also called Jenks, classifies data by emphasizing how the data are naturally grouped together. The method searches for natural breaks in data values and selects class intervals based on the natural breaks. As a result, similar values are grouped together, as there is minimal difference between data values in the same class. As you can see in figure 6.15, Wisconsin, Utah, Iowa, Indiana, Pennsylvania, West Virginia, and Connecticut with a high percentage of owner-occupied housing are in one class, and California, Nevada, and New York with a low percentage of owner-occupied housing are in another class.

The equal interval method classifies data in such a manner that each class has equal class interval, as the name suggests. The method takes the range of data values, calculated using the highest and lowest data values, and divides the range by total number of classes. This gives the interval for each class and is used as the dividing point between each class. For instance, in the owner-occupied housing map, the data is divided into four classes and the range goes from the state or district with the lowest percentage of owner-occupied housing to the state or district with the highest percentage of owner-occupied housing. In this case, the District of Columbia has the lowest percentage of owner-occupied housing with 0.38% and Iowa has the highest percentage of owner-occupied housing with 0.66%, giving a range value of 28%. Dividing the range by the total number of classes gives a class interval of 7% and is used to define each class. The first class represents states that have a 7% higher percentage of owner-occupied housing from the lowest value, that is, all states that have a percentage of owner-occupied housing from 38% to 45%. The equal interval method simply classifies data based on the data range and fails to consider natural clusters of data or data distribution. As you can see in the equal interval map shown in figure 6.15, the first class only has the District of Columbia, while most of the states have been grouped into the last class.

The third classification method shown in figure 6.15 is quantiles. This method takes the total number of data values and divides them into a number of classes in such a way that each class has an equal number of values in it. In the example of percentage of owner-occupied housing, we have 51 states (50 states plus the District of Columbia) and four classes, so each class will have data values for around 13 states. Since the quantiles method places an equal number of data values in each class, each class will occupy approximately the same area on the map. Hence, the map tends to display an even distribution of values. Like the equal interval method, this method does not take into account the distribution of data values.

The fourth method is the standard deviation method. Standard deviation is the measure of the average distance a single data value from the mean of data values. In the standard deviation classification method, class breaks are calculated by adding or subtracting standard deviation from the mean of the data. This method is particularly good at highlighting which data values are closer to the average and which data values are the outliers. The standard deviation classification method works well for a normally distributed data. In figure 6.15, ArcGIS classified the data into six classes by adding and subtracting the standard deviation of the data from the mean of the data.

As you can see from figure 6.15, choropleth maps created using the same data will look very different depending on the classification method used. The map will therefore convey a different message to the map reader. When choosing a classification method, it helps to know how your data values are distributed: whether it is a normal distribution or a skewed distribution and whether the data is positively skewed or negatively skewed. The choice of a classification method will depend on the type of data and purpose of the map. Ultimately the goal must be to place similar data values in the same class. Different classification methods will work differently with your data. The natural breaks (Jenks) method helps to display natural patterns in data values, whereas quantiles places the same percentage of data values in each class, presenting an even distribution of data values. The method is particularly useful in displaying groups of data of interest to statisticians.

Choropleth maps are also influenced by the type of data values being mapped. If the data being mapped represents a data collection area or unit, the size and shape of the area can influence the appearance of the map and the message it conveys. If the sizes of the areas being mapped are very different, very large areas will overshadow smaller areas and dominate the map. As a result, smaller areas with significant values will grab little attention on the map. Such a map may not be successful in effectively conveying any emerging patterns. For instance, if you are mapping the number of owner-occupied houses in the United States, a very large state like Texas will probably have a larger number of houses (9,977,436 total housing units with 5,685,353 owner-occupied units, according to the 2010 Census) compared to a smaller state like Connecticut (1,487,891 total housing units with 925,286 owner-occupied units, according to the 2010 Census). So if you make a choropleth map that shows the total number of owner-occupied housing in each state, Texas will show many more owner-occupied houses than Connecticut. Moreover, larger states like Texas and California will be prominent on the map due to their sheer size and will overshadow smaller states like Connecticut and Rhode Island.

In order to avoid this scenario where the size of polygons or total number of features within a data collection area influence the appearance of your map, you must always take into consideration the size of your data collection area by *normalizing* the data. **Normalization** is the process of transforming data values into a ratio with the original data value as the numerator to minimize the differences in values caused by the size of areas or the total number of features in each area. In the example of owner-occupied housing units, a better way to map the distribution would be create a choropleth map that displays the percentage of housing units that are owner occupied. This way the big difference in the total housing units between Texas and Connecticut will not factor to influence your map. When you map the percentage of owner-occupied housing units, Texas has 56% of owner-occupied housing units while Connecticut has 62% of owner-occupied housing units. Thus your choropleth map now truly showcases the distribution of owner-occupied housing in US states.

However, not all data must be normalized for choropleth mapping. Ask yourself, will the data values be influenced by the size of data collection areas? If not, then normalizing is not needed. For instance, if you are to map the median income or average rent of all states in the United States, do you need to normalize the data to create your choropleth maps? Will median income or rent be influenced by the size of polygons representing states?

Imagery in Maps

It is common practice to create maps that include aerial photography, satellite imagery, or other remotely sensed data. We have become accustomed to seeing such imageries on our maps. Take the example of Google Earth, Google Maps, Bing Maps, or maps that you see on the ArcGIS Living Atlas of the World. The maps on all of these applications use remotely sensed data as basemap or as background information. Using orthophotos or aerial photos as basemap to digitize vector data, as discussed in chapter 5, is common practice. GIS softwares like ArcGIS software

FIGURE 6.16 World Imagery added from ArcGIS Online to a US states map in ArcMap. States are shown using states shapefile with symbology set at 50% transparent to visualize world imagery.

Data source: ArcGIS Living Atlas of the World. Attribution: Esri, TomTom, US Department of Commerce, US Census Bureau, Maxar, Earthstar Geographics, CNES/Airbus DS, and the GIS User Community. ArcMap interface courtesy of Esri, Inc.

provide users the option to add imageries as basemap directly from ArcGIS Online. This adds a layer of rich background information to the map.

Figure 6.16 shows a World Imagery layer added to a states map from ArcGIS Online. The state layer is shown by using shapefile. The World Imagery is a very high-quality imagery that provides a very detailed view of the environment. As you zoom into the map, you can visualize features on the imagery, providing rich contextual information. The symbology for the states shapefile has been made 50% transparent so that a user can visualize the rich information from the World Imagery under the shapefile. A user can use this information to digitize and create a vector representation of features shown on the World Imagery, if needed.

Questions and Activities

1. Browse the web and select any five maps of your choice. Give a brief description of the maps that you have selected in terms of (a) the purpose of the map and (b) characteristics of the map chosen. For each of maps, explain whether it is a reference map or a thematic map. For thematic maps, what is the theme of the map?
2. For this question we will use the ArcGIS Living Atlas of the World. Go to https://livingatlas.arcgis.com/en/home. Click "Browse." Here you can browse or search for geospatial data. In the search box type *USA States*. We will use the first dataset from the list, *Usa States Generalized Boundaries* (Feature Service by esri_dm). Click on "USA States Generalized Boundaries." This will take you to a page where you will find detailed information about the data and different options to view the data. You will see different options on the right-hand side of the window to open the dataset. Select the option for *Open in Map Viewer*.

   Map Viewer allows you to create and interact with online maps in ArcGIS Online. Just like you have created GIS maps using ArcGIS software in the previous chapters, here you will design and create an online map. (Software versions and online resources may change rapidly. This exercise was designed using the most recently available version of ArcGIS Online at the time of this writing.)

   A map of USA states opens in the map viewer (figure 6.17). When the map opens, you will see a map of USA states symbolized using a single color. We will now create a map to show other attributes of USA states. At the right-hand corner top of the map, click on "Open in Map Viewer Classic."

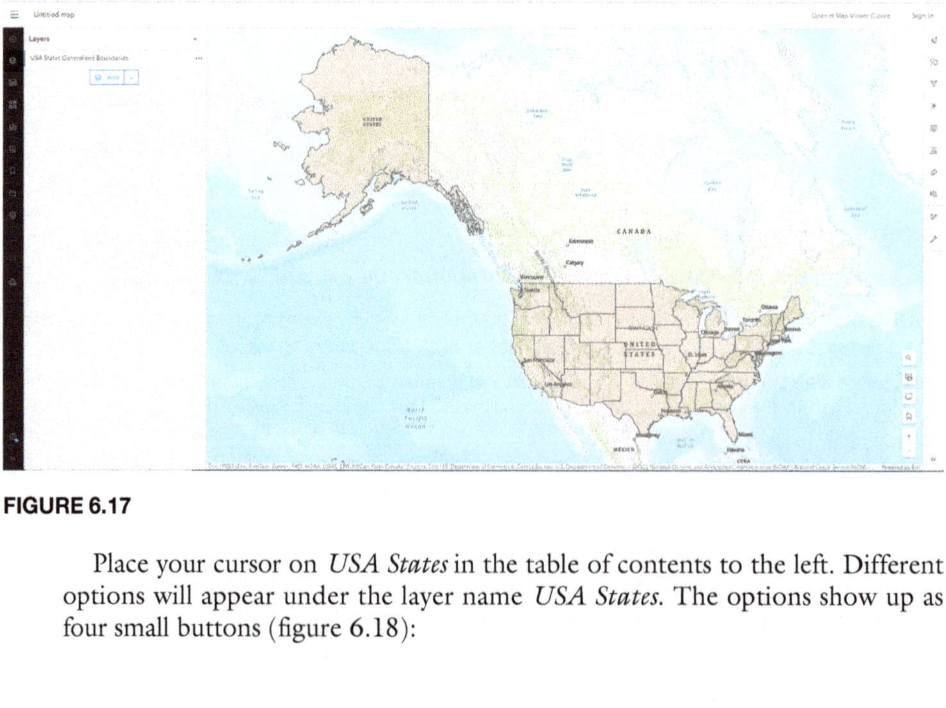

FIGURE 6.17

   Place your cursor on *USA States* in the table of contents to the left. Different options will appear under the layer name *USA States*. The options show up as four small buttons (figure 6.18):

a. The first button, "Show Legend," is for showing legends for the layers shown on the map.
b. The second button, "Show Table," is for viewing the attribute table.
c. The third button, "Change Style," gives a user different options to design a map using different symbology, classification, and so forth.
d. The fourth button, "Filter," allows a user to conduct attribute queries.

FIGURE 6.18

We will be using the first three buttons. Click on the second button, "Show Table." The attribute table of USA states will open under the map. For each column in the table, when you click on a column name you will get two or three options. If the column stores qualitative data, you will see two options: Sort Ascending and Sort Descending. If the column stores quantitative data, you will see a third option when you click on the column name: Statistics. How many columns does the USA states attribute table have?

For the column POP10_SQMI, arrange it ascending. Which state had the highest population per square mile in 2010 and which state had the lowest?

Close the attribute table. There is an X sign on the upper right-hand corner of the table to close/exit the attribute table. We will now create a map to show the 2010 population and population density.

Select the third button, "Change Style." For *1 Choose an attribute to show*, select POP10_SQMI. For *2 Select a drawing style*, you will see four different options to symbolize POP10_SQMI. We will select the first drawing style, *Counts and Amounts (Color)*, if it is not already selected. This style will create a map that shows quantitative data using colored symbology. Click on the "Options" button. Keep the Theme at *High to Low*. We will symbolize the states using a color gradient that shows population density gradually increasing from less than 1 to greater than 1,000.

Click on "Symbols" and select a color ramp of your choice. Once you have chosen your color ramp, change the values by sliding the number slider. Put the lowest slider at 1.2. You can click on the number and type in the value. Put the second slider at 1,000. Click "OK," then click "Done" (figure 6.19).

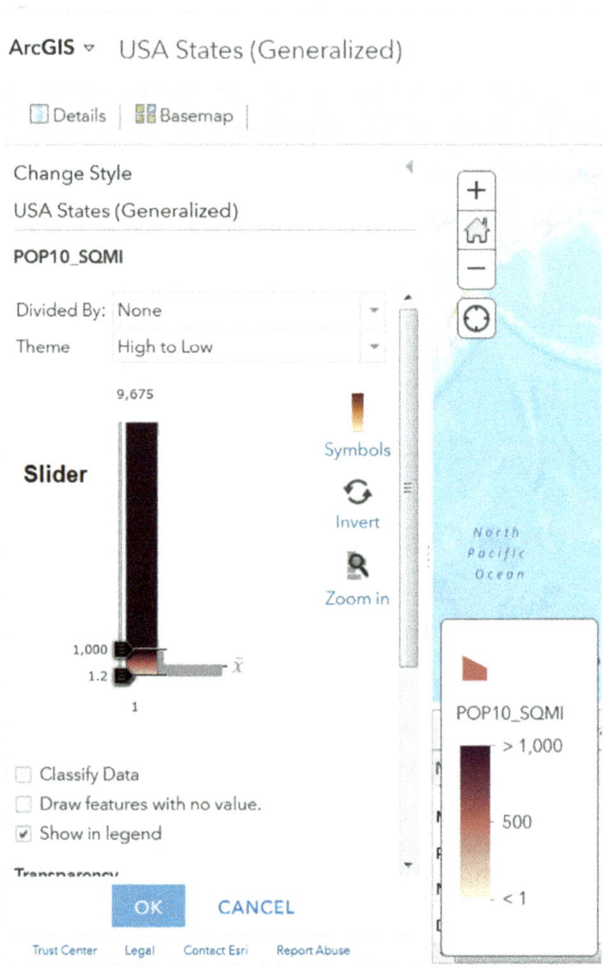

FIGURE 6.19

What is the population density in your state? Which other state has a similar population density?

3. Next we will create a proportional symbols map where each state will be represented by varying-sized point symbols. The points will be symbolized by proportionally scaling the symbol size to population density.

   Select the third button, "Change Style." For *1 Choose an attribute to show*, select POP10_SQMI. For *2 Select a drawing style*, we will now select the second style, *Counts and Amounts (Size)*. After you have selected the style, click on "Options." You can click on "Symbols" and change the symbol color. Click on "Symbols." The symbols options window opens. We will not change the symbol shape and keep the circle. Click on "Fill" and choose a color of your choice. We have selected light brown color. Click "OK."

   For values, you can change the values represented by each symbol size by using the slider. Set the lowest value at 10 and the highest value at 1,500. Make sure that *Adjust size range automatically* is selected. Click "OK," then click "Done" (figure 6.20).

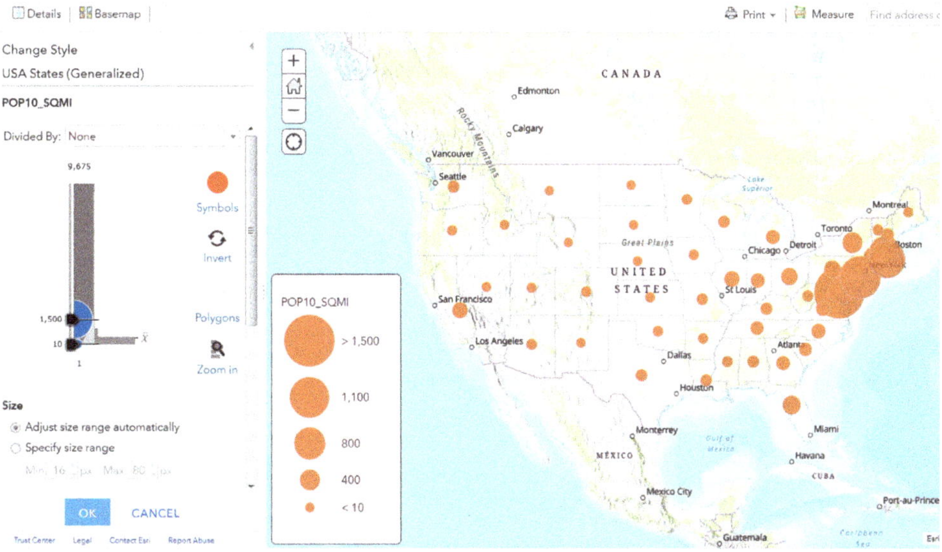

FIGURE 6.20

Of the two maps, the one created in question 2 and the one created in question 3, which one according to you does a better job at representing population density of states? Why? Explain your answer in a one-paragraph write-up.

4. Next we will create a choropleth map by classifying the data. Select the third button, "Change Style." For *1 Choose an attribute to show*, select POP10_SQMI. For *2 Select a drawing style*, you will again select the drawing style *Counts and Amounts (Color)*. After selecting the drawing style, click on "Options."

Check the box to "Classify Data." In the "Using" drop-down menu, select/keep *Natural Breaks*. For number of classes, select *5* for the "With __ classes" option. Click "OK" (figure 6.21).

How were the states classified in this method? What can you say by looking at the map legend?

Next we will change the classification method to *Equal Interval*. Change the number of classes to *4*. Click "OK." What change do you see in your map? Do all the states fall into one class? Zoom into the map so that you are able to see Maryland and Delaware. What do you notice? The District of Columbia falls into one class.

Next we will change the classification method to *Quantile*. How does the appearance of your map change now? Does your map look different when you have 4 classes versus 5 classes? Why?

Now we will select *Manual Breaks* and you will select your own class breaks. Change the number of back classes to *5*. You will select your class breaks by using the slider. For the first class, set the slider at 25. For the second class, set the slider at 75. For the third class, set the slider at 250. For the fourth class, set the slider at 7,500. You can choose your own class breaks. What class does your state fall into?

Write a paragraph explaining the four classification methods that you have used to create your choropleth map. Which method according to you did a good job at representing population density of USA states?

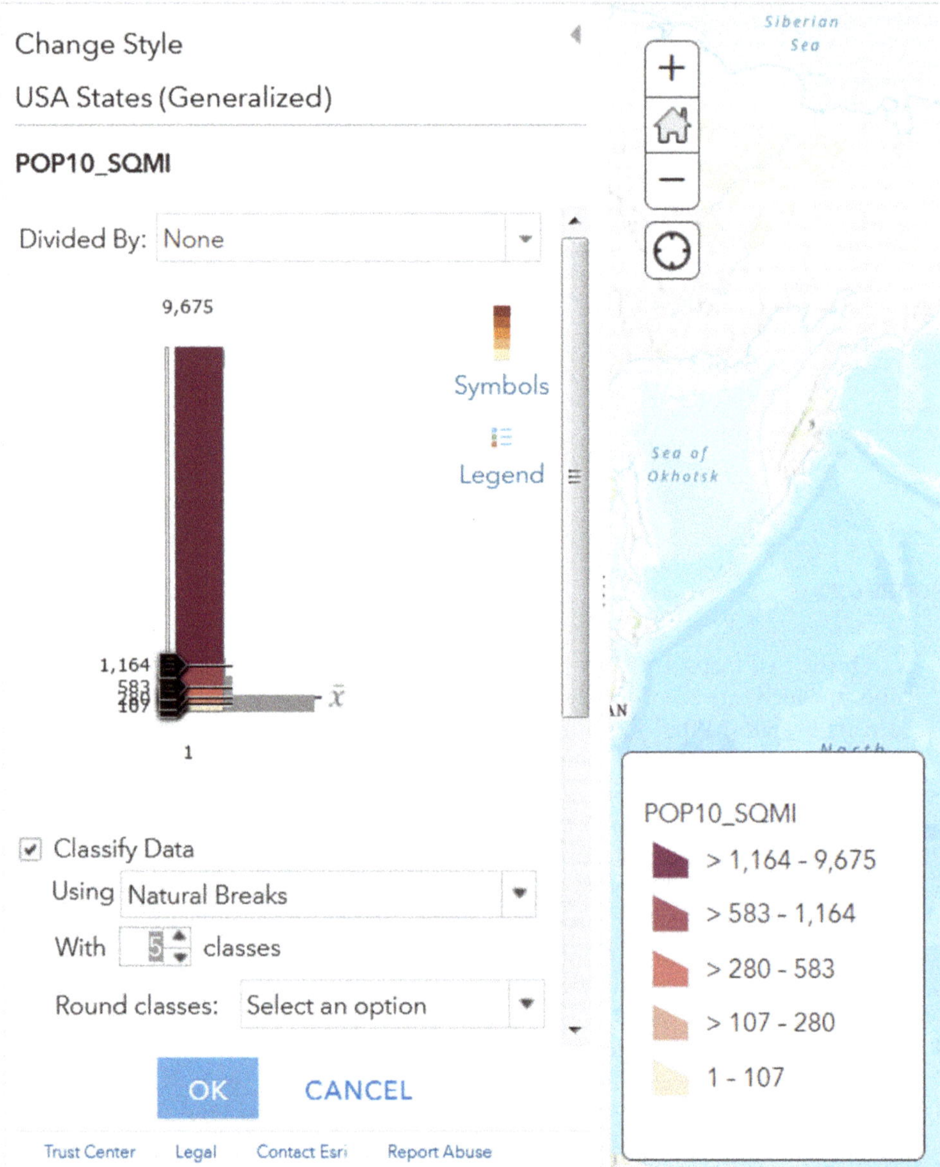

FIGURE 6.21

Key Terms

bathymetric isobaths (p. 147)
cartographic design process (p. 138)
data classification (p. 150)
flow map (p. 147)
graduated symbols map (p. 144)
isarithmic map (p. 147)
isobars (p. 147)
isolines (p. 147)
isotherms (p. 147)
legend (p. 140)
normalization (p. 154)
north arrow (p. 140)
picture symbol map (p. 143)
proportional symbols map (p. 144)
reference map (p. 142)
thematic map (p. 142)
topographic contours (p. 147)
typography (p. 140)

References

ArcGIS Living Atlas of the World. Accessed February 2, 2022. https://livingatlas.arcgis.com/en/home.

Buckley, Aileen, and Kenneth Field. 2011. "Making a Meaningful Map: A Checklist for Compiling More Effective Maps." *ArcUser* 14, no. 4: 40–43.

Jensen, John, and Ryan Jensen. 2013. *Introductory Geographic Information Systems*. Boston: Pearson.

Kimberling, Jon A., Aileen R. Buckley, Phillip C. Muehrcke, and Juliana Muehrcke. 2009. *Map Use*. Redlands, CA: Esri Press.

Mukherjee, Falguni, and Deepika Singh. 2020. "Assessing Land Use–Land Cover Change and Its Impact on Land Surface Temperature Using LANDSAT Data: A Comparison of Two Urban Areas in India." *Earth Systems and Environment* 4: 385–407. https://doi.org/10.1007/s41748-020-00155-9.

Price, Maribeth. 2016. *Mastering ArcGIS*. New York: McGraw Hill Education.

Shellito, Bradley. 2018. *Introduction to Geospatial Technologies*. New York: W. H. Freeman–Macmillan Learning.

USGS. 2005. *Topographic Map Symbols*. Washington, DC: US Geological Survey. Accessed February 2, 2022. https://pubs.usgs.gov/gip/TopographicMapSymbols/topomapsymbols.pdf.

7 GIS Softwares

Learning Objectives

- Explain different GIS softwares
- Describe the features of ArcGIS software including ArcGIS Desktop, ArcGIS Pro, and ArcGIS Online
- Differentiate the three ArcGIS softwares based on their key features
- Describe the features of QGIS software
- Describe the features of GRASS GIS and gvSIG
- Explain key considerations for GIS software selection
- Use QGIS software for basic GIS operations
- Use ArcMap for basic GIS operations
- Download GIS data freely available from the ArcGIS Living Atlas of the World

GIS Softwares and Their Capabilities

So far in this book we have covered different aspects of GIS and geospatial data. An important component of any working GIS consists of GIS software. A lot of what you can or cannot do with your geospatial data depends on the capabilities of the GIS software used. Most of the activities and exercises in this book are based on ArcGIS software. However, ArcGIS is not the only GIS software available on the market.

Several GIS software packages are available today, each with its own unique capabilities. GIS softwares range from freely available data visualization softwares to open-source GIS softwares to highly expensive, proprietary commercial softwares. The fate of any GIS project depends largely on the software selected to perform the GIS tasks. Hence, before committing to any GIS software, careful consideration of the software capabilities and drawbacks must be made. The software's capabilities and functions must match the overall goals of the organization and what an organization plans to achieve through its GIS projects. Such a decision is not based on one project or a few projects. The decision to invest in GIS software and a specific GIS software must fit within the overall goals, current and future, of the organization. This involves long-term planning and decision-making.

In this chapter we will cover a few of the important GIS softwares that you may encounter and discuss their capabilities in brief.

ArcGIS

ArcGIS is a suite of GIS softwares created and distributed by Esri, a Redlands, California–based company. Esri was founded by Jack and Laura Dangermond in 1969 and today is the market leader in GIS softwares. Esri's GIS programs are used by academic institutions, government agencies, the military, and private corporations with applications in different fields such as planning, environment, hydrology, and geology. Esri released its first GIS software, Arc/Info, in 1982. Over the years Esri's GIS software has gone through many changes and has evolved from a command-based system to an extremely powerful but user-friendly graphical user interface (GUI)–based system.

ArcGIS, Esri's latest GIS program, is a highly versatile suite of softwares that can be used on multiple platforms: on a desktop, on the web, or as a mobile app. Irrespective of the platform on which the ArcGIS program is used, a user can avail the same techniques and conduct the same GIS tasks and analyses. ArcGIS includes functionalities to conduct a variety of sophisticated vector data analysis as well as preliminary raster data analysis. Advanced capabilities targeting specific specialty areas can be added to the program by purchasing extensions to the ArcGIS suite. Additionally, it also has capabilities to import data in different formats and perform GIS tasks and analyses. A benefit of using Esri's ArcGIS product is the availability of good-quality geospatial data that you will have access to to get you started with exploring and using GIS. Esri's suite of ArcGIS programs fall into three main categories: ArcGIS Desktop, ArcGIS Pro, and ArcGIS Online.

ArcGIS Desktop

ArcGIS Desktop was first released in 2001. It runs on personal computers and was designed to operate on a Windows operating system. ArcGIS Desktop was first released as ArcGIS Desktop version 10.8. At the time of this writing, the most recent version is ArcGIS Desktop 10.8.2 and this would be the last release of ArcGIS Desktop. ArcGIS Desktop consists of five integrated applications: ArcMap, ArcCatalog, ArcToolbox, ArcGlobe, and ArcScene.

ArcMap is the main component of ArcGIS Desktop and is used for displaying, viewing, and analyzing geospatial data and creating maps. You are already familiar with ArcMap, having used it in exercises in the previous chapters. When you browse data on the Esri Living Atlas of the World and want to view the data on ArcGIS Desktop, it opens the data in ArcMap. An ArcMap consists of two main sections: the table of contents and the map display area. The table of contents shows all the data

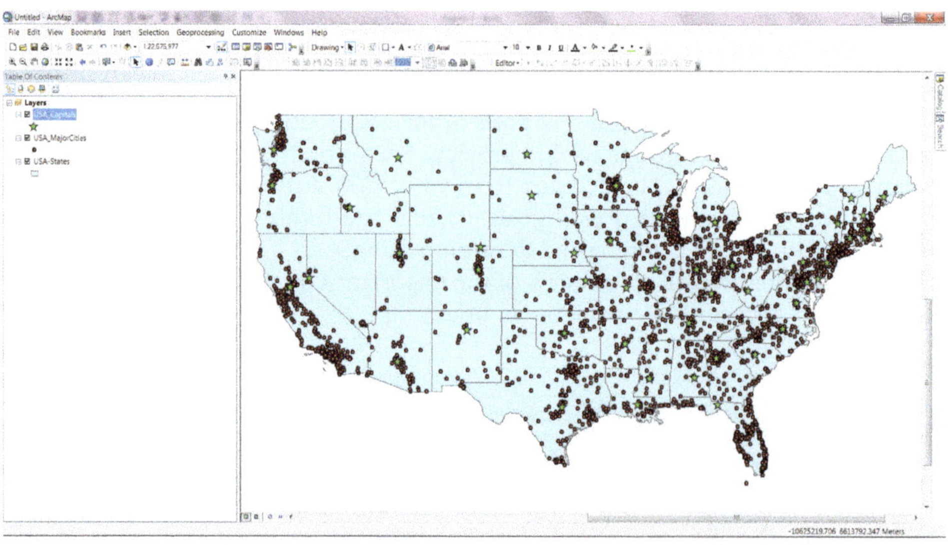

FIGURE 7.1 ArcMap interface.

ArcMap interface courtesy of Esri, Inc. Data Sources: Esri; U.S. Department of Commerce, Census Bureau; U.S. Department of Commerce (DOC), National Oceanic and Atmospheric Administration (NOAA), National Ocean Service (NOS), National Geodetic Survey (NGS)

that have been added to ArcMap, while the map display area allows a user to visualize the data as map layers. ArcMap provides its users multiple geoprocessing tools for data analysis, visualization, and manipulation.

ArcCatalog is mainly used for data management purposes in addition to browsing data and moving data from one location to another. It is also used to manage metadata of geospatial datasets. ArcToolbox contains all the analysis tools used for geospatial data. ArcGlobe and ArcScene add 3D capabilities to ArcGIS. ArcGlobe is used to view geospatial data on a globe surface, while ArcScene is used for three-dimensional visualization.

ArcGIS Desktop is available at three licensing levels: ArcGIS Desktop Basic, ArcGIS Desktop Standard, and ArcGIS Desktop Advanced. Each licensing level is available with different functionalities, with each successive license level adding more functional capabilities to ArcGIS Desktop. ArcGIS Desktop Basic has the least number of functional capabilities. This license provides tools for basic map creation and interactive visualization. ArcGIS Desktop Standard has all the capabilities of ArcGIS Desktop Basic and additionally provides functional capabilities for multiuser editing and data management. ArcGIS Desktop Advanced includes all the functional capabilities of ArcGIS Desktop Standard plus capabilities for sophisticated geospatial analysis, advanced cartography, and database management.

The functional capabilities of ArcGIS Desktop can be further enhanced by adding extensions such as spatial analyst, geostatistical analyst, and survey analyst. These extensions enhance the functional capabilities by adding tools to ArcToolbox.

ArcGIS Pro

A new addition to the ArcGIS Desktop family is ArcGIS Pro. It is the next generation of ArcGIS Desktop and will ultimately replace ArcMap. ArcGIS Pro is most notable for its advanced architecture and enhanced visualization. It has a 64-bit architecture. ArcGIS Pro is a standalone GIS program that can run side by side with ArcMap. It has a modern and sleek interface and works seamlessly with web GIS data and maps. ArcGIS Pro has enhanced three-dimensional visualization

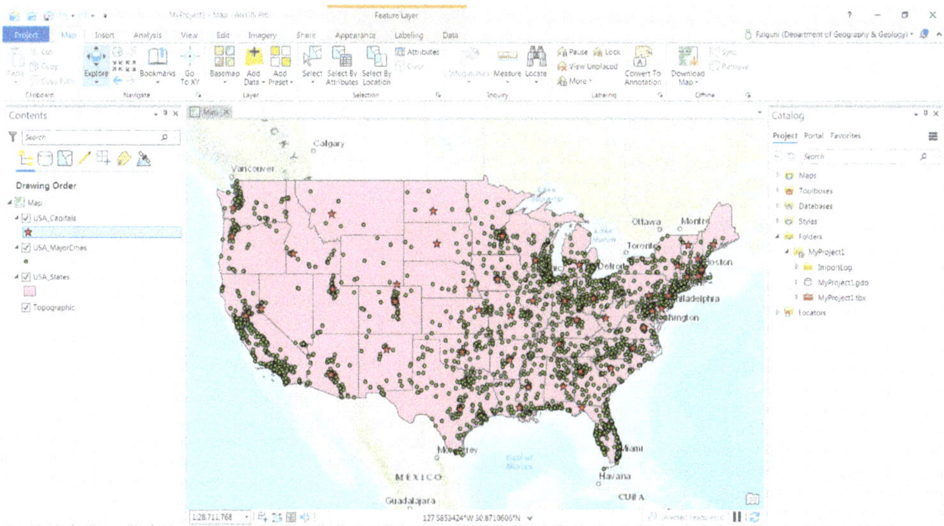

FIGURE 7.2 ArcGIS Pro interface.

ArcGIS Pro interface courtesy of Esri, Inc. Data Sources: Esri; U.S. Department of Commerce, Census Bureau; U.S. Department of Commerce (DOC), National Oceanic and Atmospheric Administration (NOAA), National Ocean Service (NOS), National Geodetic Survey (NGS)

capabilities that are integrated into the program and provide users upgraded geospatial functional and analytical capabilities. Nonetheless, it uses the same basic GIS concepts of maps and layers that all GIS users are familiar with.

ArcGIS Online

ArcGIS Online is Esri's cloud-based solution to creating and analyzing geospatial data and maps. It provides users the capability to create web maps, three-dimensional web scenes, and apps using a web browser. Users can analyze geospatial data or create maps using a web browser and share with others. You don't need to install a software on your own machine. All you need is a web browser and access to the internet. When using the ArcGIS Living Atlas of the World, when you open your data in Map Viewer, you are utilizing ArcGIS Online. You have, in fact, used ArcGIS Online for your chapter 6 exercise.

QGIS

Another widely used GIS software is QGIS. QGIS is an open-source desktop GIS software that is available for free. The software has a unique capability to run on Windows, Linux, Unix, and Mac operating systems. QGIS was created by Gary Sherman, and the QGIS project became the initial project for the Open Source Geospatial Foundation (OSGeo). OSGeo is a nonprofit nongovernmental organization established with the goal of providing organizational, legal, and financial resources and support to promote the advancement of the open-source geospatial community. The development of QGIS software is an important step in that direction.

QGIS version 1.0 was released in 2009. The GIS software is primarily written in C++ and makes extensive use of the Qt library, a tool kit to develop open-source

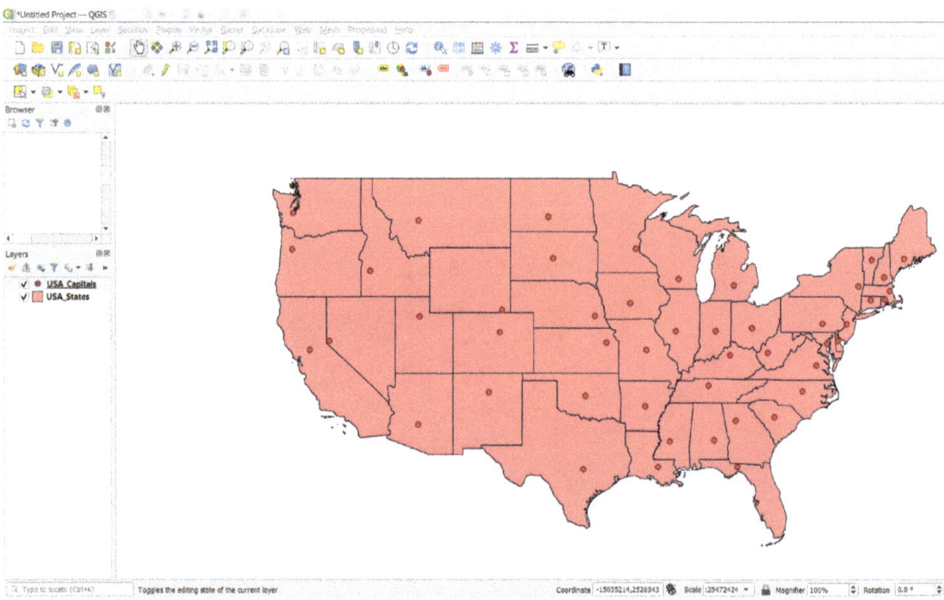

FIGURE 7.3 QGIS interface.

QGIS interface courtesy of QGIS

softwares. QGIS is developed and maintained by a community of volunteer software developers and organizations. The software is available in 48 different languages and has been adopted globally by private and public organizations. It is used in a variety of different fields, including engineering, planning, military, education, and research.

QGIS Desktop is used to create, manipulate, display, and publish geospatial data. The software supports both vector and raster data formats. It supports different vector data file formats such as shapefiles, geodatabases, and DXF. QGIS also integrates with other open-source GIS softwares such as PostGIS, GRASS, and MapServer. It provides users the flexibility to customize the software to specific individual needs. The software is available to download for free at https://qgis.org/en/site.

GRASS GIS

Geographic Resources Analysis Support System (GRASS) is another open-source GIS software. GRASS GIS was a founding member of OSGeo and is affiliated with the OSGeo like QGIS. The software was developed by the US Army Corps of Engineers' Construction Engineering Research Laboratory (USA/CERL). The development of this software is also a global collaboration between software developers and scientists from different disciplines hosted through GitHub.

Just like QGIS, GRASS GIS runs on Microsoft Windows, Linux, and Mac operating systems. The software is available with an interactive graphical user interface (GUI) and a command line interface. It is used for managing, modeling, analyzing, and viewing both vector and raster datasets. The software has a global user base within academic and business communities. GRASS GIS is available to download for free at https://grass.osgeo.org.

gvSIG

gvSIG is another open-source GIS software developed to work with geospatial data. Similar to QGIS and GRASS GIS, gvSIG is affiliated with OSGeo. It is capable of working with both vector and raster datasets supporting different file formats. The software is available in different languages and works on Windows, Mac, and Linux operating systems. gvSIG is available for desktop, web, and mobile applications at http://www.gvsig.com/en.

Many such GIS softwares are available that are either open-source or proprietary commercial softwares. Some work exclusively with vector or raster data, and others have the capability to work with both. Some only work on Windows operating systems, while others have the flexibility to work with different operating systems. Table 7.1 summarizes some popular GIS softwares.

Software Selection

As you can see, many GIS softwares are available for you to choose from. Some are open source, while others are proprietary commercial softwares. Some like ArcGIS work exclusively on a Windows operating system, while softwares like QGIS and

TABLE 7.1 Open-Source and Proprietary Commercial GIS Softwares

Name	Vector/ Raster Data	Operating System	Open Source/ Proprietary	Website
ArcGIS Desktop	Vector	Windows	Proprietary commercial	https://esri.com
ArcGIS Online	Vector	Windows	Proprietary	https://esri.com
AutoCAD Map 3D	Vector	Windows, Unix	Proprietary commercial	https://www.autodesk.com/products/autocad/included-toolsets/autocad-map-3d
Bentley Map	Vector	Windows	Proprietary commercial	https://www.bentley.com/en/products/product-line/asset-performance/opencities-map
Cadcorp	Vector	Windows	Proprietary commercial	https://www.cadcorp.com
Caliper Maptitude	Vector	Windows	Proprietary commercial	https://www.caliper.com
ERDAS Imagine	Raster	Windows, Unix	Proprietary commercial	https://www.hexagongeospatial.com/products/power-portfolio/erdas-imagine
GE Smallworld	Vector	Windows	Proprietary commercial	https://www.ge.com/digital/applications/smallworld-gis-geospatial-asset-management
GRASS GIS	Vector and raster	Windows, Linux, Mac	Open source	https://grass.osgeo.org
gvSIG	Vector and raster	Windows, Linux, Mac	Open source	http://www.gvsig.com/en
IDRISI	Raster and vector	Windows	Proprietary commercial	http://www.clarklabs.org
Manifold GIS	Vector and raster	Windows	Proprietary commercial	http://www.manifold.net
MapInfo Pro	Vector and raster	Windows, Unix	Proprietary commercial	https://www.precisely.com
QGIS	Vector and raster	Windows, Linux, Unix, Mac	Open source	https://qgis.org/en/site
TatukGIS	Vector	Windows	Proprietary commercial	https://www.tatukgis.com

MapInfo Pro work on Windows or Unix or even a Mac operating system. At some point in your career you may be faced with the decision of choosing an appropriate GIS software for your organization. How does one do that? How do you determine which GIS software is best suited for you and your organizational needs? The success of any project will depend on that decision. A lot of factors are involved when deciding which GIS software to use.

One of the biggest considerations is cost. Over the years GIS software costs have gone down considerably. Nonetheless, when you consider other costs associated with software implementation, cost can be a major factor. When investing in a GIS software, one must understand that this is not a one-time investment. Most GIS software companies charge for things like software maintenance, upgrades, and support. Software licenses are not cheap. One must take into consideration the costs involved in a single-use license versus multiple or unlimited licenses. Moreover, no organization, whether public, private, or academic, has unlimited financial resources, so the decision about how much funds to allocate to purchase and maintain a GIS software must be made very carefully.

The cost of GIS software varies depending on the type of organization. For instance, the cost is considerably higher for commercial use by private companies compared to academic institutions. When considering software costs, one must also take into consideration the frequency of software updates, particularly when working on long-term projects. Many software companies compel you to upgrade to a newer version by discontinuing support to older versions of the software. This is an important consideration that can have serious repercussions on the success of a project and must be considered in the budget.

As shown in table 7.1, different GIS softwares are compatible with different operating systems. Some only work with Windows, while others work with Mac, Linux, and Unix as well. One must choose a software that works on an operating system that they are comfortable with. Within an organizational context it is important that the GIS software works with the organizational operating system. In addition to having an appropriate operating system, when installing a GIS software the device must have an adequate processing capacity, memory, and disk space for the software to operate efficiently.

Another key consideration is functional capabilities and data format. What GIS functions the software is capable of performing and what kinds of digital data the software will be able to handle and support are two of the questions that must be considered. If your main focus will be raster data analysis, you must select a software that is capable of working with raster data such as ERDAS Imagine or ENVI. Conversely, if you are mainly focusing on vector data analysis, a software like ArcGIS Desktop or Pro would be more appropriate. Also, the level of licensing one pays for must be commensurate with the required functional capabilities. GIS software may have overcapacity. For instance, if the purpose of investing in a GIS software is to use the software for editing geospatial data, a standard license would serve this purpose instead of an advanced license. This would make a big difference in the cost of the software.

A software must have the capability to handle different data file formats, particularly common formats such as shapefile, geodatabase, TIFF, JPEG, and IMG. A software must be able to handle and support geospatial data, whether vector or raster, available in the public domain such as US Census Bureau TIGER/Line files, USGS DEM files, and DOQs. A GIS software should be able to open these data files with ease. The software must also be able to export geospatial data and database files in formats that other GIS softwares can read.

Software support is an important consideration when investing in a GIS software. Support is provided in different ways, in the form of trainings, online support, customer service, help line, training workshops, manuals, and software user and support community, among others.

There is no doubt that a steep learning curve is involved in GIS use. Few GIS applications can be used with little to no prior training. Many GIS software vendors provide technical training to their users. This may be a one-day training workshop, an online training seminar, or a weeklong training endeavor. For instance, Esri provides a variety of training to ArcGIS users through Esri Academy. The trainings can be instructor-led or self-paced. Some are free trainings, while others require fees. The trainings vary from 30 minutes to eight hours to three days in duration. Trainings can thus be time-consuming and at times expensive.

Irrespective of the technical training provided by the software vendor, all GIS users will find themselves in need of technical support at some point in the project. At that time one generally resorts to a software manual. GIS users heavily depend on software manuals. Does the software have a detailed user-friendly manual? Is the manual available online? Is the manual interface easy to navigate? Is the manual integrated with the software? These are some pertinent questions to consider.

Before committing to a major investment in a specific GIS software, users must always examine the support documentation that accompanies the software. Figure 7.4 shows the online help interface of ArcMap. Here information about printing a map is requested using the "Help" menu in ArcMap. The table of contents of the "Help" menu provides a list of topics a user can use to conduct specific searches. Here the online manual opens in a new window while the software runs in another window. Thus a GIS user can use the manual while working on the software simultaneously and also have the manual available for reference at a later time. Similarly, open-source GIS softwares like QGIS and gvSIG have detailed online documentation.

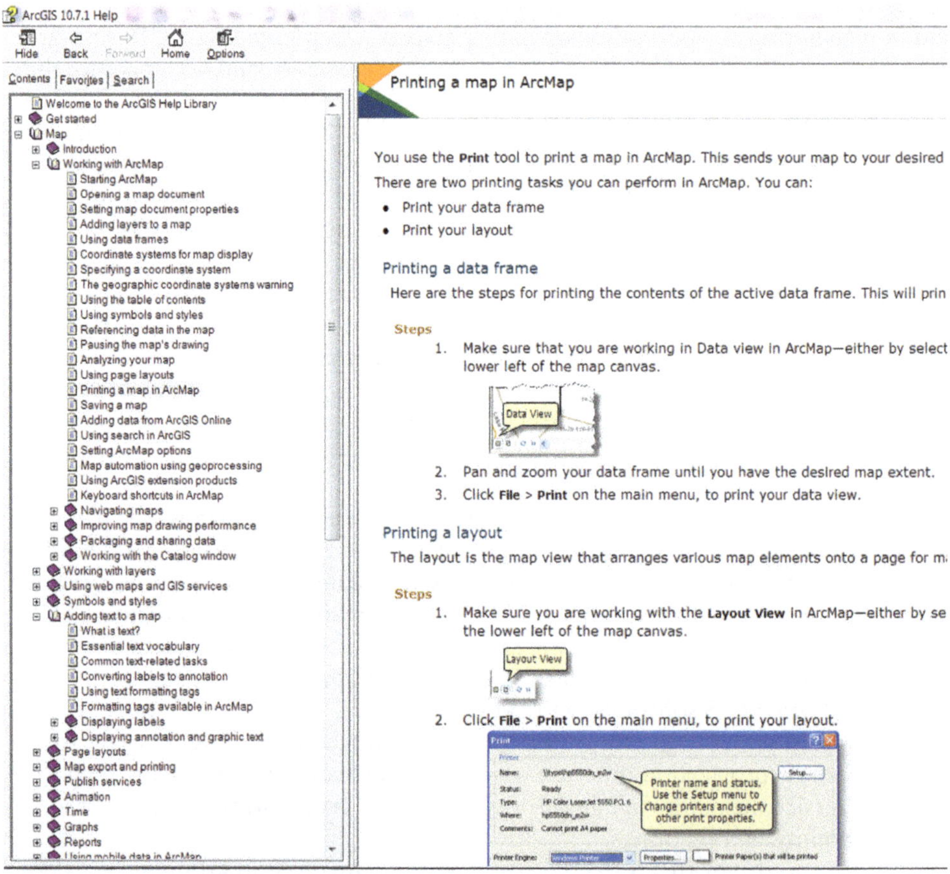

FIGURE 7.4 ArcMap Help interface.

ArcMap interface courtesy of Esri, Inc.

Eventually a need arises when a GIS user must contact customer support through a phone call, chat, or email to answer specific questions. Queries may vary from software installation questions to software bugs or errors. The ability to talk, email, or chat with someone can solve the problem quickly and is of immense value. Blogs and software user groups and communities are also a big help in solving issues. If a software has a large user base, it will be easy to find solutions to software-related issues by posting the question on a user site. Multiple users will respond with answers and possible solutions and help solve the issue at hand.

Software maintenance is also an important consideration. Most software vendors offer maintenance plans that users can purchase at the time of software installation. This is particularly pertinent for large-enterprise systems involving multiple departments and users. Softwares must be maintained regularly to keep them updated to the latest version. As an organization grows and has diverse needs, a GIS software must have the capability to grow with the organization and accommodate its growing needs.

It is therefore important that a user takes into consideration all the requirements of their organization, research different softwares, and examine thoroughly and question before committing to a specific software. The GIS software must be budget-friendly and user-friendly and have functional capabilities that meet the diverse needs of the organization over a long period of time.

Questions and Activities

1. Esri is no doubt the market leader when it comes to GIS softwares. Its ArcGIS Desktop and ArcGIS Pro are used globally in many different industries. Explore the different uses of ArcGIS. Go to https://www.esri.com/en-us/industries/index and explore the different industries in which GIS is used and how it is used.
2. Pick any one industry from a discipline of your choice. Explore how GIS is used in that specific industry and write a one-page narrative describing the uses of GIS in the industry.
3. How are the developments at Esri, particularly the future direction of its software development, fueling development in the industry of your choice?
4. The field of GIS software is constantly evolving. Esri keeps its users constantly updated about its products and future directions. It does this through its publications: ArcNews, ArcUser, and ArcWatch. Choose any one of the publications and read about Esri's future direction. In what direction do you think Esri is headed?
5. How was software development at Esri influenced by the COVID-19 pandemic? What is Esri's contribution in the fight against COVID?
6. Visit the computer laboratory in your department. What operating systems are available for students in your school's computer lab? What GIS softwares are available in the computer lab? Do you have access to GIS software in buildings other than your department?
7. What is the Open Source Geospatial Foundation (OSGeo)? What is its role in the GIS world? Write a one-page narrative explaining the importance of OSGeo and the significance of OSGeo for GIS software development and OSGeo projects.
8. In this chapter we discussed a few open-source GIS softwares. Do your own research to find other open-source GIS softwares and create a table comparing them. What operating systems are supported by the softwares, and what kind of

support documentation is available for users? Which one according to you has a more user-friendly software support available for users and why?

9. Why do you think a large base of user community and support groups is important for any GIS software user?

Key Terms

ArcCatalog (p. 165)
ArcGIS (p. 163)
ArcGIS Desktop (p. 164)
ArcGIS Desktop Advanced (p. 165)
ArcGIS Desktop Basic (p. 165)
ArcGIS Desktop Standard (p. 165)
ArcGIS Online (p. 166)
ArcGIS Pro (p. 165)
ArcGlobe (p. 165)
ArcMap (p. 164)
ArcScene (p. 165)
ArcToolbox (p. 165)
Linux (p. 166)
Open Source Geospatial Foundation (OSGeo) (p. 166)
operating system (p. 164)
platform (p. 164)
QGIS (p. 166)

References

GRASS GIS. Accessed March 5, 2022. https://www.osgeo.org/projects/grass-gis.

gvSIG. Accessed March 5, 2021. https://www.osgeo.org/projects/gvsig.

Jensen, John, and Ryan Jensen. 2013. *Introductory Geographic Information Systems.* Boston: Pearson.

Open Source Geospatial Foundation (OSGeo). Accessed March 5, 2021. https://www.osgeo.org.

QGIS. Accessed March 5, 2021. https://www.qgis.org/en/site.

Shellito, Bradley. 2018. *Introduction to Geospatial Technologies.* New York: W. H. Freeman–Macmillan Learning.

8 GIS and Society

Learning Objectives

- Understand the relation between GIS and society
- Discuss the principles of GIS project management
- Explain how the principles of project management influence GIS use in organizations
- Describe the factors necessary for effective GIS use in an organization
- Explain the impacts of GIS use

GIS and Society: Why?

Consider a situation where you are visiting a new city and as you embark from the plane, the first thing you must do is install a GIS-based application that will monitor your every movement and post your location to a server. Or a situation where you are required to take a selfie and then upload the photograph using the application on your phone, which will geotag the photograph, allowing officials to know of your whereabouts. Or you visit a retail store that sells sports equipment to purchase golf clubs. Soon after, you start receiving marketing material from different retailers in your mail, advertising for different golf equipment. You also start receiving marketing material from different sporting goods stores, even those that are not in your vicinity.

Now, consider a situation where you work for a neighborhood-based organization. It is a small nonprofit organization and your role is to make future plans for the neighborhood with goals to improve neighborhood quality by focusing on current issues such as rental properties or delinquent properties. You must prepare planning documents and submit them for grant funding. The funding organization tends to prefer grant applications that have GIS-based maps. Your nonprofit organization does not have the funds to invest in a GIS nor does it have technical knowledge on how to use it. But now you must figure out how to include GIS in your planning document, despite your document being very detailed and comprehensive, in order to give an "edge" to your grant application.

All of the scenarios described here involve the use of geospatial data and/or a GIS-based application. Examine each of these examples and ask yourself: Do any of these applications of geospatial data and technology have a relation with its surrounding society? Are any of the applications in any way influenced by its surroundings? Do they impact the society? Do they influence who can and who cannot participate in a specific task? You may find it difficult to answer these questions because rarely are there any discussions on these topics. Also, because we are used to thinking about GIS or geospatial data in a very different way. Outside the realm of hardware or software, we rarely discuss what leads to GIS use, why geospatial technology is used, and how those answers ultimately influence specific GIS use. Are there any political goals, profit-making goals, or societal benefits attached to GIS use? Is the use driven by a state or private corporation agenda?

In chapter 1 we discussed different definitions of geographic information system (GIS). One definition that was coined during the 1990s completely changed the field of GIS, highlighting its important role in society. GIS is more than hardware, software, and geospatial data. Two key components of any working GIS involves people and methods, bringing to the forefront other issues that shape GIS use and implementation, such as ethics, geodemographics, community involvement in decision-making, and non-Western GIS implementations, to name just a few.

Nicholas Chrisman underscored the importance of broadening our theorizations of GIS to include the social, economic, ethical, and institutional implications of operating a geographic information system (Chrisman 1997), defining GIS as an "organized activity by which people measure and represent geographic phenomena then transform these representations into other forms while interacting with social structures" (Chrisman 1999, 175). The role of social structures and social contexts in GIS use and its social impacts are therefore paramount. It is extremely important that every GIS user is aware of the multiple contexts that shape their GIS use and decision-making, whether in an organizational or an institutional context, as well as the broader impacts of its use on society.

Discussions about the implications of GIS use have mainly come from two areas of specialization. One has focused on GIS usage within organizations aimed at examining GIS implementation, adoption, and diffusion following project management protocols. The other discussion about the implications of GIS use within the discipline of geography has come from social theory specialists.

GIS and Organization: Principles of GIS Project Management

Prior to the 1990s, studies on GIS use focused on the technical and financial aspects of GIS that treated GIS as any other software to be implemented within an organization. Discussions about the merits and integrity of GIS use within an organizational setting only started in the 1990s. These discussions were motivated by the use of GIS within planning agencies and government organizations in the Western world (Obermeyer and Pinto 1994, Campbell and Masser 1995, Pinto and Onsrud 1997, Gilfoyle and Thorpe 2004). The discussions particularly focused on issues surrounding GIS adoption, implementation, diffusion, employee habits, inter-personnel relationships, and interorganizational issues. However, there was a lack of assessment criteria and evaluation frameworks to guide such GIS research (Campbell 1991, Budic 1998). Frameworks developed by Clapp et al. (1998), Antenucci et al. (1991), and Calkins and Obermeyer (1991) provided guidelines to evaluate GIS use in an organization (Budic 1999).

Clapp et al. (1998) proposed a framework consisting of four levels of assessment: (1) operational efficiency, (2) operational effectiveness, (3) program effectiveness, and (4) contribution to well-being. Based on their consulting experience, Antenucci et al. (1991) proposed five types of distinct benefits of GIS use that included measurable parameters such as efficiencies and improvements in existing practices, expansion or addition to existing capabilities, benefits that can be derived from GIS use, and sale of geospatial information. Calkins and Obermeyer (1991) introduced 24 evaluation criteria based on six categories that include uses of geographical information, effectiveness of use of geographical information, benefits of use of geographical information, measurements of benefits, characteristics of geographical data, and spatial analysis and organizational factors (Budic 1998).

Another seminal work during this time examined the success of using GIS technology using the principles identified in Everett Rogers's innovation diffusion meta theory (Rogers 1983). Rogers defines **innovation diffusion** as "the process by which an innovation is communicated through certain channels over time among the members of a social system" (15–16). Studies that used innovation diffusion theory examined whether a specific GIS use was successful or not based on the criteria of simplicity of the system, whether there were any observable benefits derived from GIS use, the relative advantages to an organization or its employees from GIS use, whether GIS use permitted the ability to make small trials, and finally the compatibility of GIS with the overall organizational culture (Masser and Onsrud 1993, Innes and Simpson 1993, Budic 1998).

A lot of research has been done on organizational GIS use. GIS use within an organization is influenced by a number a key factors such as upper management support within the organization for GIS use, staff support for GIS use, length of experience using GIS, organizational culture towards system sharing, comprehensiveness of GIS database, number of organizational applications that will use GIS, and type of GIS applications that will be used within an organization (Campbell 1991, Croswell 1991, Budic and Godschalk 1994, Onsrud and Pinto 1993, Budic 1998).

Organizational politics play an important role in GIS use, a topic that is seldom discussed. Organizational politics influence key decisions that shape GIS implementation and use and substantially shape its outcomes. Winning the support of key decision-makers and upper-level management is essential for effective GIS implementation and usage. It ensures long-term commitment for a GIS project that may continue for years and grow with the diverse needs of an organization. GIS use involves a complex mix of hardware, software, and geospatial datasets. A GIS implementation takes place within an organization and its hardware and software needs must align with the overall organizational environment. Therefore, the presence of a technical personnel who understands the needs and requirements of a GIS implementation, someone who is a GIS specialist within an organization, is essential for a smooth-functioning GIS.

Implementing GIS and its diffusion within an organization takes time. The benefits of GIS use are not attained immediately. Using GIS involves a steep learning curve. One needs to work with it for a long time before being familiar with it and in a position to use it to its full potential. Thus patience and commitment to long-term support is essential.

Other criteria are also important when it comes to GIS implementation. In an organization that is completely new to GIS, it is important that the technology is as user-friendly as possible. Moreover, it has to be meaningful to those involved in using it and making decisions about it. It is not necessary that everyone in the organization understands every minute detail of how a GIS works. A simple explanation communicating its essential capabilities will suffice. The value of adopting GIS must be evident to all. Particularly those who are responsible for its adoption must be well aware of the reasons for devoting their time and efforts to the endeavor and be able to examine and understand its value.

A GIS implementation requires investment of financial and nonfinancial resources, which typically does not align with an organization's cost-reduction strategies. And the benefits derived from GIS use take time to materialize. The initial phases of any GIS implementation project in particular offer very little to compensate for the high costs involved. The high costs are not only in terms of financial costs and manpower. GIS implementations also involve other intangible costs. For instance, introduction of GIS into an organization alters interdepartmental power structures, disrupting traditional practices, and has the potential to cause conflicts. As a result, GIS adoptions are bound to face challenges at the onset. Therefore, in order to ensure an effective GIS implementation and wide-scale adoption, it is essential that the GIS is well suited for the applications and also aligns well with the overall organizational culture.

Building an organization-wide GIS requires different departments and agencies to collaborate with one another. Sometimes this means that departments or agencies that normally do not work together must collaborate. It may also mean that departments or agencies that are at odds with one another must work together. The staff may have different professional outlooks, technical capabilities, and information needs and they must put their differences aside and work together. A lot of times differences in perspectives between project participants, differences in understanding the potential capabilities of GIS, and differences in geospatial skills and other technical capabilities may present challenges. These conditions are further exacerbated by an inability to achieve agreement and arrive at a consensus over issues related to access, leadership, technical standards, and training.

Implementing a GIS thus has the potential to have far-reaching impacts within an organization. It may necessitate organizational restructuring that can lead to rearrangement of responsibilities across organizational boundaries resulting in a new organizational philosophy, new lines of communication, and redistribution of business functions (Huxhold and Levinsohn 1995). For an organization-wide GIS implementation project to be successful, it is essential that an organization has the capability to endure such conditions.

Key project management strategies are very important here. Interorganizational issues essential for an effective GIS implementation include an assurance of long-term funding, a well-formulated project plan, realistic goals, a project that progresses in phases, and prevention of individual agendas and conflicts. Other important interorganizational issues include a clear knowledge of who is contributing what to the project, persistent communication with all project stakeholders, commitment, stability of leadership, clear protocols about data access and ownership, and clear and transparent database responsibilities. It is absolutely important for all stakeholders who are involved in the project to know every stakeholder's contribution to the project in advance. Concerns are bound to arise among members that individual contributions are commensurate with their size. It is rare that a single department will be willing to take responsibility for excessive expenses in developing any joint components of a project. Questions pertaining to each partner's roles and responsibilities and project output must be addressed as soon as possible.

Control structure must also be established. The question of which department will house the GIS and the shared database must be decided and concerns about control over the GIS and shared database must be addressed early on. One way to do this is to establish a common ground for all stakeholders involved and create a common working language for communicating and negotiating positions, solutions, and ideas throughout the life cycle of the project. Every participant in a project must feel empowered to plan, make decisions, and bring them to realization. If members feel left out or their needs are not addressed, the likelihood of partnerships failing is very high. Equally important are issues regarding shared responsibilities about project costs and duties, particularly those pertaining to geospatial data. Interorganizational relationships that clearly acknowledge member responsibilities are likely to be more effective.

One must also pay close attention to departments responsible for acquiring and maintaining data and ensure that they have the requisite resources to perform their tasks. Generally, departments that spearhead data acquisition and maintenance start incurring additional workloads, expenses, and responsibilities over time and if a perception of inequity in commitments and workload arises among project participants, there is bound to be a negative environment that will impact the progress of the project. Shared responsibilities must be laid out early in a project. It is best to refrain from practices that would address issues related to shared roles and responsibilities as the project progresses without any protocols, as it will give rise to problems in the future. Implementation of a GIS project is thus dominated by interorganizational dynamics.

Successful adoption of a GIS also depends upon how organizational members accept and utilize it. Perception of personal benefits is an important factor for individuals adopting GIS. Benefits such as salary raise, advancement in position, personal satisfaction, and improved professional prestige act as incentives for individuals adopting GIS. People who have a strong computer background and experience are more likely to adopt the technology. Similarly, individuals who are open to change, ready to take on a challenge, and willing to adapt to a new environment are more likely to adopt GIS.

Ultimately, building multidepartmental or multiorganizational partnerships has the potential to give rise to power shifts, turf wars, and conflicts over authority and distribution of power. The key to an effective GIS project is to navigate cautiously through these complex and messy issues. Also crucial is to recognize that GIS adoption is dependent upon the interaction between the technology and its users within specific cultural and organizational contexts. Several studies have highlighted the importance of differing conditions between different countries (Sahay and Walsham 1996, Hoeschele 2000, Mukherjee 2020) and its impact in shaping GIS adoption, particularly for countries in the Global South. Even if GIS vendors try their best to create a software product that is user-friendly and can be easily implemented and utilized, there are still significant social and cultural impediments to its adoption and use in many countries because geospatial softwares and applications have been traditionally created and developed to meet Western world user requirements. Moreover, a culture of low compensation and negative attitudes towards access to information, privacy, and training may not only lead to poorly educated and poorly motivated personnel but also create an overall negative environment impeding GIS adoption and use.

GIS and Society

The rise in GIS use in the 1970s and 1980s led to enthusiastic analysis that heralded the benefits of GIS. Certain geographers were particularly excited about GIS since they saw it as a technology paving the path for a scientific and quantitative methological resurgence within the discipline of geography (Dobson 1983, Openshaw 1991). This led to passionate discussions within the geography community and the formation of two camps within the discipline: the fervent proponents and the vehement opponents of GIS. Concerns were raised that those who heralded the benefits of GIS failed to address the social, political, and technological impacts of its applications. Opponents pointed out the problematic nature of the data handling and processing techniques that are a fundamental element of GIS.

The problematic nature of techniques and processes inherent in a GIS was discussed by John Pickles in his seminal work, *Ground Truth* (1995). GIS was critiqued for its antidemocratic nature due to the differential access to the technology. If you pay close attention to the theories and concepts involved in GIS use, you will see that GIS involves a steep learning curve and not everyone has the ability or the means to be able to use it to the same degree. Thus, by virtue of its nature, it gives an advantage to a certain class of users. In order to be able to use GIS, one needs to be computer savvy, have a certain level of education, have access to technology, and possess technical acumen. It facilitates differential access to the necessary equipment, knowledge, and comfort in working with GIS (Sheppard 2006). It tends to marginalize users along the axes of education, geographic region, age, economic means, gender, and race, thereby creating a digital divide within society.

GIS was also critiqued for including certain types of knowledge, representations of space, reasoning, and logic in its processes that were predominantly quantitative in nature while excluding other forms of knowledge such as qualitative information and non-Western concepts of space and reasoning. Human geographers in particular were concerned that GIS failed to accommodate nontraditional representations of geographic data such as sketch maps, mental maps, narrative and oral histories, and pictorial images. Therefore, the criticism was that endorsing GIS as an acceptable tool for all kinds of data and information processing, and making decisions based on those, reinforced a specific form of knowledge and reasoning in decision-making at the expense of other forms of knowledge. Spatial decision-making that follows any GIS use is therefore associated with power relationships and politics and is heavily influenced by disparate access to GIS and geospatial data.

GIS use does not occur in a vacuum devoid of its environment and it is neither objective nor value free. GIS use and decision-making are dependent upon the different options that users may have and decisions a user may make. These options and decisions are influenced by certain restrictions imposed by the very nature of GIS about the types of data models, attributes, data characteristics, scale, and analytical approach and outcomes that are produced from the resulting analysis.

Concerns have also been raised about the surveillant capabilities of GIS and its ability to access, integrate, and combine databases from different sources and promote widespread use and visualization of unregulated data. For instance, take a look at the use of geospatial technologies by different governments the world over during the COVID-19 pandemic. Governments have utilized a variety of different geospatial and location technologies in their so-called war against COVID. These technologies are being used for geofencing, contact tracing, tracking, monitoring, and managing crowds for strict enforcement of quarantine and lockdown norms. But have these technologies really helped to control the pandemic? Do they really help to curb the spread of the virus? Is it even ethical to surveil and monitor the movement of people, and for what purposes? What about the ethics of collecting location information from individual smartphones? Who is responsible for individual privacy and security of vast amounts of location data collected via applications installed on smartphones? Or is this an opportunity for governments to roll back privacy laws?

Human geographers and social theorists emphasized the lack of attention to ethical questions that arise as GIS organizations, processes, and methods are embedded into social, political, and economic fields of inquiry. Particularly concerning is the central role of private proprietary companies in the implementation of GIS and creating a hidden technocracy. Proprietary GIS applications are developed under certain assumptions and logic, and the closer an organization's environment is to those assumptions and logic, the more appropriate and well suited will be the implementation. For instance, urban GIS based on the southwestern United States contains assumptions and logic that may be irrelevant and will not work for a South Asian urban local body. Social theorists cautioned that early GIS applications indicated an artificial disconnect among people, society, and technology and missed the important social, cultural, economic, and political contexts that influence technology and the people who develop and use this technology. Moreover, it is argued that GIS technology facilitates an external vantage providing its users a "god's eye view." As argued in the feminist critiques of Bondi and Domosh (1992, 202):

> [GIS] implies the existence of an external vantage point from which everything on earth can be uniquely fixed. It is a "god's eye view," from which all problems are solvable provided we possess and have sufficient computer power to process all the appropriate data.

For instance, in their "war" against the COVID-19 pandemic, state and local government agencies in India have made extensive use of COVID "war rooms." Figure 8.1 shows a COVID war room in the Indian city of Jaipur. These war rooms are equipped with GIS technology to receive data from cameras stationed at road junctions, from healthcare workers in the field and on the ground, and generated from the myriad of COVID-19 applications. The war rooms thus receive massive amounts of data which is then processed and examined by government officials from the comfort of the COVID war room and used for surveillance, tracking, and tracing purposes. The COVID war rooms with their advanced technologies and data feeds have in fact provided the government officials with a "god's eye view" of the city, allowing them to govern from a distance.

This intellectual divide between GIS opponents and proponents was discussed at a meeting on "GIS and Society." The meeting sponsored by the National Center for Geographic Information and Analysis (NCGIA) was held at Friday Harbor, Washington, in November 1993. The conference brought the proponents and opponents of GIS together to the table and led to a proposal for an NCGIA Initiative 19. The goal of Initiative 19 was to study the social consequences of GIS, paving the path for developing a GIS and society research agenda (Mukherjee 2015).

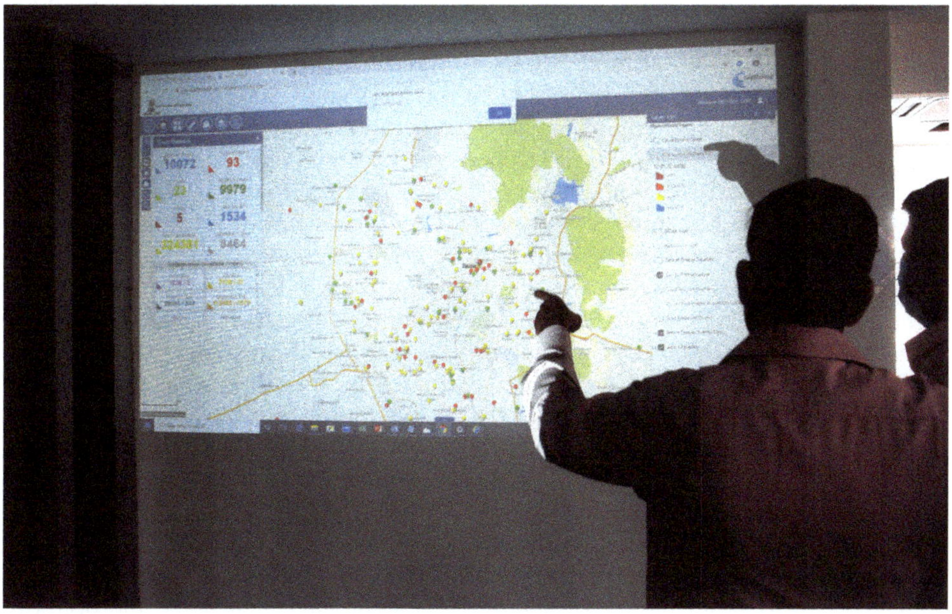

FIGURE 8.1 COVID war room in the Indian city of Jaipur.

Credit: Himanshu Vyas / Hindustan Times via Getty Images

In 1998, the University Consortium for Geographic Information Science (UCGIS) included a detailed GIS and society research agenda within its definition of GIS research consisting of critical social theory, social history of GIS, legal and ethical issues, institutional issues, and public participation GIS (Sheppard 2006). The NCGIA initiative was one of several platforms to give rise to an active GIS and society research agenda where researchers explored the relationship between GIS and society. Several themes emerged under the GIS and society research umbrella. Since the Friday Harbor meeting, those examining the relationship between GIS and society have conducted an active examination of a number of critiques raised by social theorists.

GIS, having adopted cartographic principles, represents space as a collection of points depicting places on earth associated with a set of attributes. It rarely represents space as a "relational mapping of the interdependencies between those places" (Sheppard 2006). For instance, Native American perceptions of space are inconsistent with representations of space and location used in GIS, making GIS inappropriate for representing Native life and worlds. Similarly, data classification categories developed for remote sensing techniques may not be applicable in all instances, particularly in non-Western worlds. Land use–land cover classification categories inherent within remote sensing and GIS softwares do not always align with actual land use on the ground. Because the classification categories have been developed in the Western world, they predominantly represent the Western environment, failing to take into consideration the actual setting in a non-Western environment. Simple terms such as range, urban, road, shrub, or grassland represent different things in different places and can be interpreted in different ways in different sociopolitical and institutional contexts.

For example, land use–land cover classification of a rural area in northwestern India may assign a group of pixels as barren land. However, for the locals that piece of "barren" land is sacred, as it consists of shrubs that provide nutritious berries, part of their staple diet, and the leaves serve as fodder for their goats. So, for the locals it is sacred land, not barren land. But for government officials from the forestry department who are very much influenced by scientific tools and techniques and more importantly by Western world approaches, the classification of land as barren is not problematic. Thus, the meaning assigned to geospatial data or the output from a software application will vary depending on the social and institutional contexts.

Context plays a vital role. This happens because we are all influenced by our environment and just like any other technology, GIS represents the biases that have subconsciously crept into the application during its development and symbolizes those biases. If you look into the history of GIS development, GIS was developed for use by large agencies such as the military, state agencies, universities, and private corporations in the Western world. It was not developed for use by ordinary members of the public or marginalized sections of society. Nonetheless, such members of society are consciously or subconsciously using some sort of a geospatial data or application by virtue of the digital age that we are in. At times this use may also be forced upon them, like the geolocation-based COVID-19 applications that have become part of the daily routine of citizens in many countries.

Thus, there are many challenges to accessing and using GIS. In order to examine this barrier to access, the research agenda of Public Participation GIS (PPGIS) has emerged, which is a direct result of many years of research efforts investigating the societal impacts of GIS. PPGIS is an area of research and practice that examines access to geographic information technologies and their social and political impacts. It is defined as "the process of acquisition and application of hardware, software and spatial data for GIS analysis as well as the social and political contexts in which GIS is being employed which influences its use and impacts" (Elwood and Ghose 2004, 20).

PPGIS has become one of the most developed research agendas within the GIS and society umbrella, and a number of subthemes have emerged within PPGIS research (Mukherjee 2015). Some have examined if and how GIS may be used to empower marginalized sections of society who have traditionally been excluded from the decision-making and creation process (Elwood and Leitner 1998, 2003; Elwood 2006, 2008; Ghose 2001), while others have investigated if an attempt to empower marginalized communities further consolidates power in the hands of existing dominant actors (Ghose 2005; Elwood 2002, 2008). Work within the PPGIS research umbrella has shown that facilitating GIS access to nontraditional users enables the creation and development of alternative forms of knowledge and the possibility of its inclusion in the decision-making process.

An extensive body of work has investigated GIS use by inner-city neighborhood, community-based, and nonprofit organizations to challenge dominant discourse, insert their voices into decision-making affecting their neighborhoods, and empower themselves. GIS can be used in different ways here. For instance, community-based organizations use GIS or the results of GIS analysis to produce new information; visually depict the social and economic condition of neighborhoods; investigate social, political, and economic processes influencing underlying neighborhood conditions; and utilize such information to draw attention of policymakers and state actors in an attempt to change neighborhood conditions (Ghose 2005, Elwood 2002).

PPGIS practices are shaped by organizational contexts discussed in the previous section along with the external contextual environment (Elwood and Ghose 2004). The external contextual environment includes factors such as network of relationships between organizations, ability of an organization to navigate the political environment, ability of an organization to recognize and reach out to partners who can assist it in its GIS use, leadership who see the value of investing in a GIS, and ability of an organization to identify existing opportunities for GIS use and apply it to the organization's neighborhood planning process, organizational stability, and mission. Such external contextual factors coupled with the internal organizational contexts shape access to GIS and spatial data during a participatory GIS process.

Participatory GIS practices involve multiple individuals and agencies working together. Like any collaborative activity that involves participants with varying roles, responsibilities, and expertise, there is bound to be a power structure involved. GIS use is an important component in shaping the struggles and negotiations that occur in any participatory GIS process. It provides traditionally marginalized communities opportunities to represent their interests, influence dominant opinions of their geographic space, and participate in planning and decision-making that impacts them. This allows them to gain greater control over their space and resources and renegotiate with dominant state actors (Ghose 2005). Thus, it provides an avenue to empower marginalized communities.

However, one must tread with caution when engaging in participatory GIS practices. Are marginalized communities actually empowered by using GIS? Do they in reality have control over the database creation and mapping and visualization activities, or does it make them overly dependent on outside actors? If a participatory GIS practice must empower, it must be approached following a bottom-up strategy. Such an approach is driven by demand rather than technology.

Several constraints to GIS access are particularly pronounced for marginalized communities. Such constraints include limited financial and time resources, and access to individuals with knowledge and training in geospatial data, computer, and GIS use. An important factor that shapes GIS use within such communities is the presence of institutions and resources that support or facilitate participatory GIS practices, availability of data, and availability of support to gain access and use information technologies.

The process of GIS use and its impacts does not follow a "one size fits all" model. As emphasized by Chrisman (2005, 28), "Multiple social forces interact in the process of developing a complex technology such as GIS. Implementation of any technology depends on the specific local environment."

GIS adoption and use within local governments is also shaped by multiple contextual factors that lie outside the confines of an organization. Geospatial data creation and GIS use within such agencies is not merely shaped by technical factors—it also includes managerial skills as well as geospatial data handling skills. Moreover, these agencies are uniquely shaped by administrative, institutional, political, and economic factors. Their GIS use is a reflection of their response as they react to accommodate changing political and economic environments.

The factors shaping GIS use by marginalized communities or government organizations are also evident in the case of GIS use by organizations in the non-Western world. In the past decade there has been a rise in GIS use by government agencies in countries in the Global South such as China and India. GIS use by government agencies in these countries is largely shaped by the political environment of the country. Countries like India are undergoing a huge urban transformation, and political leaders consider technological intervention a solution to manage the rapidly changing urban landscape of the country.

India has embarked on an ambitious informationization path (Mukherjee 2020), and technologies like GIS play a very important role. Local government agencies are encouraged to make use of GIS in a variety of different projects facilitated by the central government. However, GIS adoption and use by government agencies at the local level is quite variable despite similar incentives being provided by the central government to all local government agencies. This is because GIS use is not a universal process—what works in an organization in one location may not work in another organization at another location. This is due to the fact that two organizations in two different locations will have completely different circumstances surrounding them, which in turn will influence their GIS use in different ways.

GIS use in this case is a complex process and is shaped by the national political agenda, broader local political conditions, organizational cultures, and institutional structures. Factors such as the willingness to adopt a change, strong support from senior officials and political leaders, the presence of a champion who will promote GIS use internally, and the presence of a bureaucrat who has the ability to foresee the political as well as organizational benefits of GIS use and is able to navigate the institutional structures to procure the necessary resources for GIS adoption and use are crucial in these circumstances.

GIS is thus more than just a tool or technology. Its implementation and use and creation of geospatial information and knowledge in the process involve a unique relationship with the contextual environment surrounding the organization and society at large. Each GIS implementation is a distinct combination of social, political, institutional, and technological factors and hence no two GIS implementations are similar. Each instance of a GIS implementation will be as unique as the circumstances surrounding it. GIS and society share a two-directional relationship and both influence each other in unique ways (Chrisman 2005). GIS use thus does not occur in a vacuum devoid of its surrounding circumstances. It is influenced by political goals, institutional structures, administrative norms, and social contexts. Each implementation, adoption, and use of GIS is shaped by the internal organizational environment as well as the external sociopolitical contexts.

As a GIS user it is not enough to have access to the most robust hardware and sophisticated software. Of course, you need the hardware and software to be able to use GIS. But a successful implementation, use, and adoption of GIS depends on a number of other factors. Context is very important. Your organization must have the need for a GIS, and it has to be ready to accept and adapt to the changes that come with GIS implementation. Members within the organization must be ready to change their traditional ways and adapt to the changes too.

Sociopolitical and institutional contexts also shape GIS use in a myriad of ways. GIS use has political and institutional meaning. For instance, for a community organization, GIS use is a way to showcase to funding agencies and state officials the conditions in their neighborhood and campaign for resources to improve these conditions. For local government agencies in India, GIS adoption and use is a way to showcase themselves as a modern government agency. Here GIS use is driven by the political ambitions of government officials, institutional goals and objectives that put an emphasis on GIS use, the presence of bureaucrats who are able to envision the benefits of GIS use and collaborate with local political leaders to gain access to the necessary resources to drive their GIS use, and the presence of officials within the local government agency who are willing and able to drive the agenda forward.

All of these conditions must come together to form a conducive ecosystem for GIS use. The external contexts influence the internal contexts of GIS use. The site for GIS use could be either a community-based neighborhood organization in North America or a local government organization in India. As you can imagine, it is not possible at all times and in all instances to have all of the elements that form a conducive environment for GIS adoption and use present or for them to come together. Therefore, every instance of GIS use will vary, and as a GIS user it is vital that you are cognizant and acknowledge the variety of forces that come together to shape every instance of GIS use. The *S* in GIS is very important. Science and system go together. They cannot be separated, and every GIS endeavor must acknowledge this.

Questions and Activities

1. Let's revisit the scenarios we started this chapter with:
    a. You are visiting a new city and as you embark from the plane the first thing you must do is install a GIS-based application that will monitor your every movement and post your location to a server.
    b. You are required to take a selfie and then upload the photograph using the application on your phone which will geotag the photograph, allowing officials to know of your whereabouts.
    c. You visit a retail store that sells sports equipment to purchase golf clubs. Soon after, you start receiving marketing material from different retailers in your mail advertising for different golf equipment. You also start receiving marketing material from different sporting goods stores, even those that are not in your vicinity.
    d. You work for a neighborhood-based organization. It is a small nonprofit organization and your role is to make future plans for the neighborhood with the goal to improve neighborhood quality by focusing on current issues such as rental properties and delinquent properties. You must prepare planning documents and submit them for grant funding. The funding organization tends to prefer grant applications that have GIS-based maps.

    In each of the instances described here, explain the different factors that may influence GIS use.

2. Think of three activities from your daily life that involve the use of geospatial data or technology. Are there any alternate data or technology instead of the one you are using? What are some of the factors that have influenced your use of this specific data or technology?
3. What are contextual factors? List and explain in your own words the different contextual factors that may influence GIS use in an organization. Remember to discuss what kind of an organization too.
4. There are many resources for the GIS and society research agenda. One such resource is the UCGIS. Visit the UCGIS website at https://www.ucgis.org and read about the mission of UCGIS. What is the role of UCGIS in GIS education?
5. Another valuable resource is the National Center for Geographic Information and Analysis. Visit the NCGIA website at https://umaine.edu/scis/ncgia and read about the history of NCGIA. What are the three NCGIA member institutions? Visit the websites of each NCGIA center and review their roles, research areas, and projects.
6. Review the research projects of NCGIA-Maine. Pick any three projects of interest. Write a paragraph describing each of the projects. In your opinion, how do these projects fit the GIS and society research agenda? Discuss with another classmate.
7. Another valuable resource is PPGIS.net. Visit https://www.ppgis.net and thoroughly review the contents. Go over the videos posted in the "Videos" section. Write a one-page narrative explaining your understanding of PPGIS.
8. The use of technology during the COVID-19 pandemic has been covered by many news outlets. Several online news articles have discussed the use of technologies during this time for different purposes. One such article is available at https://www.cnbc.com/2020/03/27/coronavirus-surveillance-used-by-governments-to-fight-pandemic-privacy-concerns.html. Read this article.

   How did different countries in the world mobilize the use of technologies during the COVID-19 pandemic? What according to you is problematic about these technological applications?

   Can you draw comparisons between the technology use during the COVID-19 pandemic and the massive use of geospatial technology that is so well integrated into our daily lives today? What are the societal consequences of such use?

Key Terms

innovation diffusion (p. 175)
interdepartmental (p. 176)
National Center for Geographic Information and Analysis (NCGIA) (p. 179)
Public Participation GIS (PPGIS) (p. 180)
University Consortium for Geographic Information Science (UCGIS) (p. 180)

References

Antenucci John, Kay Brown, and Peter Croswell. 1991. *Geographic Information Systems: A Guide to the Technology.* New York: Van Nostrand Reinhold.

Bondi, Liz, and Mona Domosh. 1992. "Other Figures in Other Places: On Feminism, Postmodernism and Geography." *Environment and Planning D: Society and Space* 10, no. 2: 199–213.

Budic, Zorica. 1998. "The Impact of GIS Technology." *Environment and Planning B: Urban Analytics and City Science* 25, no. 5: 681–92.

———. 1999. "Evaluating the Effects of GIS Technology: Review of Methods." *Journal of Planning Literature* 13, no. 3: 284–95.
Budic, Zorica, and David Godschalk. 1994. "Implementation and Management Effectiveness in Adoption of GIS Technology in Local Governments." *Computers, Environment and Urban Systems* 18, no. 5: 285–304.
Calkins, Hugh, and Nancy Obermeyer. 1991. "Taxonomy for Surveying the Use and Value of Geographical Information." *International Journal of Geographical Information Systems* 5: 341–51.
Campbell, Heather. 1991. "Organizational Issues in Managing Geographic Information." In *Handling Geographic Information*, edited by Ian Masser and Michael Blakemore, 259–82. London: Longman.
Campbell, Heather, and Ian Masser. 1995. *GIS and Organizations: How Effective Are GIS in Practice?* London: Taylor and Francis.
Chrisman, Nicholas. 1997. *Exploring Geographic Information Systems.* New York: John Wiley & Sons.
———. 1999. "What Does 'GIS' Mean?" *Transactions in GIS* 3, no. 2: 175–86.
———. 2005. "Full Circle: More Than Just Social Implications of GIS." *Cartographica* 40, no. 4: 23–35.
Clapp, James, John McLaughlin, Jerome Sullivan, and Alan Vonderohe. 1989. "Toward a Method for the Evaluation of Multipurpose Land Information Systems." *Journal of the Urban and Regional Information Systems Association* 1: 39–45.
Croswell, Peter. 1991. "Obstacles to GIS Implementation and Guidelines to Increase the Opportunities for Success." *Journal of the Urban and Regional Information Systems Association* 3, no. 1: 43–56.
Dobson, Jerome. 1983. "Automated Geography." *The Professional Geographer* 35: 135–43.
Elwood, Sarah. 2002. "Neighborhood Revitalization through 'Collaboration': Assessing the Implications of Neoliberal Urban Policy at the Grassroots." *GeoJournal* 58: 121–30.
———. 2006. "Critical Issues in Participatory GIS: Deconstructions, Reconstructions, and New Research Directions." *Transactions in GIS* 10, no. 5: 693–708.
———. 2008. "Grassroots Groups as Stakeholders in Spatial Data Infrastructures: Challenges and Opportunities for Local Data Development and Sharing." *International Journal of Geographical Information Science* 22, no. 1: 71–90.
Elwood, Sarah, and Rina Ghose. 2004. "PPGIS in Community Development Planning: Framing the Organization Context." *Cartographica* 38, no. 3 & 4: 19–33.
Elwood, Sarah, and Helga Leitner. 1998. "GIS and Community-Based Planning: Exploring the Diversity of Neighborhood Perspectives and Needs." *Cartography and Geographic Information Systems* 25, no. 2: 77–88.
———. 2003. "GIS and Spatial Knowledge Production for Neighborhood Revitalization: Negotiating State Priorities and Neighborhood Visions." *Journal of Urban Affairs* 25, no. 2: 139–57.
Ghose, Rina. 2001. "Use of Information Technology for Community Empowerment: Transforming Geographic Information Systems into Community Information Systems." *Transactions in GIS* 5, no. 1: 141–63.
———. 2005. "The Complexities of Citizen Participation through Collaborative Governance." *Space and Polity* 9, no. 1: 61–75.
Gilfoyle, Ian, and Peter Thorpe. 2004. *Geographic Information Management in Local Government.* Boca Raton, FL: CRC Press.
Hoeschele, Wolfgang. 2000. "Geographic Information Engineering and Social Ground Truth in Attappadi, Kerala State, India." *Annals of the Association of American Geographers* 90, no. 2: 293–321.
Huxhold, William, and Allan Levinsohn, 1995. *Managing Geographic Information Systems Projects.* Oxford: Oxford University Press.
Innes, Judith, and David Simpson. 1993. "Implementing GIS for Planning: Lessons from the History of Technological Innovation." *Journal of the American Planning Association* 59, no. 2: 230–36.

Masser, Ian, and Harlan Onsrud. 1993. *Diffusion and Use of Geographic Information Technologies.* Dordrecht, Boston, and London: Kluwer Academic Publishers.
Mukherjee, Falguni. 2015. "Public Participatory GIS." *Geography Compass* 9, no. 7: 384–94.
———. 2020. "Institutional Networks of Association of GIS Use: The Case of an Urban Local Body in India." *Annals of the American Association of Geographers* 110, no. 5: 1445–63.
Obermeyer, Nancy, and Jeffrey Pinto. 1994. *Managing Geographic Information Systems.* New York: Guilford Press.
Onsrud, Harlan, and Jeffery Pinto. 1993. "Evaluating Correlates of GIS Adoption Success and the Decision Process of GIS Acquisition." *Journal of the Urban and Regional Information Systems Association* 5, no. 1: 8–39.
Openshaw, Stan. 1991. "Using GIS to Put Humpty-Dumpty Together Again." *Environment and Planning A: Economy and Space* 23, no. 5: 621–28.
Pickles, John. 1995. *Ground Truth: The Social Implications of Geographical Information Systems.* New York: Guilford Press.
Pinto, Jeffery, and Harlan Onsrud. 1997. "In Search of the Dependent Variable: Toward Synthesis in GIS Implementation Research." In *Geographic Information Research: Bridging the Atlantic*, edited by Massimo Craglia and Helen Couclelis, 129–45. London: Taylor and Francis.
Rogers, Everett. 1993. "The Diffusion of Innovations Model." In *Diffusion and Use of Geographic Information Technologies*, edited by Ian Masser and Harlan Onsrud, 9–24. Boston: Kluwer Academic Publishers.
Sahay, Sundeep, and Geoff Walsham. 1996. "Implementation of GIS in India: Organizational Issues and Implications." *International Journal of Geographical Information Systems* 10, no. 4: 385–404.
Sheppard, Eric. 2006. "Knowledge Production through Critical GIS: Genealogy and Prospects." *Cartographica* 40, no. 4: 5–21.

GIS: Possibilities and the Future

9

Learning Objectives

- Explain how the geospatial industry has evolved over time
- Discuss the types of GIS degrees offered by institutions and what one can do with a GIS degree
- Identify the typical job titles in the GIS industry
- Understand digital transformation and its role in GIS
- Explain terms like big data and internet of things (IoT)
- Understand cloud technology, types of cloud services, and API
- Articulate the role of cloud computing in GIS
- Describe different cloud services available for geospatial technology: ArcGIS Online, StoryMaps, ArcGIS Location Platform, ArcGIS Hub
- Explain how digital transformation, cloud computing, big data, and IoT have changed the mechanisms used for geospatial data capture, storage, and delivery

The World of GIS

There is no doubt that GIS has evolved dramatically over the years. Since the first use of geospatial concepts by the Canada Geographic Information System (CGIS) and the creation of SYMAP at Northwestern University, GIS has come a long way. It is a rapidly advancing field that has integrated with many other disciplines and fields such as public health, disaster management, law enforcement, environmental management, resource allocation, forestry, geology, and so on. The integration of these fields with GIS has transformed not only the field of GIS but has also other fields and disciplines. It has opened novel practices and new ways of conducting business within these fields. Thus, a complementary relation has evolved between GIS and other disciplines.

GIS has grown into a sophisticated multibillion-dollar industry. It is both a technology and a science, and the vast gamut of applications that utilize geospatial data and GIS techniques have time and again attested to this reality. It is so well integrated into our society and our daily lives that you may not even realize that you are actually utilizing an application that uses geospatial data and technology. Today geospatial technology is seamlessly used with other tools and technology.

Consider your daily activities and pay close attention to examine the practices that use geospatial data or technology. You will be surprised. Every time you look up a new coffee shop, restaurant, car service center, or grocery store or travel to a new location or go on a hiking or biking trip, do you find yourself making use of paper maps? Every time you go on a road trip, do you consider carrying a paper map with you? Are you worried that if you were lost during your trip, you must wait to meet someone to find your way back? If you were running low on gas, you will find the nearest gas station in a matter of minutes. If you were looking for a new house to buy, you can check out the neighborhood, look up home prices, and check the amenities such as parks, golf courses, and tennis courts in and around the neighborhood without leaving your home. This is possible because of the use of geospatial data and technology that is well integrated with other applications, and you are using it on a daily basis. GPS devices, satellite imageries, data analytics, and digital maps have become a part of our daily lives in today's digital world.

What Can You Do with a GIS Degree?

Growth in the GIS field has been accompanied by tremendous growth in the GIS community, both professional and academic. Today GIS is offered in different educational settings from K–12 to graduate programs. Academic programs in GIS are offered in person, online, and in **hybrid models**. These include certificates and associate's, undergraduate, and graduate degrees. GIS is also an important part of STEM education.

A tremendous amount of resources are available today for GIS education. Numerous GIS academic program databases now exist and assist prospective students in understanding the myriad of choices available to them. The University Consortium for Geographic Information Science (UCGIS) (https://www.ucgis.org), Esri's GIS Education (https://www.esri.com/en-us/industries/education/overview), and Esri's K12 GIS organization (https://k12.maps.arcgis.com/home/index.html) are a few of the key resources for GIS education. Training courses provided by Esri's virtual campus are also a useful means of developing geospatial competency.

Similarly, other agencies that create and disseminate geospatial data such as the USGS, NASA, and Google have also created lesson plans that are available for K–12 teachers and college educators to adopt in their classrooms. These lessons provide a road map to educators for integrating different geospatial data and products into their curriculum on geospatial concepts and theories. As you can imagine, this has given rise to a large body of knowledge on GIS, including UCGIS's GIS&T Body of Knowledge and GIScience/GIS/Geospatial Instructional Resources.

Today GIS is well integrated into the curricula of higher education institutions worldwide. As the number of GIS programs has expanded over the years, guidelines have been developed to assist schools and academic institutions interested in starting GIS programs. For instance, the US Department of Labor has developed the **Geospatial Technology Competency Model (GTCM)** to provide institutions a framework for geospatial skills that are vital in today's workforce. The UCGIS also provides valuable resources for integrating GIS into higher education curricula.

If you are interested in further exploring GIS after an introductory GIS class, you can do so in different ways. Take a few more intermediate and advanced undergraduate GIS classes offered in your department. See if your institution has the option of a GIS or geospatial minor or a concentration. Courses that focus on geospatial technology are mainly offered in geography departments of colleges and universities. In certain institutions you may find course offerings that focus on geospatial technology within environmental, geosciences, civil engineering, forestry, urban planning, and geology programs. GIS education is delivered primarily by following three key models: on-campus GIS certificate programs, on-campus undergraduate and graduate degree programs, and online undergraduate and graduate degree programs.

Undergraduate programs provide different options. Students may take several GIS courses as part of their major. Alternatively, students can pursue either a minor or a degree with a concentration in geospatial technology. In the latter case, students focus on their discipline-specific courses while taking a specified set of geospatial classes that give them a basic introduction to geospatial technology.

This way they can apply for jobs in their own field with a basic training to utilize geospatial technology. For instance, the Department of Environmental and Geosciences at Sam Houston State University offers a bachelor's degree in geography, in environmental science, and in geology. Students pursuing a bachelor's degree in geography may choose from one of the three-track curricula as their concentration: human geography, environmental geography, and geospatial information sciences. Students pursuing a bachelor's degree in geology or environmental science may choose a minor in geospatial information science.

A GIS certificate is one of the traditional means of imparting professional GIS education. Certificate programs give students the option of taking a limited number of classes, typically five to seven, that focus on specific skill sets or application areas. Students take the specific courses in a structured manner to get a GIS certificate and obtain a basic introduction to theory and applications. Certificate programs are offered at both undergraduate and graduate levels and vary from school to school. With geospatial technology so well integrated into other disciplines, students from programs other than geography may take a few GIS courses with the goal of obtaining a GIS certificate and also derive a basic foundation in geospatial technology. This way they can apply for jobs in their own discipline with the added advantage of having familiarity with geospatial technology.

A graduate GIS certificate may be the first step towards a more intensive GIS graduate program. For instance, the Department of Environmental and Geosciences at Sam Houston State University offers a GIS certificate at both the undergraduate and graduate levels. The graduate GIS certificate involves students taking three required GIS graduate courses: an introductory GIS class, an advanced GIS class, and a spatial analysis class. Thereafter students take two additional courses from a list of available graduate GIS classes, including visualization, image processing, Lidar and radar, GIS programming, web applications, database design, project management, and model builder.

An increasing number of community colleges are offering GIS certificates. Professional GIS certification is also available through nonacademic institutions. One such institution is the GIS Certification Institute that offers GIS Professional (GISP) certification. The GISP certification process involves a rigorous exam and a portfolio review. The portfolio must describe the applicant's background, education, professional experience, and contribution to the GIS field. Similarly, Esri offers technical certification based on an exam. Esri technical certification exams are grouped into three categories: desktop, enterprise, and specialty. Accordingly, certifications are offered at three levels: entry, associate, and professional.

The past two decades have seen a rapid growth in the number of GIS graduate programs offered at academic institutions. Traditionally, graduate programs have been offered on campus with formal in-person lectures and labs in classrooms. Students gain hands-on experience by using GIS software in a lab setting with other students and are supervised by an instructor or teaching assistant. They have access to resources available on campus and face-to-face interaction with faculty members and their peers. Students may work with faculty members on research projects and gain valuable research experience. Like the GIS certificate programs, graduate programs vary from school to school. Some programs have a thesis option for students, and some also have a non-thesis option and students can choose between both.

Recently more graduate programs have begun following a distance education model for both certificates and master's degrees in GIS. This model works well for nontraditional students who are trying to balance their professional and personal commitments. The distance education model therefore works well for institutions that would like to establish a professionally oriented graduate program. It also works well for institutions that are located in remote or rural areas, making it difficult for students who wish to pursue a master's degree in GIS but cannot commit to relocating to the campus during the duration of the degree. A good example of an online graduate program in GIS is the Penn State World Campus that offers both an online GIS certificate and an MGIS degree.

Some programs offer both the options: a traditional model and a distance education model. For instance, the Department of Environmental and Geosciences at Sam Houston State University offers a master's in GIS. Graduate students interested in pursuing a master's degree have two options: a traditional track that offers in-person classes and a professional track that is offered completely online and is geared towards working professionals. The graduate program offers courses that focus on spatial analysis, advanced GIS analysis, GIS programming, project management, GIS web applications, model builder, Lidar and radar, spatial database design, and GIS applications in law enforcement and the energy industry.

The integration and vast availability of curricula on geospatial technology in higher education has helped build a trained workforce for the burgeoning GIS industry and other allied industries. The field of GIS is not only evolving but also expanding at a rapid pace. Moreover, the seamless integration of GIS with other industries has sparked a growth in other fields in conjunction with the field of GIS. In fact, the geospatial industry is one of the fastest-growing applied technology sectors in both business and government, with projected demand for individuals trained in GIS far exceeding the supply. Based on a recent report from Research and Markets, the geospatial industry is projected to grow globally at a compound annual growth rate (CAGR) of 13.2% from 2019 to 2024. According to P & S Market research, the global GIS industry is projected to reach $25.6 billion by 2030. Thus, there continues to be a high demand for professionals with geospatial knowledge and skills.

A variety of roles in the industry are available for those with knowledge and expertise in geospatial technology. Take a look at the "People in Action" sections in this chapter and in chapter 1. This chapter features three GIS professionals, each in different phases of their career, who entered the geospatial industry at different times and in different ways.

Typical roles or job titles for someone with expertise in geospatial technology listed on the GIS Jobs Clearinghouse webpage (http://www.gjc.org) include GIS specialist, GIS analyst, GIS technician, transportation planner/analyst, geospatial technical lead, GIS developer, assistant state cartographer, GIS/IMS/server applications developer, associate GIS technician, GIS database administrator, GIS team leader, GIS database/systems coordinator, GIS/mapping technician, geospatial data coordinator, web developer, senior geospatial analyst/programmer, GPS manager, remote sensing technician, director of operations, field data collector, GIS supervisor, GIS solutions engineer, planning and GIS specialist, GIS data field technician, GIS programmer, GIS administrator, geospatial project manager, and real estate development analyst. Moreover, a large percentage of job listings for academic positions in geography and environmental science require knowledge of geospatial technology.

The salaries for GIS professionals will vary geographically. Individuals with a master's degree tend to be paid more than those with a bachelor's degree. In 2017 the Urban and Regional Information Systems Association (URISA) conducted a survey to gather data for GIS salaries. The survey is based on 3,060 respondents who were employed full-time in the geospatial industry. According to this survey, the average salary of survey respondents was $70,857, an increase of 15% over 2010. The survey results showed that GISPs on average earned $10,000 more than non-GISPs.

According to multiple web sources (Indeed, Glassdoor, Payscale, Paysa) the average salaries for GIS-related jobs in the Houston area posted in August 2024 were as follows:

GIS Programmer Analyst: $96,929
GIS Developer: $79,300
Senior GIS Analyst: $104,557
GIS Programmer: $65,992
GIS Database Analyst: $68,858
GIS Analyst: $77,972
Geospatial Analyst: $70,299

What Lies Ahead?

It would be an understatement to say that the geospatial industry is evolving at a rapid pace, adapting to growing demands and integrating with other fields and industries. Today you can gain access to geospatial data and technology from anywhere using online resources. Whether it is a satellite imagery or an aerial photograph or Google Earth, you can access a variety of geospatial datasets through the click of your mouse, thanks to the internet.

The internet has fundamentally changed the world and how we interact with one another. In today's digital world, billions of physical devices around the world are connected to the internet and are continuously generating and sharing data. This interconnection of physical devices is called the **internet of things (IoT)** and it has fundamentally transformed the world of geospatial science and technology. It has not just changed the way we access data and technology—it has also transformed how we are able to utilize data and share data and information with others, be it our colleagues within an organization or others outside an organization in the public domain. Coupled with IoT, geospatial technology is a key tool for platforms to leverage data intelligence for digital transformation of operations in the public and private sectors. So, what does the future look like?

GIS and Digital Transformation

So, what is digital transformation? Every business in the industry today is talking about it. **Digital transformation** is the process of utilizing digital technologies to create new or transform existing business processes and operations. It also involves a transformation in the work culture and modification of customer experience. Digital transformation employs a strategic use of technology, people, and processes to fundamentally change how business operations are conducted. IoT, big data, and data analytics are all key components of this process.

At the heart of this "digital revolution" is **data intelligence**, the ability to extract information in order to understand the data so that we can make intelligent decisions and accordingly take actions. If you look around, you will find that we are all constantly connected to the internet through some device or the other. Your mobile phone, laptop, tablet, television, car, and home security system, even refrigerators and doorbells, are all constantly connected to the internet. And this is just your household. Think about all the businesses and industries around you. Imagine the massive amount and variety of data that is generated by all the devices that are constantly connected to the internet every second, every minute, every hour,

365 days a year. Such large volumes of data are known as big data, data that is so large in volume and complexity that traditional methods of analysis and processing cannot handle it efficiently.

Big data is generated in all forms and can be structured or unstructured. However, this type of data contains copious amount of information that can be utilized to generate intelligent insights and make key business decisions, something businesses were not able to do before. So, there is an urgent need to make sense of the massive amount of data that is being generated constantly in our digital world. Businesses in particular have a pressing requirement to make sense of the massive amount of data generated from the devices that are well integrated into our society. Additionally, they must also generate intelligent solutions and actions for their businesses from the information contained in this data.

Traditional data processing softwares do not have the ability to generate this kind of data intelligence, but geospatial technology does. There is no doubt that GIS has unique capabilities of working with location data, thereby providing a geographic context in any analysis. Coupled with that, its capability to map and spatially analyze data makes it a potent tool for digital transformation. Again, take a look at the "People in Action" in this chapter that features three GIS professionals and their use of geospatial technology. You will see the different ways they use geospatial technology at work for different purposes and use a variety of different technologies coupled with GIS.

A GIS can extract data from physical devices such as GPS, sensors, and satellites, including connected devices, UAVs, and vehicles. A GIS can be used to extract, map, and analyze data from such physical devices on a real-time basis too. A massive amount of information contained in the data is constantly being generated from all the devices, connected to either the internet or a satellite. In its raw form the data is too enormous, and it is difficult to decipher any meaning from it. Using GIS we can map this data and perform a variety of analysis, modeling, and visualization. A GIS can be used to understand the spatial relationships between the data and its surrounding environment. Moreover, location information that is part of any geospatial data assists in understanding the contexts and examining the underlying causes responsible for specific trends, patterns, or distributions. In a GIS, data can be overlayered with other datasets, allowing for convenient comparison and visualization. GIS therefore provides a completely new approach to examine and visualize information contained in big data. These unique capabilities are extremely useful for the digital transformation of any business operation.

Consider a business that is involved with marketing or sales operations. By combining census tract level data with consumer demographic information collected by the business, GIS can be used to analyze and visualize specific habits of people. This information is very important for businesses to target specific consumers in specific regions and accordingly create business strategies. For instance, this analysis would be useful for a fast-food restaurant chain operator to determine where the next outlet should be located, for a laundry service company to determine where the next self-service laundry should be established, for a retail real estate company to help its clients determine where to build the next shopping center, for an automotive company to determine where the next car dealership should be located or which specific car model should be promoted in a particular region, or for a manufacturing company to determine which vendor will be able to provide the necessary materials most rapidly. The list is unending. Thus, there is tremendous potential for GIS use in the digital transformation of any business operation. However, only 1% of available big data is actually being used effectively by businesses to extract information and make intelligent decisions (Esri 2020). Thus, a large amount of big data remains untapped.

There is immense scope for businesses to leverage the potential of big data to their advantage. The future lies in making use of geospatial data and technology in conjunction with IoT and deciphering big data for business intelligence.

GIS and the Cloud

Geospatial technology has increasingly become a key component of cloud computing. We are moving towards a trend where you are not going to need a costly software installed on your computer or laptop in order to utilize GIS. All you will need is a reliable internet connection and a web browser to use GIS. This is because everything you will need to utilize GIS will be available to you through a web browser. It is possible because your GIS software is stored on the cloud. Just like the internet has become a part of our daily lives, the cloud too is well integrated in our lives. You may just not realize it.

So, what is the cloud? The cloud is a computing technology wherein the resources required to run an application—the infrastructure, software, and platform—are stored on a server at another location and served to a user via the internet. So your data, software, database, everything that you need to use the data is stored at another location and you access it when you need it through the web. You may not realize it, but you actually use the cloud on a daily basis. Every time you use Gmail to check your emails or YouTube, Netflix, or Amazon Prime to stream a movie or a series, you are utilizing cloud service. For instance, when you log into your Netflix or Amazon Prime account, you can see the movies and series that are available for you to stream and watch. The digital files for those movies and series are stored on the cloud.

Let's say that you want to watch the third season of *The Crown* or the fifth season of *The Great British Baking Show*. You select the series, it streams, and you watch it. So where is the actual video file? It is stored on a server at some location and when you select it, it is streamed to your television through the internet for you to watch it. Similarly, let's say you filmed a video of your friend performing in the school musical show and now want to share the video file with the friend. The file is too large for you to send it as an email attachment. One way to share the video file would be to upload it to your YouTube account and then share the link with your friend. Now your friend can watch the video anytime without downloading the actual file. How is this possible? When you uploaded the video file to YouTube, it was stored on a server at some location. Your friend uses a web browser on either their mobile phone, tablet, laptop, or computer to connect to the server using the link that you provided to view the video.

If you have ever utilized a software product like Norton to back up your files, where do you think your files are being stored for backup? They are also being stored on the cloud. Similarly, if you take photos using your smartphone, you have the option to store them on the cloud; for iPhone users, it is iCloud. And if you want to show those photos to someone, your smartphone will retrieve them from the cloud for you. Dropbox, Google Drive, and OneDrive are all cloud services like iCloud that can be used to store and share files and folders. That way you avoid using up space on your computer and also have the flexibility of sharing the files with anyone. We all use cloud services today in some way or another even if we don't realize it; for instance, checking our emails in Gmail or collaborating with others using Google Docs. All of these activities involve the use of cloud services.

There are three main types of cloud services: software as a service (SaaS), infrastructure as a service (IaaS), and platform as a service (PaaS). In SaaS, a cloud service makes software available to users over the internet. SaaS is used for storing files, documents, and folders and managing web-based emails. Here the software that you need to run an application, such as email or a file backup program, is stored and served to you from the cloud. You don't need to install any software on your personal device. The software and the tools to use the software are stored in the cloud and you access them using a web browser by logging into your account. The software can be accessed from anywhere on any device as long as you have internet connection. Dropbox, Microsoft Office 365, and Gmail are all examples of SaaS cloud service.

In **IaaS**, the infrastructure required to run software tools and applications is stored in the cloud. A user does not need to install the infrastructure such as servers, networking hardware, and other physical hardware on-site. Instead, they are stored in the cloud and users use the service over the internet when needed and pay for the service accordingly. Amazon Web Services (AWS) and Microsoft Azure are examples of IaaS. **PaaS** provides a web-based environment that includes hardware and software tools, data, and interfaces for developers to build software and applications.

Most SaaS cloud services are built on PaaS or IaaS. All three cloud services communicate with one another using what is known as an **application programming interface (API)**. APIs are how different software applications communicate with one another. Every time you use your smartphone to find a location or a route from point A to point B; access a social media website like Facebook, X, or Instagram; or visit the webpage of any business you are using an API. For instance, let's say that you use your smartphone to find a Thai restaurant in your vicinity. Your phone will connect to the internet and send the information to a server. The server will retrieve data based on your query, interpret the data, and send it to your phone. An application in your phone will interpret the data it gets from the server and present it to you. All of this communication back and forth between the application on your phone and the server occurs through API.

A few of the benefits of cloud services include cost effectiveness, productivity, flexibility, scale, performance, reliability, and security. It saves organizations the cost of investing in on-site IT infrastructure and resources. Organizations are thus able to reduce the costs involved in maintaining and managing IT systems. Everything is stored in the cloud and you access everything when you need it. Accordingly, you pay for the services as and when you use them since cloud services are subscription-based. There is no need for you or your organization to worry about costly and time-consuming system upgrades and maintenance. It provides flexibility in work practices. Since everything is stored in the cloud, you can access it from anywhere. Moreover, you will have access to your data and other IT needs no matter what. Even if your local computer or laptop is damaged, you are assured that your data, applications, and tools are securely stored in the cloud and you will have access to it.

SaaS and Geospatial Technology

In terms of geospatial technology, ArcGIS Online is a good example of SaaS. Everything you need to create a GIS map is available through a web browser. You have in fact used ArcGIS Online in chapter 6. ArcGIS Online provides users all the resources to create interactive GIS maps using a web browser. The tools and GIS software to create a web-based GIS map are stored in the cloud and users access them using their web browser. Users can upload their own datasets or use geospatial data stored in the cloud. The web maps can be shared with others within an organization or with the public.

ArcGIS Online is available in two versions. The first version is free of cost. All you need to do to use it is create a free Esri account. You can then log into ArcGIS Online and start using the tools and tons of geospatial data available to create an interactive map. The second version of ArcGIS Online is subscription-based. It has a lot more features and offers its users many more options than the free version. You or your institution must have an organizational account with Esri in order to use the added features and options available in the subscription-based version of ArcGIS Online. There are different levels of user types, each with different subscription fees. Each user type includes different sets of tools and applications. More importantly, subscription to each user type includes a certain number of credits that can be used for specific premium services such as batch geocoding, routing, and feature storage.

ArcGIS Online is also integrated with the current version of ArcGIS Desktop. You can add data from ArcGIS Online to your ArcMap. Conversely, you can also upload GIS data and maps from ArcGIS Desktop to your ArcGIS Online account. Data and maps are uploaded as services to ArcGIS Online.

The tools and resources available in ArcGIS Online can be used to create web maps and web applications. Maps are created and stored in the ArcGIS Online cloud as web maps. You can create a map using ArcMap on your desktop and then publish the map document as a service in ArcGIS Online. You can then use the service to create a web map that can be shared with everybody. The map that you created in the chapter 6 exercise to show different attributes of US states using ArcGIS Online is a web map. A web application is created from a web map that has been customized using web application templates on ArcGIS Online or configured per user preference or requirements.

A GIS web application will have specific tools and functions available to its users. An example is Esri's ArcGIS StoryMaps. StoryMaps is a web application where a user can integrate ArcGIS maps, scenes, narratives, images, geotagged images, multimedia content, and other online content to create and share "stories" on a specific theme or topic. In a StoryMap you can customize your web map using different resources available in ArcGIS Online as well as utilize other online resources. You can add pictures or images from a personal Flickr account, videos hosted on your YouTube account, YouTube video links, and descriptive texts to a StoryMap to narrate a "story" about a topic or theme to the audience. The StoryMap can be shared with a specific group within an organization, or it can be shared with the public.

StoryMaps can be created for free using an ArcGIS Online account. You would use an ArcGIS StoryMaps Story Builder to create your StoryMap. The Story Builder is preloaded with a menu of options that users can use to create their stories. The menu includes options for text; media including maps, images, video, and audio: and immersive content. Each of the options has its own individual formatting options. For instance, text can be made bold or italic, font color and style can be customized, paragraph style for narratives can be customized, and text can be included in bullet points. Similarly, the Story Builder has a variety of built-in options for images, media, and immersive content. Users can add maps to individual StoryMaps from several different sources including maps saved to their ArcGIS Online account, maps that have been shared with them within their organization or group, and maps available on Esri's Living Atlas of the World. The Story Builder also has different built-in layout options and templates to create a StoryMap. Each of the options are geared towards a specific purpose for creating a StoryMap.

Figure 9.1 is an example of a StoryMap that was created to capture and map student stories and their experience as they engage in a weeklong excursion in the Mississippi Delta region (Mukherjee 2019). The excursion is part of a field course in a cultural geography class offered by the Sam Houston State University Department of Environmental and Geosciences. The StoryMap takes the viewer through a tour of all the sites students visit during the fieldwork and provides a glimpse of student experiences at each site.

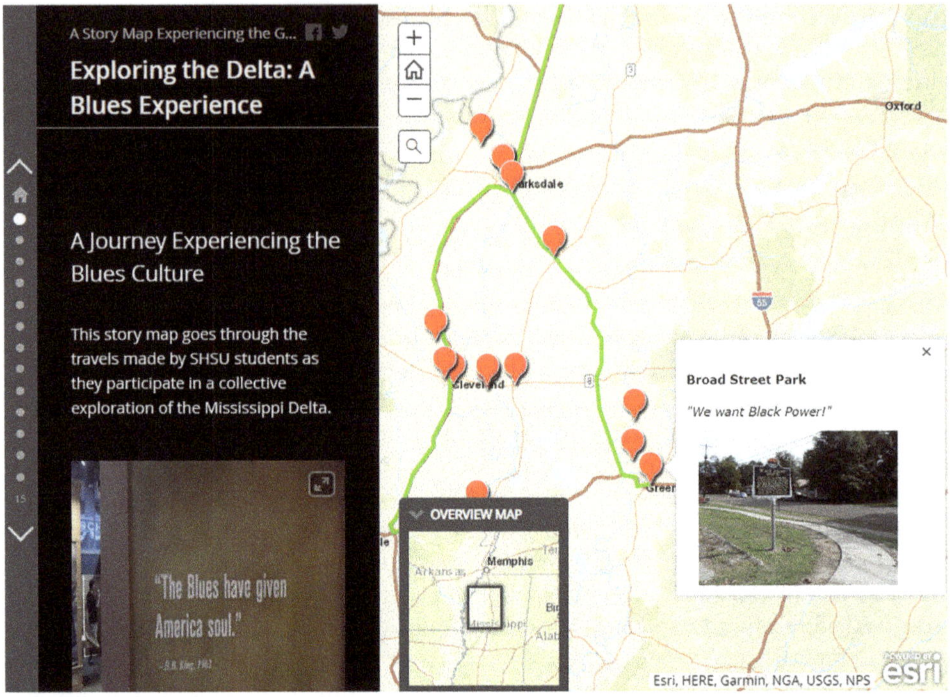

FIGURE 9.1 A StoryMap of the experiences of students enrolled in a cultural geography field class.
Credit: Mukherjee 2019

The StoryMap uses a Map Journal template consisting of different sections. Each section has a main panel and a side panel. Both panels can be used to define map actions. In this case, the main panel was used to display the associated basemap, image, video, photograph, or web page and the narrative text was provided in the side panel. The StoryMap here used data stored on the cloud maintained by different cloud vendors. The basemap was from ArcGIS Online, images from Flickr, and videos from YouTube. Thus a mix of cloud data sources were used to create the StoryMap application on ArcGIS Online. Viewers who view this StoryMap don't need to download the videos, maps, or photographs on their personal computer. They view the entire application that serves resources from different cloud sources on a web browser.

Applications like these are the wave of the future that integrate and serve software, data, tools, and resources from multiple cloud sources and can be viewed on a web browser without the need to download the data, tools, or software on your computer.

Another good example of SaaS in terms of geospatial technology is Esri's Business Analyst and Community Analyst. Both applications and the tools available with them can be accessed in a web browser without the need to install the softwares on your computer.

ArcGIS Location Platform is a PaaS cloud service for geospatial data provided by Esri. It is a location-based cloud service that provides a web environment where developers can integrate location capabilities of GIS with specific business applications, systems, tools, and resources to create location-based software and applications. It allows users to use multiple APIs including open-source resources to build customized applications and integrate those with Esri's location data, tools, and resources on the cloud. Developers can therefore build powerful location-based applications. Using ArcGIS Location Platform, developers can embed geospatial capabilities such as geocoding and routing as well as spatial analysis and data visualization into their applications. Moreover, it provides developers access to a variety of geospatial data including ready-to-use basemaps, imagery, and live-feed data hosted by Esri's cloud service.

The field of cloud computing has revolutionized and continues to revolutionize the world of geospatial technology. Gone are the days when organizations would need to install physical hardware and manage and maintain IT infrastructure or install software on computers on-site, manage software extensions, maintain software upgrades, or download data locally. Everything can now be accessed on a web browser. The same software capabilities and tools to work with geospatial data are now available more easily. Options to integrate geospatial technology with other applications and services are now readily available. Hence the possibilities of what one can do with geospatial technology are limitless. For instance, the "People in Action" interview with Bobby Saleh in this chapter underscores the use of geospatial technology and its role as a game changer in business.

Geospatial Data

Data is the heart and soul of GIS. One of the biggest hurdles to using GIS has always been access to good-quality geospatial data. Data that is current and accurate is key to effective GIS use. The types of data that are needed for any project today have evolved with time. This is because we have so many different sources and techniques for generating data. The technology to create, store, and share geospatial data has undergone a massive transformation and it continues to evolve. The possibilities are endless for new types of data, current and timely data, and higher-resolution data.

Technological innovations like digital transformation, cloud computing, big data, and IoT have completely transformed the mechanism for data capture, storage, and delivery. Good-quality authoritative geospatial data are now readily available using a web browser, making the process of data exchange non-cumbersome. As covered in chapter 5, a lot of data and maps are now available in the public domain and can be easily downloaded. GIS projects no longer have the need to create their datasets from scratch, particularly basemaps. High-quality basemaps, topographic maps, and authoritative geospatial datasets are readily available for organizations to use in their projects, so you can build your data by integrating project-specific or organization-specific data with public domain data.

New sources of geospatial data have emerged. One such source is the rising use of sensors that are embedded in pretty much everything that we use today, such as smartphones, home alarm systems, doorbells, refrigerators, sprinkler systems, vehicles, the list goes on. Sensors are an integral part of IoT and capture all kinds of data, be it temperature, pressure, humidity, speed, motion, traffic flow, and so on. IoT sensors are collecting data on a real-time basis along with location information, giving rise to important time series data. We use such data on a daily basis without even realizing it. A classic example would be our daily commute. Have you ever wondered how Google Maps provide real-time traffic information? It's due to IoT sensors and devices.

Every time you use your smartphone for directions from point A to point B, if you have installed the latest operating system on your phone, you can see real-time traffic conditions on your map. The map on your smartphone is able to do this because of multiple IoT devices and sensors that are interconnected with one another, sending and receiving data back and forth over the internet. These include IoT devices like our smartphones and road sensors. Google collects data from all smartphones running the Google Maps app. If a phone has its location feature on, the Google Maps app sends real-time GPS coordinates to Google that are used to calculate location, speed, and direction of the vehicle. Google does this for all other vehicles that are on the road at that time and have their location feature turned on. Google aggregates this data with data captured by traffic flow sensors and by local civic transportation authorities who also collect traffic data in real time. By combing all of these data, Google is able to provide an accurate measurement of how fast the traffic is moving on a specific road or highway in real time. It is also able to provide you the best path to get to your destination in the shortest amount of time. This information is then relayed back to the mapping app on your smartphone and displayed as a real-time traffic map.

Moreover, Google uses historical traffic flow data to predict what the traffic flow may be at a certain time of the day. The next time you are about to start your trip to a location that you visit regularly, notice that a message pops up on your smartphone's Google Maps app predicting the time it will take for you to reach your destination along with the traffic information. All of this is possible because of the rich and vast amounts of location-based data that are being generated and captured, in some cases on a real-time basis, by devices and sensors interconnected with one another and the internet.

Google Maps is just one example. Innumerable new applications are made possible by such sensors and devices. These devices and sensors are widely used in many different industries such as e-commerce, transportation, healthcare, retail, utilities, manufacturing, agriculture, and even in building smart cities. So a massive amount and variety of data is being generated at a very rapid pace. There is a tremendous need to analyze this massive amount of data, extract information from the data and its innumerable attributes, and visualize it. This will help businesses make intelligent decisions and take appropriate actions. GIS can not only assist organizations in analyzing the massive amounts of datasets and visualizing trends, patterns, and distributions, but also help in understanding data contexts. This is an extremely important step in recognizing what the data actually depicts in an "easy to understand" format and making key business decisions. Geospatial technology therefore has the potential to transform the way organizations are able to take advantage of and utilize these technologies of the future.

GIS Applications

There is no doubt that GIS has applications in varied fields. Its integration with cloud technology and IoT has further strengthened its applications in a myriad of fields. One such field application of GIS that particularly stands out is healthcare. The use of GIS during the COVID-19 pandemic has proved the power of geospatial technology and the scale of its impact. GIS use for public health purposes is not new, particularly for mapping purposes. However, its use during the pandemic on such a large scale and at such a rapid pace is an attestation of its power.

The use of GIS during the COVID-19 pandemic has been one of the most comprehensive use of the technology to date due to the length and breadth of its application. Organizations of all types—large and small, public and private—made use of GIS during the pandemic in some form or another. Moreover, its use has ranged from local organizations to international organizations. The scale of organizations making use of GIS has also varied dramatically. For instance, in India from the very beginning of the COVID-19 pandemic, central government agencies, state agencies, and local government agencies have all resorted to a variety of geospatial technology in their so-called war against the pandemic. Some of the GIS applications were key in providing invaluable insights about the pandemic to the world. GIS use during the pandemic has proved that it is not just about mapping a phenomenon. GIS has proved to be a key tool in spatially analyzing and visualizing the contexts and relationships as well as the underlying trends and patterns that are mostly hidden due to the complexity and volume of the data.

One invaluable tool that became the main resource for providing COVID-19 information to the world is the online GIS dashboard by the Johns Hopkins University Center for Systems Science and Engineering. The main technology used for the dashboard was the ArcGIS platform and the dashboard was delivered using Esri's cloud service. Thousands of organizations worldwide followed suit and launched their own COVID dashboard utilizing the ArcGIS platform, such as the WHO dashboard and HealthMap.

Another key resource during the pandemic was the ArcGIS Living Atlas of the World, which provides access to authoritative geospatial datasets. You have utilized the Living Atlas for many of your chapter exercises in this book. The datasets are stored in the cloud, and users can access them and carry out their analysis through a web browser without the need to download the actual datasets, making it easy to access and use geospatial data, a valuable resource during the pandemic.

Also important during this time was the capability provided by GIS to integrate massive amounts of real-time data from a variety of interconnected devices and sensors, analyze and extract meaningful information from the complex data, and then visualize it. This was a key feature in recognizing the dynamic nature of the pandemic in real time, crucial for making important decisions.

Another cloud-based GIS service that was immensely useful during the pandemic is ArcGIS Hub. ArcGIS Hub is a SaaS platform for collaboration purposes. It provides users the ability to create an information and collaboration "hub" based on specific topics such as COVID-19. You can pull together all of your topic-specific resources on the hub and start a dialogue with internal members of your organization as well as with the public. This is a very useful feature for planning purposes, especially during a pandemic or natural disaster. Such applications have been key in educating the public about the status of the pandemic in their communities.

The pandemic has thus shown the strength of integrating GIS with cloud services and IoT. GIS coupled with these technologies made it possible for the world to visualize every facet of the pandemic as it was unfolding and wreaking havoc at a rapid pace. GIS tools provide a geographic context and have helped in understanding and visualizing the interrelated causes and impacts of the pandemic on communities.

Thus, as trends like cloud computing, big data, data analytics, and IoT take over our digital world, the field of geographic information science and technology will be transformed as it gets integrated with these technologies. This will not only change the field of geospatial science and technology, but also transform the other technology fields that have become a big part of our lives. As the technologies continue to complement each other, the options for their applications are limitless.

People in Action Javier Leija

My name is Javier Leija. I grew up in the greater Houston area, in Baytown, Texas, a city about 30 minutes east of Houston. I graduated from the University of Houston–Clear Lake (UHCL) with a bachelor of science in geography, where I first discovered GIS. I was initially interested in geography and using my degree to teach, but after taking my first GIS course, I have been hooked ever since. After two semesters and three GIS courses under my belt, I began to apply for GIS internships, where I was able to demonstrate what I learned at UHCL within the professional world. I landed an internship with SetId, an organization that offers GIS data-driven products to energy organizations as solutions to growing geospatial needs. After completing the internship, I received an offer from NASA to work on research involving lava flow morphology and GIS. Currently, along with full-time work as a GIS analyst for the City of Pearland, I am also working on my master's degree in GIS from Sam Houston State University.

What does your organization/company do?

As a local government entity, the City of Pearland is responsible for serving its residents. Among some of the services the city offers are informing residents of current ongoing projects, improving city infrastructure, enforcing city ordinances, issuing residential and commercial permits, police/fire/emergency services, and water billing, to name a few. Our mission states: "To deliver unparalleled city services and responsibly manage our resources holistically to contribute to a safe and desirable Pearland community. We will be a high-performance organization focused on continuous improvement and innovation, aspiring to build an engaging, sustainable, and inclusive community of choice."

What do you use GIS/geospatial technologies for in your organization?

The uses of GIS in the organization can include anything from simple feature creation to more-advanced applications such as 3D modeling of city infrastructure or indoor mapping. We use GIS software to create feature classes, update/manage data, update our vector-tile basemaps, symbolize, publish PDF maps, and store our data in an SDE environment. Online cloud software is used to work with published projects from software on our desktop, and we can use online resources to create

(continued)

People in Action *Continued*

targeted maps, apps, and dashboards to provide city officials with solutions to problems. This is not an exhaustive list—there are many more GIS services we will offer in the future as our GIS department continues to grow.

What brought you into GIS in your field? How did your GIS career begin?
My intro to GIS began with my undergraduate studies, in an intro to GIS course. I had a great GIS professor who loved what he did and cared about the discipline. Because of his leadership and enthusiasm, I was hooked on GIS and have not put it down since. After my first semester with three GIS courses under my belt, I decided to apply to internships. Any internship I could find online that involved GIS, I applied to. My first professional experience came working 30 hours a week at an organization called Setld, which provides GIS data-driven products to oil and gas organizations. At this internship, coupled with my self-driven learning outside of work, I learned about professional cartographic principles such as labeling, color schemes, symbology, scales for layers, scales for labels, georeferencing, and applying projections and coordinate systems. My GIS skills improved dramatically by working there.

What value does GIS add to your organization? How does it impact your business?
Right now, we are still using GIS to locate assets, taking inventory of the current infrastructure the city has. This is a lengthy process and may not be done until next year. Concurrently, the city is implementing a citywide asset management program to help maintain an inventory of city assets. This new program will have the capability to integrate with our current GIS, and once the new program is implemented, we will replace any feature classes we have of inventory with the new data from asset management. This will help remove the burden on us of having to maintain/update city inventory and focus more on the user visualization/analysis of the city assets. However, that is not to say that the current systems we have in place will be rendered obsolete. Not everything we store in our database are assets, and we can also use current maps, apps, and dashboards that use city assets layers we maintain as templates for use with the new asset management software.

If you had to give advice to an incoming freshman, what would it be?
To an incoming freshman who is interested in GIS, I would suggest: (1) Begin to establish professional connections with your professors and classmates. Your colleagues and peers may have information on an open GIS position and can recommend you, as well as be a valuable reference in an application. (2) You're never done learning about GIS, especially after graduation. The field is ever-changing, and jobs expect you to take online training, take certification exams, and attend conferences outside of your normal hours to continue learning. (3) Fill electives with any classes related to GIS, AutoCAD, 3D/BIM, programming, SQL, and web application development. The field continues to involve more coding, database administration, 3D, and design of applications, something traditional GIS courses don't teach. Or, if you are interested in environmental GIS, take weather/climate, environmental, geology, and more geography courses. It will also help you decide if you would like to specialize in a specific topic within GIS.

To read the full interview with Javier Leija, visit this book's companion website at https://bloomsbury.pub/mukherjee-intro-to-gis.

People in Action Betsi Chatham

Betsi Chatham, GISP, City of Grapevine GIS Manager, has been serving in the Texas government GIS industry for the last 28 years at both the state and local levels. Betsi received a bachelor of arts in geomorphology from Arizona State University and a master's in applied geography from Southwest Texas State University.

What does your organization/company do?
Promote and coordinate the effective use and sharing of accurate and reliable geographic information through data design, development, and maintenance; spatial analysis and reporting; web map application development and maintenance; and map production.

Please tell the readers a bit about your job. What is your current role and responsibilities?
I am the GIS Manager at City of Grapevine, Texas.

- Directs and provides the support for interdepartmental GIS needs and enhances the use and development of user-friendly applications.
- Prepares and directs the quality control of interdepartmental GIS layers/geodatabases and distributes all GIS layers/geodatabases throughout the city.
- Creates an annual GIS strategic plan and modifies as needed to fit the city's organizational goals and strategic plan.

What do you use GIS/geospatial technologies for in your organization?
Event planning, property assessments, computer-aided dispatch, zoning/future land use, asset management, routing, 3D modeling, crime analysis, site selection, map production, data collection in field with GPS.

What GIS software(s) do you use?
ArcGIS Server and Desktop, ArcGIS Online, Field-Maps for ArcGIS.

How many people in your organization directly utilize geospatial technologies?
Approximately 120.

What brought you into GIS in your field? How did your GIS career begin?
A cartography class in my undergraduate work sparked my interest, and I changed my major from journalism to geography. I then pursued a master's in applied geography with a GIS emphasis. My career began by interning as a degree requirement as a GIS technician, and 27 years later I am still in the GIS profession.

Can you give an example of a typical project that your organization would do that involves GIS?
Customers request a web map so that they can view, analyze, and print data. We acquire the data, QA/QC, and then develop web maps.

Can you think of a project in your organization where geospatial technology/spatial analysis was the game changer? Why?
In most business cases, quantitative measurement is required. GIS provides these metrics.

What value does GIS add to your organization? How does it impact your business?
Decision-making is made easier as data analysis provides facts.

(continued)

People in Action *Continued*

If you had to give advice to an incoming freshman, what would it be?
Stay abreast of the technology.

What would be your advice to someone who is interested in pursuing a career in GIS?
Explore both the programming and analysis side of GIS to see what you enjoy the most.

What in your educational background prepared you for a career in GIS?
Labs with hands-on experience as well as fieldwork.

How does your company/organization use GIS for business analytics and customer and employee experience?
Citywide staff is always leaning upon statistics that are geospatially based. These are readily available after the data has been developed.

For someone who has been in the geospatial industry for such a long time, how would you say GIS has evolved over time?
It has become a tool for the end user. Once GISers develop data and applications, end users can easily access and utilize the technology through web-based applications.

To read the full interview with Betsi Chatham, visit this book's companion website at https://bloomsbury.pub/mukherjee-intro-to-gis.

People in Action Bobby Saleh

Bobby Saleh obtained a bachelor's of science in geography from the University of North Texas. He began his career serving as a GIS analyst at ConocoPhillips in Anchorage, Alaska. After three years there, he moved to Houston, Texas, to assist Shell (New Ventures of Americas) with various aspects of GIS analysis and support. Before he could get there, however, he was called to serve at the Incident Command Center in Houma, Louisiana, for six weeks of intense GIS work in response to the BP Macondo incident. There, he worked to provide many types of GIS skills and analysis, including a map that made its way to President Obama.

Working at Shell as both a geo-information analyst and a senior GIS analyst was rewarding and also challenging while Bobby earned his master of science in GIS from Sam Houston State University. Shortly after graduating, he also earned his GISP certification.

After supporting Shell for six years, Bobby took the opportunity to become the GIS manager at Guidon Energy, a start-up, private equity–backed exploration and production company based in Dallas. There, Bobby managed all aspects of GIS

(continued)

People in Action *Continued*

serving Land Operations, Development, Safety, and the leadership team. Five years in, Guidon was acquired by Diamondback Energy, a large oil and gas operator.

For the last three years, Bobby has served as the GIS manager for PureWest Energy, a Denver-based exploration and production company that focuses on advancing modern life by responsibly delivering essential energy with exceptional reliability and proven environmental stewardship. Bobby is excited to be a part of an organization that recognizes the value of GIS across most aspects of the business. Additionally, he has presented various work-related initiatives at GIS users conferences and energy resources conferences and was recently awarded the Esri Special Achievement in GIS award for an initiative to create a fully realized StoryMap from the company's annual ESG report.

In his personal life, Bobby enjoys photography, cycling, and tinkering with the creation of unique maps. His true joy comes from doing fun things like camping, traveling, and other outdoor adventures with his wife and two children (ages 18 and 10) and showing them how GIS can make a difference in the world around us.

What does your organization/company do?
PureWest Energy is an independent natural gas company focused on development in the US Rockies, with current operations in the Pinedale Anticline and Jonah Field in southwest Wyoming. We are the top natural gas producer in the state of Wyoming, where we operate more than 3,400 wells on approximately 115,000 net acres. We are dedicated to advancing modern life by responsibly delivering essential energy with exceptional reliability and proven environmental stewardship. We strive to go beyond the expected to produce natural gas in the most responsible and efficient manner possible.

Tell the readers a bit about your job. What is your current role and responsibilities?
I am the GIS manager at PureWest, which means I manage all in-house and third-party spatial data. I assist the Environmental, Social & Governance (ESG), Land, Regulatory, Development/Subsurface (reservoir engineering), and Operations departments with data analysis and analytics, mapping/cartography, and other various analyses related to both tabular and spatial data. I also manage our corporate spatial data store, where data is made available to other GIS users within the company and on our web map system.

What do you use GIS/geospatial technologies for in your organization?
We use geospatial technologies for a variety of things at PureWest. In the Land department, we use GIS to display our leasing rights relative to the subsurface or show where joint operating agreements are in effect to ensure proper procedure prior to planning new well locations.

In the Operations department, we create infrastructure maps showing pads, wells, roads, and pipelines for planning and visualization purposes, and we also create route maps to assist with trucking operations.

In the Regulatory department, we map hydrologic features and create elevation models to understand slope and drainage, species' habitat or migration areas, and other relevant cultural and environmental datasets to preserve the pristine environment where we operate. We also employ geospatial technology within this department to acquire and use drone-captured data in the form of ortho-rectified imagery as well as digital elevation model formats.

Another way we use GIS is in the creation of various geo-data analytics dashboards that we use to analyze a variety of data. Spatial data is widespread across many aspects of our business, and GIS enables us to visualize and analyze with the goal of adding value to our bottom line.

Can you think of a project in your organization where geospatial technology/spatial analysis was the game changer? Why?
One geospatial technology "game changer" project that we have been working on involves the creation of a comprehensive web map system deployed to every employee in the organization. This product is a game changer in many ways because it brings together data from many different departments across the organization, including Operations, Land, Safety, Development/Subsurface, Regulatory, and ESG. It serves as one common viewer that unites all the organization's spatial data, ensuring that no decisions are made "in a silo."

(*continued*)

People in Action *Continued*

The value that a centralized web map system provides is immense, and is also hard to specifically measure. A constant struggle within GIS is figuring out how to quantify the value it provides. While we do not have an exact dollar amount on the return-on-investment GIS will provide over the years, the entire management team stands behind the initiative and believes it is a game changer for our company and the industry as a whole. Years ago, leadership teams could only dream of having a single platform that could visualize answers to many of their questions with a few clicks, and GIS has now made this dream a reality.

What would be your advice to someone who is interested in pursuing a career in GIS?

Do as much as you can to learn about GIS and all the career possibilities that are available to you as a GIS professional. GIS is unique in that it can be applied to virtually all fields and industries. Connect with GIS professionals to get a sense of its real-world applications. Browse GIS job descriptions (on sites like LinkedIn, GIS Jobs Clearinghouse, or any reputable job search site) to get a sense of what jobs exist and the skills and education necessary to obtain and excel at them.

For someone who has been in the geospatial industry for such a long time, how would you say has GIS evolved over time?

GIS has become pervasive in every facet of the oil and gas industry. After all, everything has a location and that location matters! Although GIS technology has seen rapid advancement in the past 10-plus years, the concepts of spatial analysis have remained fundamentally the same; we are just dealing with larger volumes of data over more aspects of life. More than ever before, there are sensors (cell phones, smart home devices, etc.) all over the planet that are continuously capturing and storing all forms of data, including location. The internet of things (IoT) refers to this network of devices that is exchanging information over the internet. The sheer quantity of the data being captured, transmitted, and stored is astounding. Handling and analyzing these large datasets has become more difficult, and the industry has had to adapt to continue to stay cutting edge. GIS software has advanced into the world of 64-bit computing and requires more processing power from computers than it ever has before to be able to keep up with big data demands. We will see this trend continue in the future, which is exciting given that the more data we have, the more problems we (geographers) can help solve!

To read the full interview with Bobby Saleh, visit this book's companion website at https://bloomsbury.pub/mukherjee-intro-to-gis.

Questions and Activities

1. Explore the applications of GIS in a field/industry of your interest. A good place to start your investigation would be https://www.esri.com/en-us/industries/index. Here you will find a comprehensive list of industries that utilize geospatial technology. Pick one industry and review the applications of GIS in that industry. Write a two-page summary addressing the following:
   a. How is spatial technology utilized in this industry?
   b. Has GIS and spatial technology changed the industry? In what ways?
   c. What does the future of this industry look like, and how are geospatial technologies shaping that?
   d. Pick a case study to demonstrate the use of GIS and spatial technology in this industry.
2. Look at the geospatial course offerings at your institution. What kinds of GIS classes are being offered? Which department in your institution offers the classes? Do you see multiple departments in your university offering GIS classes? Are GIS classes offered at the undergraduate level, graduate level, or both?

3. Look at other institutions in your area. Do they also offer GIS courses or courses that focus on geospatial technology? Do any of the institutions, including your own, offer a certificate in GIS? How many GIS classes does one need to take in order to obtain a GIS certificate from one of these institutions?
4. What community colleges and technical colleges in your area offer GIS courses? What kinds of programs are available at these colleges: undergraduate, graduate, associate, or certificate?
5. The GeoTech Center National Program Locator by the National Geospatial Technology Center of Excellence is a map to locate colleges and universities offering GIS programs throughout the United States. The map is located at https://www.geotechcenter.org/geotech-center-national-program-locator.html and was created using Esri's cloud service. Open the map in the Map Viewer. You will see a map displaying GIS programs as point features. Do you see your program on the map? How many GIS programs do you have in your state? Which state has the highest number of GIS programs?
6. We have multiple resources for integrating GIS into classroom teaching. A few examples of such resources include:
   a. Google Maps Education available at https://maps.google.com/help/maps/education
   b. Esri's educational resources hub available at https://learn-arcgis-learngis.hub.arcgis.com
   c. GPS visualizer available at https://www.gpsvisualizer.com
   d. Datasets available from NASA at https://neo.sci.gsfc.nasa.gov
   e. Esri resources for K–12 GIS education at https://k12.maps.arcgis.com/home/index.html
   f. GIS educational resources available at https://www.geotechcenter.org/education.html

   Review these resources. What kinds of teaching material are available to educators for using geospatial technology in their classrooms? Try to locate other such examples of educational resources. What types of classroom activities and teaching materials are available?
7. In this chapter we discussed StoryMaps. To begin exploring StoryMaps, visit https://www.esri.com/en-us/arcgis/products/arcgis-storymaps/overview. The "Resources" tab on this page will give you helpful tips and take you through step-by-step instructions to create your own StoryMaps. Click on the "Gallery" tab. The page has a variety of interesting StoryMaps. Examine the StoryMaps.
   a. Filter the StoryMaps by topic. Pick any two StoryMaps within each of the topics. What topics are being presented with the StoryMaps that you chose? Do you find a particular StoryMap better at presenting the topic? Why?
   b. Pick two StoryMaps of a topic that interests you. What is being presented? What did you learn about the topic through the StoryMap?

Key Terms

application programming interface (API) (p. 194)
big data (p. 192)
Canada Geographic Information System (CGIS) (p. 187)
cloud (p. 193)
data intelligence (p. 191)
digital transformation (p. 191)
Geospatial Technology Competency Model (GTCM) (p. 188)
hybrid model (p. 188)
infrastructure as a service (IaaS) (p. 194)
internet of things (IoT) (p. 191)
platform as a service (PaaS) (p. 193)
software as a service (SaaS) (p. 193)

References

Esri. n.d. "Esri Education." Accessed April 27, 2021. https://www.esri.com/en-us/industries/education/overview.

———. n.d. "Esri K12 GIS Organization." Accessed April 29, 2021. https://k12.maps.arcgis.com/home/index.html.

———. 2020. *The Science of Where: Making Sense of Digital Transformation*. Redlands, CA: Esri. https://www.esri.com/content/dam/esrisites/en-us/media/pdf/making-sense-of-digital-transformation/Making_Sense_of_Digital_Transformation_x5.pdf.

GIS Jobs Clearinghouse. n.d. Accessed April 26, 2021. https://www.gjc.org.

Mukherjee, Falguni. 2019. "Exploring Cultural Geography Field Course Using Story Maps." *Journal of Geography in Higher Education* 43, no. 2: 201–23.

Pratt, Monica. 2020. "GIS Systems Lead Responses to COVID-19." *Esri Arc News*, Spring 2020. https://www.esri.com/about/newsroom/arcuser/gis-systems-lead-response-to-covid-19.

University Consortium for Geographic Information Science. n.d. GIScience/GIS/Geospatial Instructional Resources. Accessed May 5, 2021. https://www.ucgis.org/site/instructional-resources.

———. n.d. GIS&T Body of Knowledge. Accessed April 27, 2021. https://www.ucgis.org/site/gis-t-body-of-knowledge.

Geospatial Data Analysis 10

Learning Objectives

- Articulate what kinds of spatial analysis can be done in a GIS
- Discuss what types of questions spatial analysis can answer
- Explain data retrieval in GIS and the two types of data retrieval: attribute query and spatial query
- Discuss some of the basic spatial analysis using vector data such as buffer, dissolve, clip, erase, and data overlay
- Use GIS software to perform operations like buffer, dissolve, clip, erase, and data overlay
- Discuss spatial analysis using raster data such as map algebra, Boolean operators, and mathematical operators

Geospatial Analysis: What and Why?

In this book we have discussed the different components of a geospatial data (chapter 1); how location information is determined and described in a GIS (chapter 2); geospatial data structure (chapter 3); characteristics of a geospatial data (chapter 4); the use of location information in GIS and relational database management systems (chapter 4); attribute table, attribute, and spatial queries (chapter 4); geospatial data and field data collection (chapter 5); cartographic elements of GIS mapping (chapter 6); GIS software and their capabilities (chapter 7); principles of GIS project management (chapter 8); cloud computing in GIS (chapter 9); GIS and digital transformation (chapter 9); geospatial data analytics and StoryMaps (chapter 9); and what the future holds for the geospatial world (chapter 9). All of these topics have underscored the unique capabilities of GIS.

One of the biggest strengths of GIS is its capabilities of spatial analysis. This sets it apart from other mapping and database softwares. Spatial analysis gives GIS the unique capability to extract information from map layers and attribute data. This allows GIS to answer a variety of questions such as:

- What would be the most suitable location for something based on a number of criteria? This kind of question can be used to determine habitat suitability for a certain species or the location of a new restaurant, grocery store, retail chain, or supermarket.
- What is the possibility of a certain phenomenon occurring in the future based on an activity and a specific set of criteria? This kind of question can be used to determine the susceptibility of an area to natural hazards such as flooding, the environmental health of an area based on certain environmental parameters, or the rate and impact of deforestation.
- How many features of a certain type are located within a specific distance of a specific location? This type of question can be used to determine how many restaurants of a specific cuisine are located a certain distance from a location such a house or interstate, how many parks or elementary schools are located within

a certain distance from a neighborhood, how many houses that meet a certain criteria are for sale in a specific neighborhood, or how many rental properties are available within a certain school district.

One of the things that a GIS does is model real-world objects and behaviors by means of spatial features and describe spatial relationships between and within spatial features. This enables a user to extract important information from map data and attribute data. A user can model real-world features, behaviors, and spatial relationships and extract information from map data and attribute data in a GIS by means of spatial analysis. To do so the data must be transformed, and a new data may be created.

Many methods have been integrated within GIS softwares to perform these tasks. The GIS operations that perform these tasks are part of a larger suite of operations called geoprocessing. **Geoprocessing** is a term used for any GIS operation where a specific operation like buffer, clip, dissolve, or erase is performed on a geospatial dataset to create an output dataset. Geoprocessing operations are an important part of spatial analysis in GIS. In any spatial analysis project, you will perform a series of such geoprocessing operations to solve a problem and answer project questions that are being investigated.

There are many types of geoprocessing methods. In this chapter we will cover a few of the basic geospatial analysis that can done in a GIS.

Retrieving Data

An important type of analysis performed in GIS is **data retrieval**, where data is selected based on a specific criteria. In a GIS a user can select data from an attribute table or select spatial layers based on a particular criterion. Accordingly, one would perform either an attribute query or a spatial query.

Let's say you are working on a project where you are investigating the impacts of flooding on streets and roadways and your study is focused on the state of Texas. You have a shapefile of roadways for all the states in the United States, which is likely thousands of roadways. You will go into the dataset and select only those roadways that are in Texas. In this case, you can select your data from the attribute table by performing a query based on a criterion that selects all roadways in the state of Texas. Here you will be performing an attribute query. Attribute queries use any one of the **relational operators** of equal (=), not equal (< >), greater than (>), greater than or equal to (>=), less than (<), or less than or equal to (<=) to build a query. In this example, the relational operator of "equal to" would be used to build the query *'State_Name' = 'Texas'*. This is a **simple query** that tests one criterion.

Multiple simple queries can be connected to build a **compound query** that tests multiple criteria. Compound queries can be built using the Boolean connectors of AND, OR, XOR, and NOT. For example, you have a shapefile of all the counties in the state of Texas and you are interested in counties that have a population greater than 100,000 and a median income of $45,000 or greater. You will build a compound query that tests the two criteria to select the counties that satisfy both conditions. Attribute queries are conducted using Standard Query Language (SQL), a programming language, and are discussed in detail in chapter 4.

Attribute queries can only check for conditions based on data in the attribute table. They cannot be used for geographic search where the criterion is based on a relationship between two or more spatial layers, such as how many hospitals are located within a specific distance of a house or how many homes for sale are located within a specific school district or near a park. In this case you would use a spatial query. A spatial query tests specific conditions based on spatial relationship between two spatial layers or features at a time using the "Select by location" tool. Spatial queries are discussed in detail in chapter 4.

Spatial Analysis Using Vector Data

In this section we will discuss some of the basic spatial analyses that are performed in GIS. A very simple type of analysis is the buffer operation. A buffer is an area, or a polygon, built around a feature using the spatial relationship of proximity. A buffer operation identifies areas within a specified distance from a feature. Buffers can be created around a point, line, or polygon feature. For example, let's say that you want to identify how many subdivisions fall within 50 meters of a major highway. In this case, you would create a 50-meter buffer around the feature representing the major highway to determine how many subdivisions fall within the buffer. Figure 10.1 shows a 50-meter buffer created around a major highway. Buffers can be created for multiple point, line, or polygon features. Buffer is a relatively simple GIS operation. However, it can be used to extract useful information for analysis purposes.

Another simple GIS operation is the dissolve function. A dissolve operation merges features based on a common shared attribute. For example, you want to create a polygon feature for states, and you have a polygon feature of state counties. You can merge the county polygons to create a new feature representing a state polygon using the common field state name in the attribute table. Figure 10.2 is a state shapefile of Texas created by dissolving a counties shapefile. Here the counties

FIGURE 10.1 A 50-meter buffer around a major highway.

ArcMap interface courtesy of Esri, Inc. Data source: ArcGIS Living Atlas of the World, © OpenStreetMap contributors

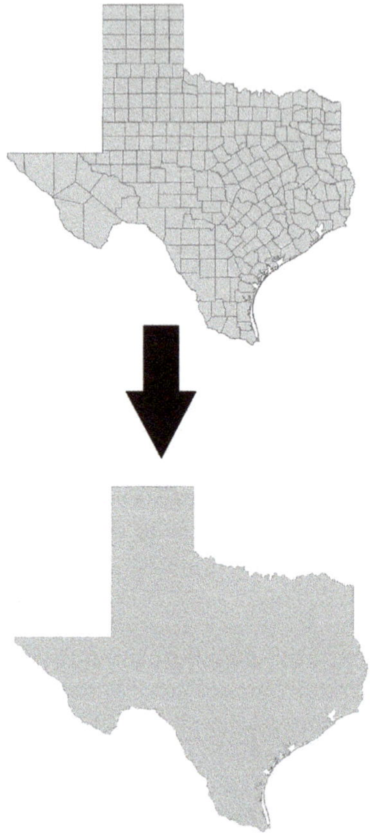

FIGURE 10.2 Dissolve operation used to create a Texas state polygon from a Texas counties polygon. The field name State_Name was used for the dissolve operation.

ArcMap interface courtesy of Esri, Inc. Data sources: ArcGIS Living Atlas of the World, Esri; U.S. Department of Commerce, Census Bureau; U.S. Department of Commerce (DOC), National Oceanic and Atmospheric Administration (NOAA), National Ocean Service (NOS), National Geodetic Survey (NGS)

shapefile had an attribute State_Name. All counties in Texas would have the same state name in the attribute table and is hence the common field that was used as the dissolve field to dissolve the county boundaries and create a new shapefile of the state of Texas. The dissolve function can be used to dissolve buffer zones created from features that are overlapping and combine the buffer zones to create one seamless buffer zone.

Many GIS projects begin with an extraction operation where the user must extract the necessary data as part of the data preparation for the project. Two GIS operations are particularly useful in such instances. One is the clip operation.

A clip operation truncates one feature based on the outline of another feature. You can therefore extract a portion of a feature based on the spatial boundary of another feature and the extracted feature is saved as a new feature. The new clipped feature will only have the attributes of the features that were extracted by clipping. In a clip operation, the input layer will be your features that must be extracted, and the clip layer is the feature that provides the spatial boundary. The clip layer will always be a polygon, whereas the feature being extracted can be a point, line, or polygon. For example, you may have a major cities point shapefile that covers the entire United States and you want to extract major cities within the state of Texas. In this case, you can use a polygon shapefile of the Texas state boundary as the clip layer to clip the major cities and extract major cities that fall within the state polygon feature to create a new shapefile of major cities in Texas (figure 10.3).

If needed, you can now use an attribute query to further extract cities from the newly created Texas cities shapefile using a specific criterion such as population or population density, provided the attribute table has a column with that information. You can also use an attribute query instead of the clip operation to extract all the cities in Texas from the major cities point shapefile, provided the attribute table of the major cities shapefile has an attribute identifying the state each city is located in, such as state name or state FIPS. A clip operation that uses a spatial boundary would be another way to arrive at the same solution.

Another useful extraction operation is the erase operation. You can think of the erase operation as the opposite of clip. An erase operation extracts features that lie outside a spatial boundary. In other words, it erases features within a spatial boundary. So, in the example of major cities in the United States, an erase operation would erase all cities within the Texas state polygon and extract major cities that lie outside the Texas state boundary, as seen in figure 10.4

Another GIS operation used frequently for geoprocessing purposes is vector data overlay. An overlay operation combines two layers based on their spatial relationship to create a new output layer that contains information from both input layers. Here the operation combines both the spatial information and the aspatial/attribute information in the new output layer. Two functions in GIS are used to perform an overlay operation: intersect and union. An intersect operation extracts common features from two layers to create a new layer that contains spatial and attribute data that is common in both of the original layers. A union operation combines all of the features in a new layer containing all the spatial and attribute data from both of the original layers.

Let's consider two polygon layers, one for land use type and the other for land cover type.

An intersect operation will create a new layer containing only those polygons that are covered by both of the input layers, land use type and land cover type. The attribute table of the new layer will merge the attributes from both input layers.

An intersect operation can be performed with point, line, or polygon features. The output layer will have a spatial dimension that is equal to or less than the smallest spatial dimension of the input layers. In other words, if the intersect operation is being conducted on polygon and line features, the output layer will be either a line or a point feature. If the input layers for the intersect operation are two polygon features, the output layer will be either a point, a line, or a polygon feature. If the input layers for the intersect operation are two line features, the output layer will be either a line or a point feature (figure 10.5).

FIGURE 10.3 Clip operation to extract major cities in Texas using the Texas state polygon feature as the spatial boundary.

ArcMap interface courtesy of Esri, Inc. Data sources: ArcGIS Living Atlas of the World, Esri; U.S. Department of Commerce, Census Bureau; U.S. Department of Commerce (DOC), National Oceanic and Atmospheric Administration (NOAA), National Ocean Service (NOS), National Geodetic Survey (NGS)

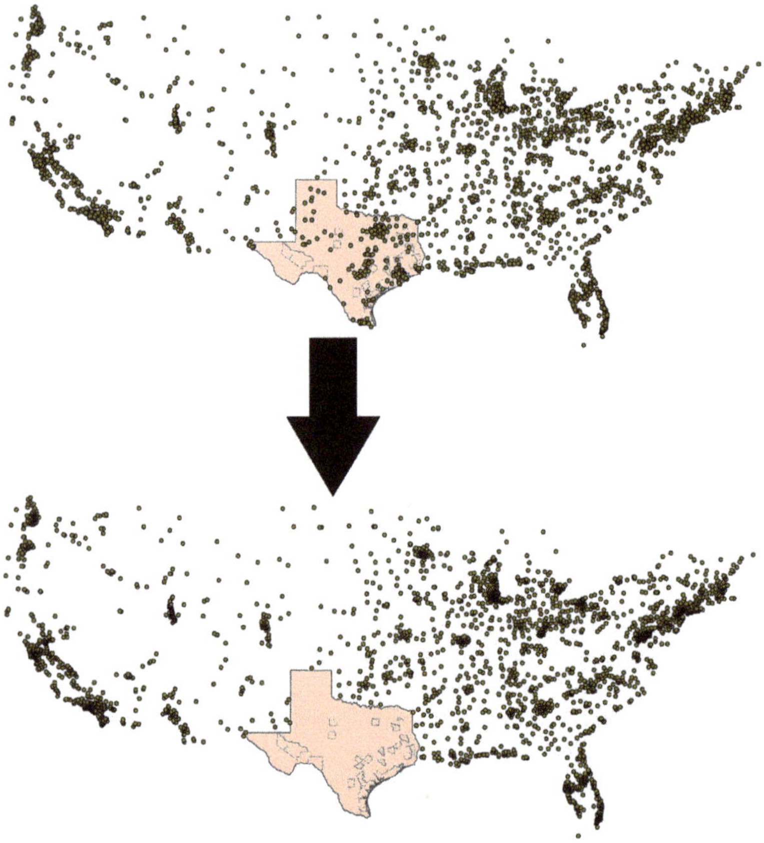

FIGURE 10.4 Erase operation that extracts all major cities outside the Texas state polygon boundary.

ArcMap interface courtesy of Esri, Inc. Data source: ArcGIS Living Atlas of the World, Esri; U.S. Department of Commerce, Census Bureau; U.S. Department of Commerce (DOC), National Oceanic and Atmospheric Administration (NOAA), National Ocean Service (NOS), National Geodetic Survey (NGS)

You would perform an intersect operation when you want to determine a solution where multiple criteria are fulfilled. An intersect uses the logical Boolean connector AND to extract common features from input layers that fulfill multiple criteria. We have discussed Boolean connectors in chapter 4. For instance, you want to purchase a house that is within five miles of a major freeway and falls inside a specific school district. In this case you would create a five-mile buffer around the freeway and then intersect the buffer layer and the school district layer to find areas that satisfy both the criteria: within five miles of the major freeway and inside the specific school district. Now you can examine which houses in the newly created intersect layer are for sale.

A union operation combines all features from the input layers to create a new layer. The new layer will be the most extensive spatial layer created by merging all the features in the input layers. The attribute table of the newly created union layer will have the attributes of all of the input features that cover them. Unlike the intersect operation, the union operation can only be performed with polygons. And it uses the logical Boolean connector OR to extract all features from both input layers.

Let's examine the example of land cover type and land use type layers. Here in the union operation, the output layer will contain features that are covered by both

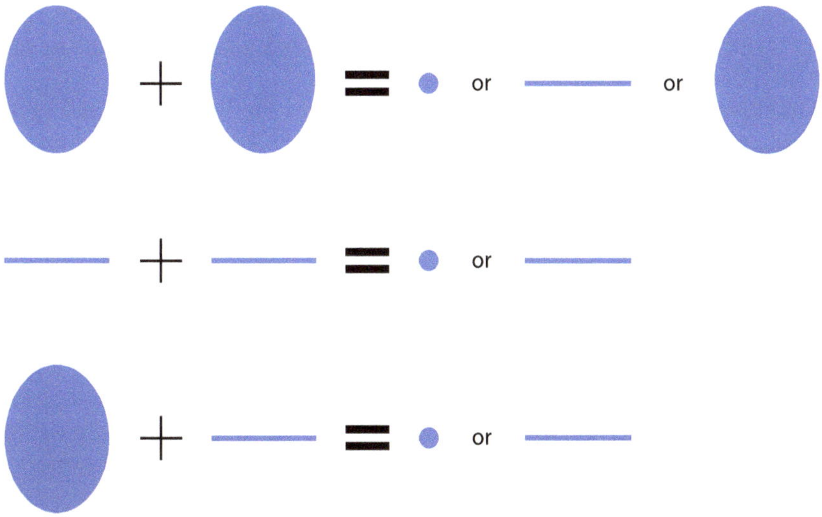

FIGURE 10.5 The output layer from an intersect operation will have a spatial dimension that is equal to or less than the smallest spatial dimension of the input layers. The output from polygons can be polygon, line, or point; output from lines can be line or point; and output from line and polygon intersect can be line or point.

layers, land use type and land cover type. Moreover, attributes from both layers will be transferred to the newly created output layer. Thus, the new layer will be a detailed layer containing features and attributes from both land cover type and land use type layers.

Spatial Analysis Using Raster Data

The overlay operations that we have discussed so far are based on vector data. Overlay can also be performed using raster data. As discussed in chapter 3, raster and vector data have very different structures and are hence processed very differently. A raster data structure consists of an array of pixels, or grid cells. Each cell is assigned a numeric value. Therefore, when a raster data is processed to extract information, the numeric value assigned to each cell in the input raster layers is combined based on the operation conducted.

Multiple raster data representing different attributes can be processed either mathematically or by using logical or Boolean operators. The operators are used in a formula for cell-by-cell analysis wherein each cell value can be mathematically analyzed and either added, subtracted, multiplied, or divided, depending on the analysis. Alternatively, cell-by-cell analysis can be conducted using logical or Boolean operators. This technique is called map algebra. A raster overlay operation is similar to a vector overlay operation, except that it uses map algebra to conduct cell-by-cell analysis. It is particularly useful when working with continuous datasets such as precipitation, elevation, and temperature. Additionally, in raster overlay multiple raster layers can be processed at a time.

Take the example of a raster layer of rainfall where rainfall data is expressed in inches. You would use map algebra and multiply the raster layer by 2.54 to create a new raster layer of rainfall in centimeters. You can use map algebra for multiple raster layers and functions; for instance, determining areas most susceptible to flooding based on different factors such as land use–land cover, elevation, slope, rainfall, and soil type. The expression on map algebra using raster layers for each of the factors would be *LULC* × *Elevation* × *Slope* × *Rainfall* × *Soil*.

Map algebra, like a vector overlay operation, uses the Boolean operators AND, OR, XOR, and NOT. The AND operator finds common areas in the input rasters and is the equivalent of the vector intersect operation. The OR operator combines areas from all the input rasters and is equivalent to the vector union operation.

Again, consider the example of determining areas most susceptible to flooding based on elevation, soil type, and land cover. The raster overlay will test the following conditions: areas most susceptible to flooding will have an elevation of less than 200 feet, soil type will be clay loamy, and land cover type will be built-up areas, roads, and settlements. Areas that satisfy all three conditions will have a value of TRUE in the output raster and the corresponding cell values will be assigned a value of 1. This can be approached either by using the Boolean operator AND or by using the mathematical operator multiply. The Boolean operator AND functions the same way as multiplication of raster layers.

Figure 10.6 shows the output that would be achieved if using the AND operator. In the first raster data, a cell value of 1 is assigned where elevation is less than 200 feet; in the second raster data, a cell value of 1 is assigned where soil type is clay loamy; and in the third raster data, a cell value of 1 is assigned where land cover type is built-up areas, roads, and settlements. Using the Boolean operator AND, if the corresponding cell values are 1 in all three rasters, the cell value in the output raster will also be 1. If any of the corresponding cell values is not equal to1, the cell value in the output raster is 0.

Now, compare the output when the three rasters are multiplied, shown in figure 10.7. You will see that the output is identical to the output when the Boolean operator AND is used.

Thus, either AND or multiplication can be used to find areas that satisfy all three conditions and are most susceptible to flooding.

Another way to approach this problem question would be to rank the areas based on their susceptibility to flooding. So, instead of finding areas that are most vulnerable to flooding, we can rank areas that are most vulnerable, moderately vulnerable, and not vulnerable at all to flooding. In this case, we would use the mathematical operator addition. Adding the three raster layers, if the corresponding cell values are 1 in all three input rasters, the cell value in the output raster will be 3. If two corresponding cell values in the input rasters are 1, the cell value in the output raster will be 2, and if one of the corresponding cell values in the input rasters is 1, the cell value

1	0	0	0	0
1	1	1	1	1
0	1	1	1	1
0	1	1	1	1
0	0	1	1	1

Cell value = 1 when Elevation< 200 feet

AND

0	0	1	1	1
0	0	1	1	1
0	0	1	1	1
0	0	1	1	1
0	0	1	1	1

Cell value = 1 when soil type is clay loamy

AND

1	0	0	0	0
1	1	1	1	1
0	1	1	1	1
0	0	1	1	1
0	0	1	1	1

Cell value = 1 when land cover type is built-up areas, roads, and settlements

0	0	0	0	0
0	0	1	1	1
0	0	1	1	1
0	0	1	1	1
0	0	1	1	1

Output raster

FIGURE 10.6 Raster overlay using the Boolean operator AND on three input rasters representing elevation less than 200 feet, soil type is clay loamy, and land cover type is built-up areas, roads, and settlements.

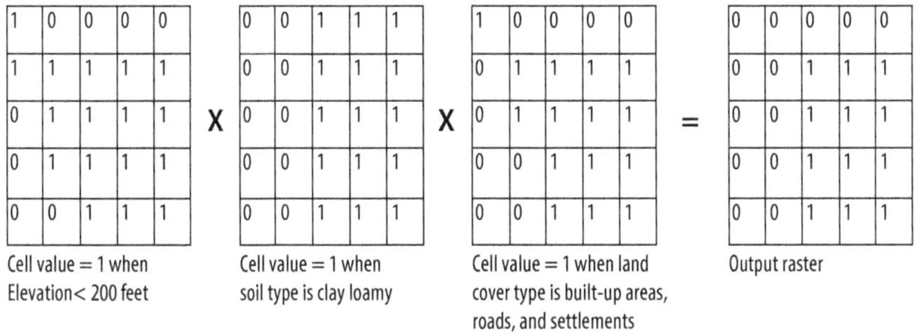

FIGURE 10.7 Raster overlay using mathematical operator of multiply to multiply three input rasters representing elevation less than 200 feet, soil type is clay loamy, and land cover type is built-up areas, roads, and settlements.

in the output raster will be 1. So now, areas where the cell value is 3 in the output raster will be most vulnerable to flooding, areas where the cell value is 2 in the output raster will be moderately vulnerable to flooding, and areas where the cell value is 1 in the output raster will be least vulnerable to flooding. Figure 10.8 shows the output raster achieved by adding the three input rasters: elevation is less than 200 feet, soil type is clay loamy, and land cover type is built-up areas, roads, and settlements.

These types of overlay operations are useful for studies involving *site suitability analysis*. There are several practical real-world examples where site suitability analysis is conducted, such as determining the most suitable habitat for a certain species of animal, determining the most suitable location for a neighborhood or urban development project, or determining the vulnerability of an area to natural hazards like flooding. If using the Boolean operator AND or mathematical operator of multiply, the output will give you results in two forms. One will be in the form of most suitable or vulnerable areas where the output raster cell value will be 1 for areas that are most vulnerable or suitable for an application, and the other will be in the form of unsuitable or vulnerable areas where the output raster cell value for unsuitable areas will be 0. Alternatively, if using the mathematical operation of addition, the output will give you results where areas in the output raster can be ranked from most suitable to least suitable.

A GIS can perform many such operations, as discussed in this chapter. Other important GIS operations used for spatial analysis include spatial autocorrelation, weighted overlay, point pattern analysis, hypothesis testing, and spatial interpolation. A GIS has the capability to process many large datasets together at once. A GIS also has the capability to combine many different conditions to determine where a

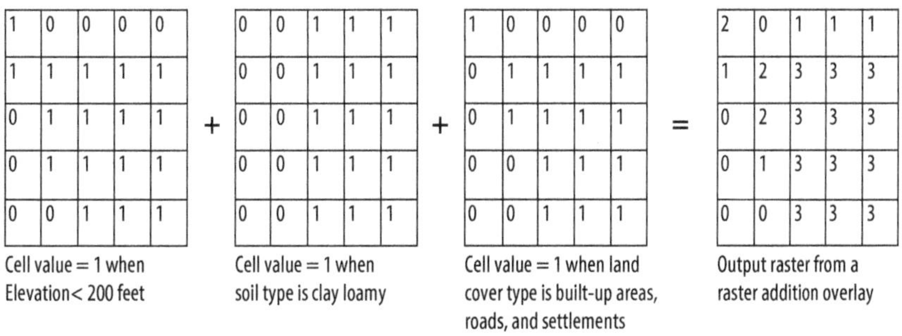

FIGURE 10.8 Raster overlay using mathematical operator of addition to add three input rasters representing elevation less than 200 feet, soil type is clay loamy, and land cover type is built-up areas, roads, and settlements.

specific phenomenon, activity, pattern, or distribution occurs or has the possibility of occurring, spatially and aspatially. The conditions that contribute to a phenomenon, activity, pattern, or distribution are constructed as layers in a GIS and combined using a variety of operations that are possible and integrated into most GIS software packages. These capabilities are key in many important decision-making processes.

Questions and Activities

1. Think of your daily routine from morning to end of day. Make a list of at least five activities from your daily routine that require spatial analysis.
2. How is a GIS able to model real-world features and relationships?
3. How would you define spatial analysis?
4. How is an attribute query similar to a spatial query and how are they different?
5. Discuss the use of an extraction operation in a GIS project. Name two GIS operations integrated into most GIS software that are used to extract data.
6. What is an overlay operation? How is an overlay operation similar to and different from an extraction operation?

Key Terms

buffer (p. 211)
clip (p. 212)
compound query (p. 210)
data retrieval (p. 210)
dissolve (p. 211)
erase (p. 213)
geoprocessing (p. 210)
intersect (p. 213)
map algebra (p. 215)
overlay (p. 213)
relational operators (p. 210)
simple query (p. 210)
spatial analysis (p. 209)
spatial query (p. 211)
union (p. 213)

Reference

Los Angeles County Department of Regional Planning (DRP). GIS-Net Public. Accessed October 11, 2021. http://rpgis.isd.lacounty.gov/Html5Viewer/index.html?viewer=GISNET_Public.GIS-NET_Public.

Glossary

absolute location: Location described with respect to a known origin and standard measurement system such as a coordinate system. (p. 12)

address matching: A form of geocoding that involves converting a combination of letters and numbers into a form of information that can be located on a map. (p. 130)

address standardization: A process in address matching where the input addresses are standardized in a consistent format so that they can be read by the address locator to match the appropriate attributes in the reference data no matter how it is represented in the input data. (p. 130)

adjacency: A topological rule that describes how one polygon is related to another polygon in its neighborhood and if they share a common boundary. (p. 58)

aerial photography: The process of taking a photograph of the earth's surface by means of a camera mounted on an aircraft. (p. 114)

almanac: Data containing information about GPS satellite constellation and the status of each satellite in the constellation. (p. 111)

application programming interface (API): How different software applications communicate with one another. (p. 194)

ArcCatalog: A component application of ArcGIS Desktop used for data management purposes in addition to browsing data and moving data from one location to another. It also used to manage metadata of geospatial datasets. (p. 165)

ArcGIS: A suite of GIS softwares created and distributed by Esri, a Redlands, California–based company. (p. 163)

ArcGIS Desktop: A GIS application that runs on personal computers and was designed to operate on a Windows operating system. (p. 164)

ArcGIS Desktop Advanced: The most advanced ArcGIS Desktop license that includes all of the functional capabilities of ArcGIS Desktop Standard plus capabilities for sophisticated geospatial analysis, advanced cartography, and database management. (p. 165)

ArcGIS Desktop Basic: A basic license of ArcGIS Desktop that has the least number of functional capabilities. It provides tools for basic map creation and interactive visualization. (p. 165)

ArcGIS Desktop Standard: The next level of ArcGIS Desktop license that has all the capabilities of ArcGIS Desktop Basic plus additional functional capabilities for multiuser editing and data management. (p. 165)

ArcGIS Online: Esri's cloud-based solution to creating and analyzing geospatial data and maps. It provides users the capability of creating web maps, three-dimensional web scenes, and apps using a web browser. Users can analyze geospatial data or create maps using a web browser and share with others. There is no need for software installation. All one needs is a web browser and internet connection. (p. 166)

ArcGIS Pro: The next generation of ArcGIS Desktop that will ultimately replace ArcMap. ArcGIS Pro is most notable for its advanced architecture and enhanced visualization. It has a 64-bit architecture. ArcGIS Pro is a standalone GIS program that can run side by side with ArcMap. It has a modern and sleek interface and works seamlessly with web GIS data and maps. It has enhanced three-dimensional visualization capabilities that are integrated into the program and provide users upgraded geospatial functional and analytical capabilities. (p. 165)

ArcGlobe: A component application of ArcGIS Desktop used to visualize geospatial data on a globe surface. (p. 165)

ArcMap: The main component of ArcGIS Desktop used for displaying, viewing, and analyzing geospatial data and creating maps. An ArcMap consists of two main sections: a table of contents and the map display area. The table of contents shows all the data that have been added to ArcMap, while the map display area allows a user to visualize the data as map layers. ArcMap provides its users multiple geoprocessing tools for data analysis, visualization, and manipulation. (p. 164)

ArcScene: A component application of ArcGIS Desktop used for three-dimensional visualization. (p. 165)

ArcToolbox: A component application of ArcGIS Desktop that contains all the analysis tools used for geospatial data. (p. 165)

area: See **polygon.**

attribute query: A query that will test a specified condition using records present in a table and give an output whose values meet the specified condition. (p. 86)

attribute table: A table that carries detailed information of the features displayed on a map. It is a table of feature attributes that exist in the background of the spatial component of the feature that one sees on a map. (p. 73)

azimuthal projection: See **true direction projection.**

band: A portion of the electromagnetic spectrum that represents electromagnetic energy of a certain wavelength. (p. 123)

bathymetric isobaths: Isolines connecting points of equal elevation below a datum. (p. 147)

big data: Data that is so large in volume and complexity that traditional methods of analysis and processing cannot handle it efficiently. Big data is generated in all forms and can be structured or unstructured. (p. 192)

brightness value: A value assigned to each pixel that corresponds to the amount of reflected energy measured by the sensor at that point on the ground. (p. 123)

buffer: An area, or polygon, built around a feature using the spatial relationship of proximity. A buffer operation identifies areas within a specified distance from a feature. Buffers can be created around a point, line, or polygon feature. (p. 211)

C/A code: Coarse acquisition code transmitted at L1 frequency. (p. 110)

Canada Geographic Information System (CGIS): An early geographic information system developed for the government of Canada beginning in the early 1960s. (p. 187)

cardinality: The relationship between two separate tables. (p. 80)

cartographic design process: A process that includes certain steps, starting from envisioning the objective of the map and concluding with the final creation of a map, that are followed to design and create maps. (p. 138)

cartography: The science of mapmaking. It is the theory behind map construction and use. (p. 23)

clip: A GIS operation that truncates one feature based on the outline of another feature. A portion of a feature is extracted based on the spatial boundary of another feature and the extracted feature is saved as a new feature. (p. 212)

cloud: A computing technology wherein the resources required to run an application—the infrastructure, software, and platform—are stored on a server at another location and served to a user via the internet. (p. 193)

color composite image: A color image where each pixel's color is determined by a combination of the primary colors of different brightness values. (p. 124)

column header: In an attribute table, attributes are stored as column headers. (p. 79)

common key: When the primary key and the foreign key have the same field name in each database. (p. 78)

compound query: A query consisting of multiple simple queries that are connected to test multiple criteria. It is built using the Boolean connectors AND, OR, XOR, and NOT. (p. 210)

conformal projection: Projection that preserves the property of shape. (p. 35)

conic projection: The globe is projected on a cone where the cone either touches the globe along a small line or slices through the globe. (p. 33)

connectivity: A topological rule that establishes how lines are formed by connecting points at the starting point and at the end point describing arc-node topology. (p. 59)

containment: A topological rule that describes how spatial objects are enclosed within a polygon. (p. 59)

continuous surface view: A view where objects or features are not separate entities with fixed location and definable boundaries. Instead, the world is viewed as made of infinite points and each point has a value representing a constantly varying surface. In this world view, values are available at every point. Thus, the world is conceptualized as a surface with values at every point that vary. (p. 53)

Continuously Operating Reference Stations (CORS): An augmentation system program operated by the National Geodetic Survey and managed by the National Oceanic and Atmospheric Administration and the National Geodetic Survey. (p. 112)

control segment: One of the three components of a GPS system that consists of a global network of ground stations that track and monitor GPS satellites and their transmissions. (p. 108)

coordinate system: The system to represent and define locations on the earth's surface. Coordinate systems are reference systems for defining real-world locations of features within a standardized common framework. (p. 41)

cylindrical projection: The globe is projected on a cylinder where the cylinder touches the globe along a small circle or slices through the globe. (p. 33)

data: Any material that is used to extract information after careful evaluation. Data is the building block of information. (p. 3)

data classification: The process of dividing data into groups or categories. (p. 150)

data intelligence: The ability to extract information in order to understand the data so that intelligent decisions can be made and actions taken accordingly. (p. 191)

data model: An abstract method for representing real-world features, phenomena, and their characteristics. It determines how data is stored, analyzed, extracted, retrieved, and displayed. (p. 53)

data retrieval: An important type of analysis performed in GIS where data is selected based on a specific criterion. In a GIS, a user can select data from an attribute table or select spatial layers based on a particular criterion. Accordingly, one would perform either attribute query or spatial query. (p. 210)

database: A standardized collection of data stored in a computer system. A database consists of a collection or series of files that store data. (p. 75)

database management system (DBMS): A specialized information system for managing multiple databases. It is a versatile system designed to create, store, manipulate, retrieve, and manage data in a database. (p. 74)

datum: An abstract system that provides a reference framework to measure locations on the surface of the earth relative to the surface of an ellipsoidal. (p. 30)

datum transformation: The process of changing measurements from one datum to another. This involves changing ellipsoids and a complex mathematical computation using transformation algorithms. (p. 32)

DEM: A raster data file format produced and distributed by the US Geological Survey. (p. 68)

destination table: The table that receives the information. (p. 82)

developable surface: The flat surface on which the earth's features and their locations are projected. (p. 33)

differential GPS (DGPS): The use of two GPS receivers. One receiver is positioned at a location that receives satellite signals and its x, y, and z coordinates are known. This is the base or reference station receiver. The second GPS receiver is a mobile or roaming receiver. Both receivers receive signals transmitted from at least four visible satellites. Because the precise location of the base station receiver is known, it can calculate corrections to any errors that cause a miscalculation of position in a roaming receiver. (p. 111)

digital number: Individual cell value of a raster data. (p. 64)

digital orthophoto quadrangle (DOQ): Orthophotos that cover an area of 3.75 minutes of latitude by 3.75 minutes of longitude on the ground, an area equivalent to one-fourth of a 7.5-minute USGS topographic map. DOQs are georeferenced so that every pixel in the photo can be accurately located. A DOQ is therefore a digital image of an aerial photo that incorporates the image characteristics of a photo with the georeferenced qualities of a map. (p. 100)

digital transformation: The process of utilizing digital technologies to create new or transform existing business processes and operations. It also involves a transformation in the work culture and modification of customer experience. Digital transformation employs a strategic use of technology, people, and processes to fundamentally change how business operations are conducted. IoT, big data, and data analytics are all key components of this process. (p. 191)

digitizing: A form of geocoding that involves capturing map data by tracing over the top of a map or an image to model the features represented on the image or map in a GIS. (p. 129)

discrete object view: Objects or features that have a definite location, a fixed starting point and end point, and clear definable boundaries. In a discrete object view, real-world features and phenomena are conceptualized as separate objects. (p. 53)

dissolve: A GIS operation that merges features based on a common shared attribute. (p. 211)

DLG: A vector data file format created by the US Geological Survey. (p. 68)

Earth Explorer: A USGS data portal that provides access to a vast number and variety of land-related digital geospatial data that includes USGS DOQs produced through 2006; satellite image data such as Landsat, Sentinel, IKONOS, and OrbView3 imageries; radar data; digital line graphs; digital elevation models; aerial photos; land cover data; and other datasets. (p. 102)

electromagnetic spectrum: Wavelength of electromagnetic energy that covers a broad range. (p. 122)

ellipsoid: A three-dimensional shape created by rotating a two-dimensional ellipse, whose equatorial radius is longer than its polar radius. (p. 27)

ephemeris: Data containing information about a GPS satellite's orbital position. (p. 109)

equal-area projection: See **equivalent projection.**

equidistance projection: Projection that preserves distance. (p. 38)

equivalent projection: A projection where the relative size of regions on the earth is preserved everywhere on the map. Equivalent projections preserve areal relationship. (p. 38)

erase: A GIS operation that erases features within a spatial boundary and extracts features that lie outside a spatial boundary. It is the opposite of clip. (p. 213)

false color composite: An imagery produced by assigning the red band to either the green or blue channel of the display device, the green band to either the red or blue channel of the display device, and the blue band to either the red or green channel of the display device. (p. 125)

flow map: A map showing changes in magnitude along a line. In a flow map, the width of the line representing the line feature is proportionally changed to match the change in magnitude. The line feature may represent changes in stream discharge, air traffic volume, or flow of goods and services. (p. 147)

foreign key: A specific column in another table that the primary key corresponds to. (p. 77)

FORTRAN: A programming language used to write programs to draw maps. Printing and plotting were done using archaic printers and plotters. (p. 16)

geocoding: The process of transforming analog data into digital format. (p. 97)

geographic coordinate system (GCS): A system that defines where the feature or the data is located on the earth using a three-dimensional spherical or ellipsoidal surface. In a GCS, locations are defined using latitudes and longitudes. A GCS consists of a datum that includes an ellipsoid, prime meridian, and units in degrees. (p. 43)

geographic information system (GIS): An automated system to process, analyze, and extract geographic information from geographic data. It is a combination of hardware, software, data, people, and methods used for processing spatial or geographic data to obtain geographic information and make decisions. (p. 4)

geoid: An irregularly shaped surface which defines zero elevation. The USGS defines a geoid as "an imaginary sea level surface that undulates or has a wavy surface extending through the land masses over all of the earth including oceanic areas." (p. 28)

geometric datum: A system for collecting and measuring positions relative to an ellipsoidal model of the earth. (p. 30)

geoprocessing: Any GIS operation where a specific operation is performed on a geospatial dataset to create an output dataset. Geoprocessing operations are an important part of spatial analysis in GIS. In any spatial analysis project, a series of such geoprocessing operations will be performed to solve a problem and answer project questions that are being investigated. (p. 210)

georelational data model: A vector data model structure where spatial and attribute information are stored in separate files. The spatial data is stored in a graphics file and the attribute information is in stored in a relational database, and the two are linked by a common ID. The spatial features are linked to their respective record in the database by means of an ObjectID. (p. 61)

geospatial data: Any data that has location information assigned to it. Location information or spatial data is associated with some kind of a geographic reference that aids in identifying where it is located on the earth's surface. (p. 8)

Geospatial Technology Competency Model (GTCM): Model developed by the US Department of Labor to provide institutions a framework for geospatial skills that are vital in today's workforce. (p. 188)

geostationary satellites: Satellites that follow an orbit where the satellite rotates at the same speed as earth. Since they take 24 hours to complete one orbit, they will be at the same location at the same time every day. (p. 126)

GeoTIFF: Raster data file format. (p. 68)

global navigation satellite system (GNSS): Any satellite constellation that provides positioning, navigation, and timing (PNT) services on a global or regional basis. (p. 107)

Global Positioning System (GPS): A type of a global navigation satellite system. GPS is a US government–owned utility operated and maintained by the US Air Force. A GPS system consists of three components that must work in sync with one another to pinpoint locations on earth: a space segment, a control segment, and a user segment. (p. 107)

Global Visualization Viewer (GloVis): USGS portal used to access publicly available remote sensing data created by federal agencies. (p. 101)

GPX: A vector data file format derived from a GPS device. (p. 68)

graduated symbols map: A map created to represent quantitative data at point locations where point data is shown with varying-sized symbols used to represent ranges of values. Here, the point data may represent either true point measurements or conceptual point measurements. A reader will not be able to tell the exact value at every point location on the map. The value will fall within a certain range representing either numeric values or ordinal level ranking. (p. 144)

graticule: The system of parallels and meridians. (p. 41)

Historical Topographic Map Collection (HTMC): USGS's map collection that includes topographic maps of the United States dating back to 1882. (p. 99)

horizontal datum: A system for collecting and measuring positions on the surface of the earth. (p. 30)

hybrid model: A model where a mix of in-person and online classes are offered. (p. 188)

hyperspectral imagery: Imagery produced by sensors that can sense and capture over 200 bands. (p. 124)

hyperspectral sensor: A sensor that is capable of measuring hundreds of bands of the electromagnetic spectrum. (p. 126)

IMG: Proprietary raster data file format used by ERDAS Imagine. (p. 68)

information: Data that has been prepared and structured in a meaningful manner that can be utilized to make informed decisions. (p. 4)

information system: An automated system used to gather, store, organize, process, and analyze data to extract information. An information system is a combination of hardware, software, data, people, and methods to transform data into information. (p. 4)

infrastructure as a service (IaaS): Where the infrastructure required to run software tools and applications is stored in the cloud, so a user does not need to install infrastructure such as servers, networking hardware, and other physical hardware on-site. Instead, they are stored in the cloud and users use the service over the internet when needed and pay for the service accordingly. (p. 194)

innovation diffusion: The process by which an innovation is communicated through certain channels over time among the members of a social system. (p. 175)

interdepartmental: Between departments. (p. 176)

internet of things (IoT): The interconnection of physical devices around the world to the internet, continuously generating and sharing data back and forth. It has fundamentally transformed the world of geospatial science and technology. (p. 191)

intersect: A GIS operation that extracts common features from two layers to create a new layer that contains spatial and attribute data that is common in both of the original layers. (p. 213)

interval data: Numeric data measured on a regular scale with no fixed zero point. (p. 84)

isarithmic map: A map that makes use of a specialized cartographic symbolization called isolines to represent a continuous phenomenon that changes values smoothly. (p. 147)

isobars: Isolines connecting points of equal pressure. (p. 147)

isolines: Specialized lines drawn on a map connecting data points of the same value. (p. 147)
isotherms: Isolines connecting data points of equal temperature. (p. 147)
join: A temporary relationship between two tables in a GIS using RDBMS. (p. 82)
KML: A vector data file format used by earth browsers such as Google Earth, Google Earth Pro, and Google Maps. (p. 68)
large-scale map: A map that covers a small area with great detail. (p. 25)
large-scale photograph: An aerial photographs taken in an aircraft flying at a lower elevation. (p. 120)
latitudes: Lines that run in an east–west direction around the globe. (p. 42)
legend: A guide for a map reader to understand what the symbols and colors shown on the map represent. (p. 140)
line: A one-dimensional vector object that has length and no width and is formed by connecting a sequence of points with a set of x and y coordinates. (p. 54)
line of tangency: The place where the globe touches the projection surface. (p. 33)
linear interpolation: The method of plotting a point at an approximate location along a line segment. (p. 131)
Linux: An open-source operating system. (p. 166)
longitudes: Lines that run in a north–south direction from the North Pole to the South Pole. (p. 42)
many to many relationship: Multiple records in the destination table match multiple records in the source table. (p. 82)
many to one relationship: Multiple records in the destination table match only one record in the source table. (p. 81)
map: A graphical representation of the earth used to represent features, patterns, activities, and phenomena on the earth's surface using different symbols. A map is an abstraction that does not show every element of the earth. Rather, it provides a generalized view of the earth and the environment. Maps present a generalized view of the earth representing selective features and phenomena. (p. 23)
map algebra: Used in a raster data overlay operation to conduct cell-by-cell analysis using mathematical, logical, or Boolean operators. (p. 215)
map generalization: The decline in scale and corresponding reduction in detail as we move from a large-scale to a small-scale map. (p. 26)
map projection: The transformation of real-world features and their locations from their places on the earth's surface onto a flat surface. It involves complex geometrical calculations and is a must in order to be able to work with geospatial data. (p. 32)
map scale: The ratio of a distance on the map to the same distance on the ground. A map scale tells you the relationship between these two distances. (p. 23)
meridians: See **longitudes.**
metadata: Data about data that carries detailed information about the specific dataset such as when the dataset was created or procured and technical specifications of the data, datum, coordinates, and other information. (p. 102)
Military Grid Reference System (MGRS): The military grid, adopted for use by the US Army in 1947 and used by many organizations and countries worldwide. It is used in conjunction with the UTM and UPS systems. (p. 45)
multipath: A cause of error in position calculation that happens when signals, instead of following a straight path from a satellite to the receiver, are reflected off buildings or walls before they reach a GPS receiver. (p. 111)
multispectral imagery: Imagery produced by sensors that can sense and capture multiple bands of the electromagnetic spectrum. (p. 123)
multispectral sensor: A sensor that measures several bands of the electromagnetic spectrum. (p. 126)
nadir: The point on the ground that is directly under the camera. (p. 119)
National Agriculture Imagery Program (NAIP): Program that acquires aerial imagery during the agricultural growing seasons in the United States. A primary goal of NAIP is to make digital ortho photography available to governmental agencies and the public within a year of acquisition. (p. 99)

National Center for Geographic Information and Analysis (NCGIA): An independent research consortium dedicated to basic research and education in geographic information science and its related technologies, including geographic information systems. The three member institutions are the University of California, Santa Barbara; the University at Buffalo; and the University of Maine. (p. 179)

The National Map: A collaborative effort between the USGS and other federal, state, and local government agencies to disseminate topographic information and geospatial data of the United States to the public. (p. 98)

near-polar orbit: Orbital path taken by satellites where satellites move close to the North and South Poles while completing multiple orbits in a day. (p. 126)

NOAA CORS Network (NCN): Provides Global Navigation Satellite System (GNSS) data, supporting three-dimensional positioning, meteorology, space weather, and geophysical applications. It improves the accuracy of position measurements throughout the United States. (p. 112)

nominal data: Any data that can identify or label an object. (p. 84)

normalization: The process of transforming data values into a ratio with the original data value as the numerator to minimize the differences in values caused by the size of areas or the total number of features in each area. (p. 154)

north arrow: A map element that helps to orient the map and understand cardinal directions on the map. (p. 140)

object-based vector model: A newer structure for a vector data model where geospatial data are treated as objects and features and attributes are stored as a single unit. This minimizes the overhead in matching features to attributes. (p. 62)

oblique photographs: Aerial photographs captured with a camera tilted at an angle of more than 3 degrees. (p. 119)

one to many relationship: Each record in the destination table matches multiple records in the source table. (p. 81)

one to one relationship: One record in the destination table corresponds to only one record in the source table. (p. 81)

Open Source Geospatial Foundation (OSGeo): A nonprofit nongovernmental organization established with the goal of providing organizational, legal, and financial resources and support to promote the advancement of the open-source geospatial community. (p. 166)

operating system: A software system that handles the connection between a hardware and a software, allowing the two to communicate. (p. 164)

ordinal data: Any data that is ranked on a numeric scale or hierarchy. (p. 84)

orthophoto: A geometrically corrected aerial photo produced using a process known as orthorectification. (p. 99)

orthorectification: The process of removing the effect of terrain and relief displacement from an aerial photo. (p. 121)

overlay: A GIS operation that combines two layers based on their spatial relationship to create a new output layer that contains information from both input layers. The operation combines both the spatial information and the aspatial/attribute information in the new output layer. (p. 213)

P code: Precise code transmitted on L1 and L2 frequencies. (p. 110)

panchromatic sensor: A sensor that measures the visible band and combines it into a single band. This produces a grayscale panchromatic image. (p. 126)

parallels: See **latitudes.**

parsing: The process of breaking up an address into individual components. (p. 130)

picture symbol map: A map where point data is symbolized using a picture. (p. 143)

pixel value: See **digital number.**

planar projection: The globe is projected on a flat plane that either touches the globe at a point or slices through the globe. (p. 33)

platform: An IT term for a hardware or software on which an application runs. (p. 164)

platform as a service (PaaS): Where hardware and software tools, data, and interfaces to develop software applications are stored in the cloud. (p. 194)

point: A zero-dimensional vector object that represents a singular, distinct feature location on the ground. (p. 54)

polygon: A two-dimensional vector object that has length and width. Polygons are closed features that are formed by connecting a series of lines in a sequence going in a specific direction. (p. 55)

primary key: A unique identifier used to link one database with another. This will be a specific column in a table that uniquely identifies records in each row of the table to link it with another table. (p. 77)

projected coordinate system (PCS): A type of coordinate system defined on a flat, two-dimensional surface, with location on the flat surface defined using Cartesian coordinates. Location is described by using x, y coordinates. A PCS is always based on a GCS. Additionally, a PCS consists of a map projection, projection parameters specific to a particular location, and linear units of measurement. (p. 43)

proportional symbols map: A map created to represent quantitative data at point locations where point data is shown with varying-sized symbols and the size of the symbols is scaled in proportion to the magnitude of values at point locations. The point locations can be true points or conceptual points. A proportional symbols map uses a specific symbol size for each numeric value rather than a range of values. (p. 144)

Public Participation GIS (PPGIS): An area of research and practice examining access to geographic information technologies and their social and political impacts. PPGIS has become one of the most developed research agendas within the GIS and society umbrella, and a number of subthemes have emerged within PPGIS research. (p. 180)

QGIS: An open-source desktop GIS software that is available for free. The software has a unique capability to run on Windows, Linux, Unix, and Mac operating systems. (p. 166)

qualitative data: Any data that is described using text. (p. 84)

quantitative data: Any data that is described using numbers. (p. 84)

raster data model: A data model that represents the world as a collection of evenly spaced tiny squares of pixels called grid cells. A raster data consists of grid cells arranged in rows and columns. Cell size in a raster data is determined by measuring one side of a square cell and represents the spatial resolution of the raster data. (p. 63)

ratio data: Numeric data that has a meaningful fixed zero point. (p. 84)

reference map: A map used to show location information. Reference maps are useful for providing geographic reference of locations of features. On a reference map no one particular feature is emphasized over another feature. All features on a reference map are shown with equal importance. (p. 142)

relate function: A relate function is used to associate two tables in a RDBMS when the rule of joining is not satisfied. (p. 83)

relational database: A database consisting of multiple tables where each table shares a common link with another table by means of a common attribute or an identifier. (p. 75)

relational database management system (RDBMS): A system designed to manage a relational database. (p. 75)

relational operators: Operators used to build attribute queries. (p. 210)

relative location: Location described in reference to another location such as an address, postal code, census tract name, or road name. (p. 12)

relief displacement: A shift in the position of a feature on an image due to its height. When an aircraft mounted with a camera flies over a terrain, some features will be closer to the camera, such as tall buildings, mountains, and towers, and will have a larger scale, while other features will be farther away from the camera and will have a smaller scale. As a result, tall features in the photo appear to be leaning away. (p. 121)

remote sensing: The process of collecting, measuring, and recording information pertaining to electromagnetic energy reflected off the earth's surface using a device that is not in physical contact with the object or phenomenon, such as an instrument mounted on an aircraft or satellite. (p. 113)

representative fraction (RF): RF is the ratio between map and ground distance. It is written as 1/x or 1:x. The unit of measurement is same for the numerator and the denominator. 1 is the distance on the map and x is the distance on the ground. One of the most common ways to express scale is by using RF. (p. 23)

satellite-based augmentation system: An augmentation system to improve accuracy of position measurements based on corrections transmitted from a satellite. An example of a satellite-based augmentation system is the Wide Area Augmentation System (WAAS). (p. 112)
secant projection: The projection surface will intersect or cut or slice through the globe. (p. 33)
simple query: A query that tests one condition or criterion. (p. 210)
small-scale map: A map that covers a large area with less detail. (p. 25)
small-scale photograph: An aerial photographs captured in an aircraft flying at a higher elevation. (p. 120)
software as a service (SaaS): A cloud service used for storing files and documents, backing up files and folders, and managing web-based emails. SaaS makes software available to users over the internet. (p. 193)
source table: The table that contains the information that will be appended. (p. 82)
space segment: One of the three components of a GPS system that consists of a constellation of GPS satellites orbiting the earth in fixed paths. These satellites transmit radio signals to users on earth. (p. 108)
spaghetti data model: First generation of simple structured vector data model. It has a very simple structure consisting of lines with arbitrary starting points and end points. Points, lines, and polygons lack any structure and are represented by a string of x and y coordinates. The spaghetti data model does not store feature relations. Features are stored independent of each other as geometrical objects. (p. 58)
spatial analysis: The ability to model real-world features, behaviors, spatial relationships and extract information from map data and attribute data in a GIS. (p. 209)
spatial query: A data retrieval operation in GIS used for geographic search where the criterion is based on a relationship between two or more spatial layers or features at a time using the "Select by location" tool. (p. 211)
spatial resolution: The area on the ground covered by one pixel. (p. 125)
spheroid: See **ellipsoid.**
standalone table: A table that is not associated with geographic data. Standalone tables must be linked with an attribute table for spatial analysis and visualization purposes. (p. 79)
State Plane Coordinate System (SPCS): A coordinate system created in the 1930s to define parcel boundaries in the United States with the goal of simplifying land parcel perimeter and area calculations and achieving accuracy within a foot of the true measurement. In order to achieve this level of accuracy, each state was divided into zones with an individual grid for each zone. SPCS is widely used by city and county governments as well as surveying, engineering, and utility companies that require highly accurate land surveying information. It is a projected coordinate system like UTM and covers the entire United States. Each state has its own map in the SPCS system. (p. 46)
structured query language (SQL): Used to access a database and communicate with data. (p. 79)
sun-synchronous orbit: Orbital path taken by satellites where satellites cover an area on the earth's surface at the same local time. (p. 126)
table: A file with data stored in the form of rows and columns. (p. 75)
tangent projection: The projection surface will touch the globe either at a point or along a line. (p. 33)
temporal resolution: The time it takes for a satellite to return to the same location to capture an image of the same spot on the earth's surface. (p. 125)
thematic map: A map that emphasizes geographical distribution of a specific theme and that particular theme is given prominence on the map. (p. 142)
topographic contours: Isolines connecting points of equal elevation above a datum. (p. 147)
topological data model: A vector data model structure that logically determines how and where nodes connect points and lines on a map; determines how points, lines, and polygons share geometry; and models vector object relationships. (p. 58)
Topologically Integrated Geographic Encoding Referencing (TIGER): An important dataset created by the US Census Bureau to modernize their mapping process for the decennial census and create a single source of good-quality geographic data. The TIGER system provided the public the first free nationwide digital map of roads and boundaries covering the entire United States at the street level and played a key role in the GIS world. (p. 103)
topology: The ability to understand the environment, how features connect with other features in their surroundings and behave, and understand feature relations. The concept of topology is

derived from geometrical mathematics and is used in GIS to record, understand, and describe logical spatial relationship between features. (p. 58)

trilateration: A mathematical process of pinpointing positions from three distances on a two-dimensional map. (p. 110)

true color composite: A color image where the red band is assigned to the red channel of the display device, the green band is assigned to the green channel of the display device, and the blue band is assigned to the blue channel of the display device. Features in a true color composite image will look exactly the way they appear on the earth's surface, like the human eye would see it. (p. 125)

true direction projection: A projections that preserves direction from the center of the map to any other location. (p. 37)

typography: The art of lettering or type. (p. 140)

union: A GIS operation that combines all of the features in a new layer containing all of the spatial and attribute data from both of the original layers. (p. 213)

United States Geological Survey (USGS): Part of the Department of the Interior, the USGS is a key government agency that supplies geospatial data covering land features, topographical information, and man-made features of the United States. (p. 97)

United States National Grid (USNG): A coordinate system used for all referencing and mapping applications in the United States, so that users of various devices that utilize geospatial data and require accurate ground coordinate information have a uniform reference system nationwide. This system is used widely by the military and by organizations such as the USGS, Federal Emergency Management Agency (FEMA), and the National Guard. The USNG is based on the UTM coordinate system and the alphanumeric referencing system used in the MGRS. (p. 46)

Universal Polar Stereographic (UPS) system: A rectangular coordinate system that consists of a north zone and a south zone. The north zone covers the area from 84° N to the North Pole. Similarly, the south zone extends from the pole to 80°30' S. It is based on the stereographic projection customized for each pole. All maps of the polar regions are based on the UPS system. (p. 44)

Universal Transverse Mercator (UTM) coordinate system: A planar coordinate grid system based on a secant transverse Mercator projection, hence the name. The UTM system divides the earth into 60 north–south zones and location is determined by x and y coordinates in those zones. (p. 43)

University Consortium for Geographic Information Science (UCGIS): A nonprofit organization that creates and supports communities of practice for GIS science research, education, and policy endeavors in higher education and with allied institutions. (p. 180)

unmanned aircraft systems (UAS): Aircraft that are driven from the ground using a remote control. (p. 116)

user segment: One of the three components of a GPS system that consists of a GPS receiver device that receives signals from GPS satellites and uses the information to calculate its three-dimensional position on earth. (p. 109)

VCT: A proprietary vector data file format used by IDRISI. (p. 68)

vector data model: A data model that represents real-world features and phenomena as points, lines, or polygons. (p. 53)

vector objects: Points, lines, and polygons that form the foundation of a vector data model. (p. 53)

vertical datum: A system created using a collection of points on the earth whose height has been accurately determined with respect to a nationally defined reference surface such as mean sea level. Vertical datums are used to measure and describe land elevations. (p. 30)

vertical photographs: Aerial photographs captured with a vertical camera axis. (p. 119)

visible light spectrum: The portion of the electromagnetic spectrum that the human eye can see. The human eye is sensitive to reflected light that has a wavelength ranging between 0.4 and 0.7 micrometer. (p. 122)

Wide Area Augmentation System (WAAS): An air navigation aid developed by the Federal Aviation Administration to augment the Global Positioning System, with the goal of improving its accuracy, integrity, and availability. It is a type of satellite-based augmentation system. (p. 112)

Index

Page numbers in italics refer to figures and tables.

About the Author

Falguni Mukherjee is professor of environmental and geosciences and associate dean for graduate student success in the Graduate and Professional School at Sam Houston State University. Dr. Mukherjee's research centers on the interaction between science, society, and technology. She particularly focuses on the process of technology use for urban governance in a non-Western context, and the focal point of her research are urban local bodies in South Asia. She also utilizes remote sensing technology to understand earth-environment interactions in her research.

Series: Exploring Geography

Series Editor: David H. Kaplan

An Introduction to the Geography of Tourism, Third Edition
By Velvet Nelson
An Introduction to Transportation Geography: Transport, Mobility, and Place
By Julie Cidell
Food Geographies: Social, Political, and Ecological Connections
By Pascale Joassart-Marcelli
Introduction to Geographic Information Systems
By Falguni Mukherjee
Biogeography in a Human-Influenced World
By Kenneth R. Young
Political Geography
By Corey Johnson
Economic Geography: Capitalism, Globalization, and Landscapes
By Barney Warf
An Introduction to Health and Medical Geography
By Sue C. Grady

www.ingramcontent.com/pod-product-compliance
Ingram Content Group UK Ltd.
Pitfield, Milton Keynes, MK11 3LW, UK
UKHW051932150926
475560UK00005B/352

9 781538 152003